Heterogeneous Catalysis

Heterogeneous Catalysis

Selected American Histories

Burtron H. Davis, Editor
University of Kentucky

William P. Hettinger, Jr., Editor
Ashland Petroleum Company

Based on a symposium sponsored
by the Division of History of Chemistry
at the 183rd Meeting
of the American Chemical Society,
Las Vegas, Nevada,
March 28–April 2, 1982

ACS SYMPOSIUM SERIES 222

AMERICAN CHEMICAL SOCIETY
WASHINGTON, D.C. 1983

Library of Congress Cataloging in Publication Data

Heterogeneous catalysis.
(ACS symposium series, ISSN 0097-6156; 222)

"Based on a symposium sponsored by the Division of History of Chemistry, at the 183rd Meeting of the American Chemical Society, Las Vegas, Nevada, March 28-April 2, 1982."

Includes bibliographies and index.

1. Heterogeneous catalysis—Congresses.
I. Davis, Burtron H., 1934- . II. Hettinger, William P., 1922- . III. American Chemical Society. Division of the History of Chemistry. IV. American Chemical Society. National Meeting (183rd: 1982: Las Vegas, Nev.) V. Series.

QD505.H47 1983 541.3'95 83-8745
ISBN 0-8412-0778-X

PRINTED IN THE UNITED STATES OF AMERICA

ACS Symposium Series

M. Joan Comstock, *Series Editor*

FOREWORD

The ACS SYMPOSIUM SERIES was founded in 1974 to provide a medium for publishing symposia quickly in book form. The format of the Series parallels that of the continuing ADVANCES IN CHEMISTRY SERIES except that in order to save time the papers are not typeset but are reproduced as they are submitted by the authors in camera-ready form. Papers are reviewed under the supervision of the Editors with the assistance of the Series Advisory Board and are selected to maintain the integrity of the symposia; however, verbatim reproductions of previously published papers are not accepted. Both reviews and reports of research are acceptable since symposia may embrace both types of presentation.

CONTENTS

Preface ... xi

EARLY PIONEERS

1. **Heteroheneous Catalysis Before 1934** ... 3
   Robert L. Burwell, Jr.

2. **Insiders, Outsiders, and Surfaces: Irving Langmuir's Contribution to Catalysis** ... 13
   George L. Gaines, Jr. and George Wise

3. **V. N. Ipatieff: As I Knew Him** ... 23
   Herman Pines

4. **Sir Hugh Taylor: The Man** ... 33
   John E. Benson and M. Boudart

5. **Paul H. Emmett: Six Decades of Contributions to Catalysis** ... 45
   Robert L. Garten

6. **The Contributions of Eugene J. Houdry to the Development of Catalytic Cracking** ... 61
   Alex G. Oblad

7. **Herman Pines and Organic Heterogeneous Catalysis** ... 77
   Norman E. Hoffman

8. **Early Catalysis Research with Para-Hydrogen and Heavy Hydrogen** ... 89
   Adalbert Farkas

LATER RESEARCHERS

9. **Allan T. Gwathmey and His Contributions to the Science of Catalysis** 121
   Henry Leidheiser, Jr.

10. **Contributions of P. W. Selwood to Catalysis Research** ... 131
    R. P. Eischens

11. **The Development of the Platforming Process—Some Personal and Catalytic Recollections** ... 141
    Vladimir Haensel

12. **Otto Beeck and His Colleagues in Catalysis** ... 153
    J. Norton Wilson

13. **Ernest W. Thiele: A Pioneer in Defining the Role of Diffusion in Heterogeneous Catalysis** ... 173
    Ravi Randhava

14. **Ahlborn Wheeler—Catalytic Scientist** 179
G. Alex Mills

15. **Early Infrared Studies of Adsorbed Molecules at Beacon** 183
R. P. Eischens

16. **The Fixed Nitrogen Research Laboratory** 195
Paul H. Emmett

17. **Einstein in the U.S. Navy** 217
Stephen Brunauer

18. **The History of the BET Paper** 227
Edward Teller

CATALYTIC CRACKING

19. **The Carbonium Ion Mechanism of Catalytic Cracking** 235
Hervey H. Voge

20. **A History of Early Catalytic Cracking Research at Universal Oil Products Company** 241
Charles L. Thomas

21. **Development of the Theory of Catalytic Cracking** 247
Rowland C. Hansford

22. **The Invention of Zeolite Cracking Catalysts—A Personal Viewpoint** 253
Charles J. Plank

23. **The Development of Fluid Catalytic Cracking** 273
C. E. Jahnig, H. Z. Martin, and D. L. Campbell

24. **The Development of Hydrocracking** 293
Richard F. Sullivan and John W. Scott

RESEARCH DEVELOPMENTS

25. **Selective Oxidation by Heterogeneous Catalysis** 317
Robert K. Grasselli

26. **Methanol: Bright Past—Brilliant Future?** 349
Alvin B. Stiles

27. **Applications of Magnetic Resonance in Catalytic Research** 375
Wallace S. Brey

28. **Chain Growth and Iron Nitrides in the Fischer–Tropsch Synthesis** . . 389
Robert B. Anderson

29. **Discovery and Development of Olefin Disproportionation (Metathesis)** 403
Robert L. Banks

30. **The Development of Automotive Exhaust Catalysts** 415
George R. Lester

31. **Attempts to Measure the Number of Active Sites** 435
Russell W. Maatman

MORE PIONEERS

32. Leonard C. Drake—His Contributions to the Development of Mercury Porosimetry for Catalyst Characterization 451
Burtron H. Davis

33. Catalysis at Princeton University Chemistry Department 463
John Turkevich

34. Murray Raney—Pioneer Catalyst Producer 491
Raymond B. Seymour and Stewart R. Montgomery

35. Frank G. Ciapetta—Pioneer in Catalytic Chemistry 505
W. P. Hettinger, Jr.

36. A Society of Catalytic Chemists and Engineers 519
Heinz Heinemann

Index 523

PREFACE

HETEROGENEOUS CATALYSIS spans a broad range of scientific disciplines including such seemingly unrelated topics as solid state physics, materials sintering, and organic reaction mechanisms. While much of the early work in catalysis was a result of academic curiosity, the need for improved performance in industrial applications had a strong influence in focusing the many research disciplines onto heterogeneous catalysis.

During the 1800s much of the research in catalysis occurred in Europe and heterogeneous catalysis did not become firmly established in the United States until the 1920s. Because extensive activity in this area has produced an enormous amount of literature documenting many important advances, only a limited number of benchmark works and scientists could be included in a symposium even if the coverage was limited to U.S. contributions.

This seems an appropriate time to pause and survey the nearly 50 years of progress in U.S. heterogeneous catalysis research. In this volume, some of the U.S. pioneers from the 1920s and 1930s provide a personal view of their work, and, in addition, several experts provide a personal, historical account of a research area or pioneer. Much of this early research had an Edisonian character. The unique, creative personality of the investigator or unanticipated events were, in many cases, important features leading to a major advance. Frequently, these facets of science are not recorded in detail.

Advances in instrumentation have also occurred, so that present day investigators can obtain a detailed structure of a catalyst surface layer. Information can also be obtained about the concentration and chemical state of chemisorbed species. Therefore, even though Edisonian approaches will remain an important aspect of this research, we are entering a new era of research in heterogeneous catalysis.

Nearly all of the papers presented at the Las Vegas symposium are included in this book. In addition, some early researchers who were unable to attend the meeting have made written contributions. Our intention is to convey to you many fascinating aspects of catalysis and its broad scope, ranging all the way from alchemy to Edisonian research to high technology, and to provide insight into those most unusual, stimulating, and intellectual pioneers that gave it life. We are especially grateful to Dr. Robert Burwell for providing, in the very limited space allotted to him, such an outstanding and stimulating view of some events in European catalytic history prior to the 1920 period.

We heartily thank the following companies for a donation to provide travel funds for some of the retired pioneers: Air Products and Chemicals, Inc., Celanese Research Company, Chemical Data Systems, Inc., General Electric Company, Monsanto Company, Phillips Petroleum Company, Rohm and Haas Company, Texaco, Inc., The P.Q. Corp., Union Carbide Corporation, and W. R. Grace & Co.

Many assisted with organizing the symposium and preparing this book. We acknowledge the outstanding assistance of the IMMR Publications Section (Rhonda Pettit, Krina Cummins, Kathleen Dumler, Susan Hudak, Martha Moore, Kathie Sauer, Vicki Schumacher, and Camille Weber), Joyce Hieger, Janet Litteral, and the encouragement of our respective spouses, Nancy and Alice.

BURTRON H. DAVIS
University of Kentucky

WILLIAM P. HETTINGER, JR.
Ashland Petroleum Company

January 25, 1983

ACKNOWLEDGMENT

Xytel Corporation assisted in the preparation of this volume with a generous grant to defray a portion of the publication expenses. In addition, the editors are grateful to Serge Randhava, Ravi Randhava, and David Goltermann of Xytel Corporation for hosting the social hour at the Las Vegas ACS meeting in March, 1982.

EARLY PIONEERS

Heterogeneous Catalysis Before 1934

ROBERT L. BURWELL, JR.

Northwestern University, The Ipatieff Catalytic Laboratory, Department of Chemistry, Evanston, IL 60201

This article was prepared as an introduction to this symposium. Its aim was to provide a brief survey of heterogeneous catalysis before American contributions began. It makes no pretence to any contribution to scholarship (1).

We shall then go back to some suitably remote time in the past and proceed to the terminal date rather rapidly. In going back in time, let us, however, make a stop at 1909. In his Nobel Award address of that year, Wilhelm Ostwald said: "The employment of the concept of catalysis has served hitherto as an indication of scientific backwardness."

We may hope that this is no longer true and presumably Ostwald felt that he had made catalysis respectable. But what led Ostwald to make this statement?

During the nineteenth century, all kinds of interpretations of catalysis were advanced and most had a closer relationship with metaphysics than with science. Here is a example from Stohmann in 1894 (2).

> "Catalysis is a process involving the motion of atoms in molecules of labile compounds which results from the presence of a <u>force emitted by another compound</u> and which leads to the formation of a more stable compound and the liberation of energy." (Emphasis added).

In the same year Ostwald gave a definition of catalysis which looks rather better (3): "Catalysis is the acceleration of a slow reaction by the presence of a foreign material."

0097-6156/83/0222-0003$06.00/0

Let us now see what led to some of the curious theories of catalysis, although we should keep in mind that even theories of uncatalyzed reactions were necessarily very inadequate and sometimes rather metaphysical.

The philosophers' stone of the alchemist was of course a catalyst and it was not until nearly 1800 that chemical theory - the ideas of chemical elements and the nature of chemical change - made transmutation suspect. The lingering idea of the philosophers' stone may have influenced the development of the idea of catalysis. The following clearly describes a catalytic experiment although the text is unclear as to whether the experiment involves homogeneous or heterogeneous catalysis (4).

> "Paracelsus sent his waiting man to deliver a piece of paper containing a small amount of a blood-red powder with the command that it be poured into molten lead and stirred well.... The master of the mint paid several thousand guilders for the resulting gold."

The experiment was not reproducible, but we have no record that the work was formally withdrawn.

During the eighteenth century various reports of what we would call transmutations appeared in the literature. But, in the absence of the theory of chemical elements, there was no way to distinguish between what we would now call a chemical change and a transmutation. Although most educated people came to believe that most alchemists were charlatans, there was no scientific reason why transmutation was impossible and another master of the mint, Sir Isaac Newton, took it seriously. But, in the nineteenth century, alchemy disappeared from the scientific literature. The discovery of radioactive transmutations revivified "alchemy" for a short period in the earlier 1900's and the philosophers' stone made its last appearance in the form of a Pd/asbestos catalyst.

In the Berichte der Deutschen Chemisches Gesellshaft for 1926 (59B, 2039), the eminent radiochemist Fritz Paneth and K. Peters published a paper entitled "On the Transmutation (Verwandlung) of Hydrogen Into Helium." Paneth had developed a method for the measurement of very small quantities of helium. In this paper, he reported that 1 g of a 50% Pd/asbestos led to the formation of 10^{-7} cm^3 of helium per day. The authors were well aware of the energetics of the process. Certainly, this was the most significant heterogeneous catalytic reaction ever to be reported. Unfortunately, the authors had to withdraw the paper very shortly after its appearance. Despite the statements of the authors in the communications withdrawing the work, my suspicion is that the helium collected had diffused through the glass walls of the apparatus.

In the second half of the eighteenth century there were various reports of what we would recognize today as catalytic reactions. However, the idea of a catalytic reaction could not have been understood before chemical elements and chemical change had appeared. This early work is a historical curiosity which had no influence on the development of catalysis. The idea of catalysis can be taken as starting in 1814 when Kirchoff published his work on the hydrolysis of starch to glucose by acids. A number of people had investigated the hydrolysis, but Kirchoff was the first clearly to understand what was going on.

The second event in the development of catalysis came in 1817 when Sir Humphry Davy discovered that the introduction of a hot platinum wire into a mixture of air and coal gas led the platinum to become white hot. Davy considered that there was oxidation but no flame and that the platinum was unchanged.

His cousin, Edmund Davy, continued the work and in 1820 he discovered that the platinum could be introduced at room temperature provided that it was finely divided. Döbereiner continued this work and in 1823, he found that, in the presence of platinum, vapors of ethanol reacted with oxygen to form acetic acid. Work on selective oxidation continues to this day but Döbereiner did it first. He also noticed, in line with the work of Edmund Davy, that divided platinum became red hot in the presence of hydrogen and oxygen. Döbereiner did more than notice. He developed Döbereiner's Tinder Box (Döbereiners Feuerzeug) (5). This involved a small Kipp generator containing zinc and dilute sulfuric acid. When a valve was opened a jet of hydrogen emerged, the acid rose into the zinc and the jet continued. The jet fell upon spongy platinum and burst into flame. One then lighted his fire or his pipe. Over a million Tinder Boxes were sold. Döbereiner's Tinder Box represented the first technological application of heterogeneous catalysis.

In 1824, Henry reported the first example of poisoning. Ethylene inhibited the reaction between hydrogen and oxygen on platinum. He also noted selective oxidation in the reaction between oxygen and a mixture of hydrogen, carbon monoxide, and methane.

Continuing the line of catalytic oxidation on platinum, Peregrine Philips (1831, British Patent No. 6096) patented the oxidation of SO_2 to SO_3 on platinum, but he must have died before the first contact process plant for the production of sulfuric acid went on stream. And finally, along this line of work, Schweigger in the same year discovered that hydrogen sulfide poisoned platinum.

Another line of catalytic work was started by Thenard (1818). He discovered hydrogen peroxide and he who discovers H_2O_2 is apt also to discover its catalytic decomposition. Thenard did, and he investigated the matter carefully using both homogeneous and heterogeneous catalysts.

Today, someone might possibly write an annual review of the coordination chemistry of chromium. But, Baron Jacob Berzelius, the "Mr. Chemistry" of his day, wrote an annual review of chemistry every year for twenty-eight years. In 1835, he surveyed Kirchoff's work on acid hydrolysis, the catalytic oxidations on platinum and Thenard's work on the decomposition of H_2O_2. He discerned a phenomenon common to the three areas, and he invented catalysis and catalyst. Unfortunately, he coined another term, catalytic force (6).

"Catalytic force actually means that substances are able to awaken affinities which are asleep at this temperature by their mere presence and not by their own affinity." (Emphasis added).

During the rest of the century, much fruitless effort was devoted to trying to develop explanations for catalytic force. It was this that gave catalysis something of a bad reputation and led to Ostwald's unfriendly remark. However, all through the century there were those who advanced chemical theories of catalysis. For example, de la Rive (7) in 1838 proposed that platinum catalyzed the oxidation of hydrogen by a cycle of alternate oxidations of platinum followed by reduction of the surface oxide. He said:

> "Ce n'est pas necessaire de recourir á une force mystérieuse telle que celle que Berzelius a admise sous le nom de force catalytique. [It is unnecessary to resort to a mysterious force like that which Berzelius has proposed under the name, catalytic force]."

However, believers in a chemical explanation were in the minority up into the twentieth century. Berzelius did not refer to Faraday's paper of 1834 (8), but we should. It was an excellent study of the catalytic characteristics of platinum foils for the oxidation of hydrogen: effect of pretreatment, rates of reaction in a primitive fashion, deactivation, reactivation and poisoning. For example, ethylene inhibited the reaction of a H_2+O_2 mixture, but after some hours vigorous action commenced. H_2S and PH_3 were, however, permanent poisons. It would be interesting to rewrite Faraday's article in modern form and terminology, to add a few imaginary experiments with XPS and EXAFS, and to submit the result to J.Catal. It would not be wise, however, to insert Faraday's mechanism in such a paper. He thought that hydrogen and oxygen were condensed (one might say, physisorbed) on the surface of the platinum and that reaction to form water resulted from the mere proximity of H_2 and O_2.

"The platinum is not considered as causing the combination of any particles with itself, but only by associating them closely around it."

One cannot fault Faraday for not inventing dissociative chemisorption. Hydrogen and oxygen were not known to be diatomic in 1834. Anyway, Faraday did not believe in atoms and assigned such ideas to metaphysics.

The concept of atoms "did not afford me the least help in my endeavour to form an idea of a particle of matter...with experience, I outgrew the idea of atoms.... Such ideas are mere hindrance to the progress of science (9)."

Following Faraday, there was a long, rather dry spell in academic heterogeneous catalysis. Most of the earlier work involved catalysis of oxidations or the like. The world was not ready for the oxidation of naphthalene to phthalic anhydride - either academically or technologically. Organic heterogeneous catalysis remained unexplored for many years after 1834. However, the beginning of the technology of heterogeneous catalysis came in this period (still oxidations). The Deacon process was developed in the 1860's - the oxidation of hydrogen chloride to chlorine on copper chloride - and Messel started the first plant for the oxidation of sulfur dioxide to sulfur trioxide in 1875, 44 years after the Peregrine Philips patent.

Things began to pick up towards the end of the century. In 1888, Ludwig Mond discovered steam reforming of hydrocarbons on nickel/pumice to give carbon monoxide and hydrogen. Sir James Dewar noted that oxygen adsorbed in large amounts on charcoal at the temperature of liquid air and that it desorbed as such on warming. However, oxygen adsorbed at 25°C could be desorbed only at high temperatures and as oxides of carbon. These observations can be taken as the beginning of physisorption and chemisorption.

The development of thermodynamics led to the recognition that a catalyst could only promote a rate in the direction of the position of equilibrium and that a catalyst could not change the position of equilibrium. Further, starting with Nernst, it became likely that one would need to worry about diffusional problems in heterogeneous catalysis.

In 1909, Ostwald was awarded the Nobel Prize in chemistry for his work in catalysis. My suspicion is that the committee decided to award him the prize - he was the "Mr. Physical Chemistry" of his day - and they chose his work in catalysis as providing as good a basis as any other. In fact, not much of his career had been devoted to catalysis. In 1884, he reported a study of the acid-catalyzed hydrolysis of methyl acetate which introduced kinetics, in the modern sense, into catalysis. He also tied the concept of catalytic activity to rate. Both of these items were important. Then in 1901-1904 he and his former student, Brauer, developed the Ostwald process for the oxidation on platinum of ammonia to nitric oxide. The first plant went on stream in Bochum in 1906 at a level of 300 kg of nitric acid per day. In 1908, the production was 3000 kg per day. The process actually goes back to Kuhlmann in 1838, but there had been no industrial interest in such a process, because Chili saltpeter was cheaper source of nitric acid than ammonia. However, at the beginning of the twentieth century, the ease with which the British fleet could sever the sea lane between Chili and Germany had become a stimulus to the development of the Ostwald process.

German writers tend to make Ostwald the giant of catalysis as the following quotation (1) from G.-M. Schwab illustrates.

> "It is a very wide ranging undertaking to talk on Ostwald's work on catalysis. It is similar to that which would be involved in talking on Newton's contributions to mechanics or Planck's to quantum theory."

I doubt that many French, English, Russian or American workers would write so strongly. Personally, I would put the influence of Sabatier who did not get the Nobel Prize until three years after Ostwald; Ipatieff who never got it, and perhaps Haber and Mittasch as great or greater.

However, in the years after Ostwald, his former students and collaborators dominated catalysis in Germany. We can mention Bodenstein who put the study of the kinetics of heterogeneous catalysis essentially in its modern state. The Ostwald school made little contribution to mechanism, but it adhered to a view surprisingly close to that of Faraday. The famous paper of Bodenstein and Fink of 1907 interpreted the kinetics of the oxidation of sulfur dioxide on platinum in terms of the diffusion of sulfur dioxide or oxygen through a polymolecular layer of adsorbed material. In extreme cases, accord with the observed inhibition by sulfur trioxide would have required adsorbed layers so thick that they could have been pared with a razor.

Langmuir's work put a final quietus to such ideas, but independently support for chemical theories of catalysis came from such workers as Ipatieff and Sabatier. The latter said in an article written at the end of his career (10):

> "Like my illustrious master, Marcellin Berthelot, I always assumed that the fundamental cause of all types of catalysis is the formation of a temporary and very rapid combination (he meant a chemical combination) of one of the reactants with a body called the catalyst.... This theory has been much discussed. Other theories more or less complex and based on modern concepts of the atom, have been proposed. I have tenaciously held to my theory of a temporary combination. It has guided my work both in hydrogenations and in dehydrations."

He had won the battle completely, but he was still rather defensive.

By 1934, the only remaining trace of catalytic force lay in the low temperature catalysis of the interconversion of the newly discovered ortho- and parahydrogen by paramagnetic surface sites.

Starting about 1900, Sabatier (10) and Ipatieff opened up the area of organic heterogeneous catalysis. The impact of Sabatier's work was more immediate. Anyone, in a day or two could set up the apparatus needed to duplicate Sabatier's discovery of the hydrogenation of olefins and benzene. Further, before Sabatier, the conversion of an alkene to an alkane was an operation of such difficulty and of such low yield that it was rarely carried out. Sabatier's discovery created something of a sensation. Furthermore, it was rapidly put to hydrogenating vegetable oils to make margarine.

As an artillery officer, Ipatieff started with autoclaves and high pressures. His work was not so readily applied in the usual laboratory. In the long run, however, Ipatieff's work was as influential as Sabatier's and it was more important in technological applications of catalysis to the modern industries of petroleum refining and the production of petroleum-based chemicals. Further, as will have been noticed, catalysts before Ipatieff were largely Group VIII metals. He introduced oxides like alumina into the catalytic repertory. Overall, the work of Sabatier and Ipatieff led to the development of organic heterogeneous catalysis, an area which had hardly existed in the nineteenth century and which was to provide a powerful stimulus to heterogeneous catalysis both academically, and industrially.

The work of Haber and Mittasch during the first decade of the twentieth century in developing the synthesis of ammonia by the hydrogenation of nitrogen was of major importance although more limited in scope. The industrial application of heterogeneous catalysis was further extended by the development of the commercial hydrogenation of carbon monoxide to methanol by the Badische Anilin- und Soda-fabrik. Commercial production started at Merseburg in 1923 and the plant was producing 10-20 tons of methanol per day at the end of that year (11). The catalyst was not a Group VIII metal (we now know that they could have used palladium) but $ZnO{\bullet}Cr_2O_3$.

Finally, I should note that the early thirties saw the beginnings of detailed mechanistic proposals based on the intermediate compound theory but in advance of the rather vague ideas of Ipatieff and Sabatier. The first two mechanistic proposals in the modern sense were the Bonhoeffer-Farkas (12) and the Horiuti-Polanyi (13) mechanisms. In the Bonhoeffer-Farkas mechanism (1931), the ortho- para-hydrogen conversion at high temperatures was assumed to proceed via dissociative adsorption of hydrogen followed by associative desorption and the mechanism could be immediately extended to $H_2 + D_2 \rightarrow 2HD$. The Horiuti-Polanyi mechanism (1934) for the hydrogenation of ethylene could also be readily extended to other hydrogenation reactions. Both mechanisms are still in use.

We have now reached into the period in which American work in catalysis had started and we have mentioned the European work of two scientists whose work continued in the United States, V.N. Ipatieff and A. Farkas.

Industrial applications of heterogeneous catalysis were well underway in the chemical industry: The contact process for sulfuric acid, the Haber process for the synthesis of ammonia, the BASF process for the synthesis of methanol, the catalytic hydrogenation of vegetable oils, and the catalytic water-gas shift process for producing the hydrogen needed in the preceding processes. The applications of heterogeneous catalysis to the petroleum industry had hardly begun. That was a development which was largely to occur in the United States.

Acknowledgment

This article was written while the author was enjoying the hospitality of Prof. J. Fraissard and the Laboratoire de Chimie des Surfaces of the Université Pierre et Marie Curie, Paris.

Literature Cited

1. The articles on W. Ostwald by G.-M. Schwab, Z. für Elektrochem. (1953) 25, 878, and by W. Schirmer, Sitzungsber. Akad. Wiss. DDR, Math., Naturwiss. Tech. (1979) 33, (13N) were particularly helpful to the author and he wishes to thank Professors Schwab and Schirmer for providing him with copies of their papers. Among other secondary sources, the work on Berzelius by J.E. Jorpes ("Jac. Berzelius, His Life and Work," Uppsala, 1966) and the excel-

lent articles by W.D. Mogerman in the Inco Reporter for 1965 should receive particular mention.
2. Stohmann, Z. Biol. (1894), 31, 364.
3. Ostwald, W., Z. physik. Chem. (1894) 15, 705.
4. Schwab, G.-M., "Catalysis", translated by Taylor, H.S., and Spence, R., van Nostrand, New York, NY, (1937).
5. Collins, P., Educ. Chem., (1977) 14, 14.
6. Jorpes, J.E. "Jac. Berzelius, His Life and Work", Uppsala, 1966, p. 112. "Power in Jorpes has been changed to "force".
7. de la Rive, Compt. rend. (1838) 7, 1061.
8. Faraday, M., Phil. Trans. Roy. Soc. (1834) 124, 55.
9. Mogerman, W.D., Inco Reporter, (1965) No.5, July.
10. Sabatier, P., Bull. Soc. Chim. France (1939) V6, 1261. This paper provides a fascinating account of how Sabatier came to discover hydrogenation.
11. Lormand, C., Ind. Eng. Chem. (1925) 17, 430.
12. Farkas, A., Z. physik. Chem. (1931), B14, 371.
13. Horiuti, J., and Polanyi, M., Trans. Faraday Soc. (1934) 30, 1164.

RECEIVED November 17, 1982

2

Insiders, Outsiders, and Surfaces: Irving Langmuir's Contribution to Catalysis

GEORGE L. GAINES, JR. and GEORGE WISE
General Electric Research and Development Center, Schenectady, NY 12305

Irving Langmuir's contributions to the advancement of catalysis research were many and varied, ranging from important theoretical concepts to the invention of vacuum pumps and gauges which facilitated experiment. In this paper, we review this work and attempt to place it in the historical context of the science of the time and the laboratory in which he worked, and describe how Langmuir took full advantage of his opportunities to act as both an insider and an outsider in the field of catalysis.

Experiments by Dobereiner, Dulong, and Thenard in 1823, and by Faraday a decade later, first identified the catalytic action of clean platinum on reactions between gases. Following up on Faraday's conjecture that "the effect is produced by most, if not all, solid bodies,"(1) scientists extended those pioneering findings over the next 80 years into the subject of heterogenous catalysis, or catalysis involving more than one phase of matter. But over that period, no one proposed an acceptable mechanism for it. Then, in 1915, chemist Irving Langmuir described heterogenous catalysis as something that occurred in a single layer of gas molecules held on a solid surface by the same forces that held the atoms of the solid together.(2) Other scientists have since shown Langmuir's model to be an oversimplification. But it proved an immensely powerful oversimplification. It revitalized the field, shaped research and theory for a generation, and remains a vital practical tool for describing catalysis today.

This path-breaking idea arose in the mind of a scientist who, in 1915, had published only one previous paper even partly concerned with catalytic reactions; who did not work in a university, and who, indeed, worked for an industrial organization that specialized not in chemistry at all, but in electrical manufacturing. So the episode appears to exemplify the way a scientific field can be revolutionized by an outsider to the field. "The role of Irving Langmuir in the development of the theory of catalysis and its kinetics," wrote chemist S.Z. Roginsky, "is a striking example of a fruitful and deep influence exerted by an 'outsider' on one of the most important branches of chemistry."(3)

0097–6156/83/0222–0013$06.00/0

But this, too, is an oversimplification. Langmuir's position in 1910-1915 gave him a happy combination of the perspectives of an insider and an outsider. He had been educated by the insiders. He'd kept in touch with their literature. And he worked in 1915 at a unique industrial laboratory that encouraged him to use the insiders' methods to attack problems broad enough to open up important areas of chemistry to an outsider's perspective. In the fall of 1914, a key visit by an insider in a different field would provide Langmuir with the finishing touches to his new synthesis.

Kinetics, Checkerboard, and Chemical Forces

The contribution Langmuir made in 1915 can be illustrated by considering one example of heterogenous catalysis; describing what chemists knew about it before 1915; indicating how some of them had interpreted it before 1915; and then summarizing the essence of Langmuir's new interpretation.

Carbon monoxide (CO) and oxygen (O_2) combine to form carbon dioxide (CO_2). Carrying out the reaction near a clean platinum surface greatly speeds up the reaction at ordinary temperatures and pressures. Students of the platinum-catalyzed reaction, notably Bodenstein and Fink in 1907, had shown the rate of the reaction to be directly proportional to the oxygen concentration, and inversely proportional to the carbon monoxide concentration. This contrasts with the uncatalyzed reaction. Its rate is directly proportional to both O_2 and CO concentrations.(4)

Why did adding CO slow up the reaction? Bodenstein and Fink drew on earlier work by Walther Nernst to suggest an answer. They assumed that CO and O_2 would only react rapidly at the platinum surface. The CO, they suggested, formed a film several molecules thick on that surface – the more CO, the thicker the film. And the thicker the film, the longer it took the O_2 to diffuse through and get to the platinum, so the slower the reaction.

Langmuir proposed a sharply different model, based on three key ideas: kinetics, the checkerboard, and chemical forces. Where previous physical chemists had dealt with concentrations of the reacting gases, Langmuir considered individual atoms or molecules and their motions near the surface. When they hit the surface, would they bounce off, or stick to it? Here he proposed viewing the surface as a sort of atomic checkerboard – an orderly array of atoms, presenting an array of squares for a particle from the gas to hit. And he suggested that those surface squares, unlike the ones below them, held unsaturated chemical bonds – short range tentacles of attractive chemical force sticking out into space – that caught and held the atoms or molecules that hit the squares. As a result, the particles that hit all stuck, forming a layer, at most one atom or molecule thick (no kings allowed on the atomic checkerboard!). They might vibrate loose a moment later, but for an instant at least, they stuck.

In the specific case of CO molecules and 2 oxygen atoms (supplied by the O_2 molecules) on a clean platinum checkerboard, both species stuck to the surface. But while the stuck-on oxygen atoms could react with CO molecules that hit them from the gas, the stuck-on CO molecules merely took up space on the checkerboard without reacting. So the higher the pressure of CO, the more reaction sites on the checkerboard occupied by unreacting CO molecules, and the slower the reaction.

Langmuir didn't just suggest this model. He provided a quantitative version of it and supported his view by experimental results. But where did the model come from? Where did Langmuir get the ideas of the kinetic approach, the checkerboard, and those unsaturated, short-range, chemical forces? And how did he put them together?

From Insider to Outsider

Irving Langmuir started his scientific career in 1907 with attributes that should have eventually landed him an insider's spot such as a professorship at a major university. He possessed talent and dedication. He came from a solid middle class background, the son of an insurance executive. He got a first-rate education – private schools in Paris and Philadelphia, undergraduate work at Columbia, and graduate work under Walther Nernst at Gottingen culminating in a Ph.D. degree. He landed, in 1907, in the chemistry department of Stevens Tech at Hoboken, N.J., ready to combine teaching and research and move up in the academic chemistry profession.(5)

Then something went wrong. He found himself overloaded with teaching responsibilities, deprived of time, equipment, or assistance for research, and, finally, at odds about his value to the school with the new chairman of its chemistry department. In 1909, he learned from Columbia classmate Colin G. Fink (co-author of that 1907 paper with Bodenstein on catalysis) about a summer job opportunity with Fink's employer, General Electric. Here was an escape route. "On Monday I go to Schenectady and in all probability will do good enough work so that Whitney (the GE laboratory director) will offer me salary anywhere from $1200 to $1400 for the next year," he wrote his mother.

> If I like the work I shall accept this ... while at Schenectady, I will be looking around for a really good position in a university.(6)

A quartet of GE leaders had founded the General Electric Research Lab to invent better light bulbs. But the director they chose for that laboratory, Willis R. Whitney, believed that encouraging first-rate research offered the best path to that better light bulb. He set up a laboratory with many of the academic amenities, from the latest numbers of the German journals to a weekly colloquium. A Ph.D. in physical chemistry himself, Whitney encouraged Langmuir to take a physical chemist's approach to analyzing light bulb performance. So to determine the influence of the degree of vacuum on the life and efficiency of light bulbs, and particularly on why they turned black in use, Langmuir set out to find out what happened to small quantities of gas deliberately introduced into the bulb. Whitney and others had already showed that the vacuum of a lamp seemed to improve over time. They'd named this the "clean up" effect. But how did it work?

Whitney supplied Langmuir with a skilled toolmaker named Samuel Sweetser as a personal assistant, and let Langmuir take things from there. Langmuir immediately discovered two outsider's advantages the electrical industry provided. Lamp experts taught him how to make high vacua using such techniques as liquid air traps and potassium peroxide "getters." And his GE colleague William D. Coolidge supplied him with the extremely heat resistant filament material ductile tungsten. It could conveniently be heated for many hours to drive off all the gas trapped within it before beginning high vacuum experiments.

Using these tools, Langmuir specified and Sweetser built a system that Langmuir initially described as "marvellous in its complexity."(7) Later, after Sweetser had carried out experiments with it on nearly every working day for five years, Langmuir would view it as "relatively simple." As he described it:

> A bulb containing one or more short filaments, usually of tungsten, was sealed to an apparatus consisting essentially of a mercury Gaede pump, a sensitive McLeod gauge for reading the pressures, and an apparatus for introducing small quantities of various gases into the system and for analyzing the gas residues obtained in the course of the experiments. By means of the pump, the pressure in the system could be lowered to 0.00002 mm of mercury ... By means of the apparatus for analyzing gas, a quantitative analysis of a single cubic mm (at atmospheric pressure) of gas could be carried out, determining the following constituents: hydrogen, oxygen, carbon dioxide, carbon monoxide, nitrogen and the inert gases.(2)

So in moving from the position of a Gottingen insider to an outsider located at an electrical manufacturing plant in Schenectady, New York, Langmuir had secured some outsider's advantages. He worked at a unique industrial laboratory that combined the focus and urgency of industrial problems some of with the atmosphere of academic science – a library, a colloquium, and the encouragement of research. He had tools his academic colleagues lacked. And, inspired by the needs of light bulbs, he was doing physical chemistry with an apparatus that few if any of his academic contemporaries saw fit to adopt.

New Experimental Methods

In order to carry forward this work, Langmuir had himself before 1915, devised new techniques which enhanced measurement capabilities greatly. His scheme for quantitative analysis of gases, as already noted, permitted the determination of some six different gases with a precision of 5% in only one cubic millimeter of a gas mixture. In 1913, he had devised two new vacuum gauges, both more sensitive than any previously in use. And within a year after his 1915 paper, he invented the mercury condensation vacuum pump, a major – and still generally used – tool of vacuum technology.(8, 9, 10) Even had he not been the originator of major conceptual advances, it seems fair to say that, on the strength of his experimental innovations alone, he would rank as a major pioneer in the science of catalysis.

Kinetics

Langmuir hadn't been trained as an atomist, or as a devotee of the kinetic theory of matter. The school of physical chemistry that gave him his training, the "ionist" school, remained indifferent, or in some cases even hostile, to the atomic viewpoint. But Langmuir, interpreting the behavior of his low pressures of gases in the evacuated neighborhood of a hot filament, found himself driven to a view that focused on the motion of individual particles. As he would put it in 1915:

> Very few experimenters, however, have realized that by working with gases at extremely low pressures, the experimental conditions may be enormously simplified and the velocity of the reaction is then much more intimately related to the behavior of the individual molecule than it is at higher pressures. In fact, by working continually with gases at these low pressures, one soon acquires an entirely new view point and sees almost daily fresh evidences of the atomic and molecular structure of matter. The kinetic theory of gases then becomes the great guiding principle. According to this viewpoint, the velocity of a reaction is a matter of statistics. The question becomes: Out of all the gas molecules which strike the surface of a heated filament, what fraction enter into reaction with it?(2)

In the papers Langmuir published from 1911 through 1914, one can see him account for his observations of the "clean up" of gases in a glass bulb by models that relied increasingly on the literal existence of atoms and molecules. The energy lost from the hot filament during the clean up of hydrogen far exceeded the expected value. "This is explained," Langmuir wrote, "by assuming the hydrogen at very high temperatures is dissociated into atoms." As oxygen cleaned up at above 350 degrees C, a straw-colored film formed on the filament. Above 1200 degrees C, the film disappeared. This he explained by a mechanism that he'd later abandon, but that shows how literally he took the kinetic picture:

> The oxygen molecules that strike the filament do not react directly with tungsten atoms, but as a first step become negatively charged by taking up an electron from the metal. The oxygen molecule, after taking up a negative charge, is held by electrostatic forces to the positively charged tungsten atoms and soon, by secondary reactions, combines with the tungsten and with oxygen atoms to form WO_3.(11)

Nitrogen cleaned up far more slowly than either hydrogen or oxygen. Langmuir had a kinetic model for this, too: tungsten atoms evaporated off the hot filament, and reacted with the nitrogen molecules near the filament to form tungsten nitride (WN_2). These studies of nitrogen cleanup led to his invention of the gas-filled incandescent lamp, now used throughout the world.(12)

In the course of developing these kinetic pictures, Langmuir reawakened his interest in catalysis, a subject he'd touched on briefly only once before, in his doctoral work. He began, in 1912, to focus on the question: what fraction of atoms or molecules in a near-vacuum that strike a hot metal surface stick to it? He found this fraction to depend very little on temperature. This led him to consider the Bodenstein and Fink picture of catalysis, and the formation of relatively thick films on the catalyst surface. This idea he rejected. "Undoubtedly this sort of film plays an important part in gas reactions between solids and gases at atmospheric pressure," he wrote, "but it seems extremely improbable that such would be the case at very low pressures and high temperatures."(11)

His observations in light bulbs had never indicated the presence of such thick films. Had thick films of, for example, oxygen, formed on the filament they would have cut down on light emission. In October, 1914, considering in his notebook the cleanup of one more gas, carbon monoxide, by tungsten, Langmuir built on this experience. "Let us assume," he wrote "that the amount of CO required to form a film of WCO on the surface is negligible compared to the amount in the bulb."(14) He went on to assumine that this thin film covered a certain fraction of the surface. A portion of that covered fraction evaporated off the surface in any given amount of time. But, at equilibrium, the same amount of the surface got covered by new CO molecules that hit and stuck. Expressing this mathematically:

$$s = ap\,(1-s)$$

where s is the fraction of the surface covered
p is the pressure
and ap is the ratio of CO molecules leaving the surface to molecules striking the surface
solving for s gives the now-familiar "Langmuir adsorption Isotherm"

$$s = \frac{ap}{1+ap}$$

"The theory seems to agree splendidly with experiments," he concluded. So, by October, 1914, Langmuir had assembled the kinetic theory tools to attack the problem of heterogenous catalysis. But he lacked the checkerboard and the idea of chemical forces. He hadn't specified the structure of the metal surface, or indicated the sites it offered for occupancy of atoms or molecules. And he hadn't explained how these particles stuck to the surface. The best he'd been able to do was his weak suggestion of electrostatic forces, and a vague mention of secondary reactions. "The present experiments," he'd admitted, "do not throw much light on the nature of these secondary reactions."

Checkerboard and Chemical Forces

In October, 1914, as Langmuir's notebook indicates, the problem of surfaces and forces remained unsolved. But by the time Langmuir arose to address the American Chemical Society on March 5, 1915, his uncertainties had been completely cleared up. In a masterful landmark paper entitled "Chemical Reactions at Low Pressures," he surveyed his "clean up" experiments; introduced his theory of adsorption, based on the adsorption isotherm; and described the reaction between carbon monoxide and oxygen in contact with platinum as a catalytic reaction. "The theory here outlined," he concluded, "would seem to be generally applicable to all heterogenous reactions, even at atmospheric pressures."(2)

How had he made this sudden leap to a complete theory? His notebooks enable one to pin down the timing, but make the theory appear even more to have sprung unaided from the brow of Zeus. Through the winter of 1914-1915, the entries deal with wireless telephony, or radio, Langmuir's other main technical effort of the period. Then, suddenly, on February 2, 1915, there's the heading "Theory of the Catalytic Action of Pt on the Reaction 2 CO + O_2 = $2CO_2$." Under it Langmuir mentions some experiments he and Sweetser have done on the reaction. They replicate the findings of Bodenstein and Fink. But Langmuir again rejects the Bodenstein and Fink theory. "I today thought up the following which seems much more probable," he writes:

> The CO molecules striking the clean Pt surface adhere to it (perhaps not all but say a fraction a_1) probably largely because of the unsaturated character of their carbon atoms. This action may be more rapid (that is a_1 may be larger) at low temps than at high (for example, action of CO at low temps on W) and probably results in the formation of a very unstable compound.(14)

He goes on to describe formation of oxygen and CO layers on the platinum, concluding that the "layers are never more than one molecule thick."

The key new idea is "the unsaturated character of the carbon atoms." By April 21, 1915, Langmuir has recognized its importance and had added the checkerboard to it, and is ready to put in his notebook the "Fundamental Laws of Heterogenous Reactions."

> 1. The surface of a metal contains atoms spaced according to a surface lattice (no. per sq. cm.).
> 2. Adsorption films consist of atoms or molecules held to the atoms forming the surface lattice by chemical forces.(15)

And so on through the rest of the by-now familiar picture.

So, sometime between early Fall, 1914, and Spring, 1915, Langmuir had come to assert that the adhesion of molecules to surfaces was governed not by electrostatics, or secondary reactions, but by the same short range chemical bonds that created the crystal lattice of the surface itself.

Langmuir never states where he got the clue for this leap. But comments he makes in a later paper, coupled with some strong circumstantial evidence, allows one to infer a likely source of the idea. That source is the work on crystal structure of William Henry and William Lawrence Bragg.

In 1912, Von Laue, Friedrichs, and Knipping had detected the scattering of X-rays by crystals. Between 1912 and 1914, the Braggs had applied those ideas to describe the crystal structure of solids. They'd made it clear that the atoms of a crystal are arranged in an orderly lattice of dimensions comparable to the wave length of X-rays. If so, the atoms on the surface of that lattice must present unsaturated surface chemical bonds, ready to combine with unsaturated bonds of species that alight on the surface.

In 1916 and 1918, Langmuir referred to his debt to Bragg. "The structure of crystals as found by the Braggs leads to a new and more definite conception as to the nature of chemical forces," he wrote. "The writer has consistently endeavored to apply this new conception in his work on heterogeneous reactions."(16) But Langmuir's landmark 1915 paper, to the American Chemical Society, doesn't refer to Bragg's work. How do we know that the work influenced Langmuir directly in late 1914? Here the evidence is circumstantial. In the guest book of the GE Research Laboratory, under the date November 15, 1914 – a month after Langmuir's first notebook entry on his isotherm, but three months before his theory of catalysis – we see the entry: "W.H. Bragg, University of Leeds, England, X-Ray Spectra."(17) The reminiscences of Langmuir's colleague Albert W. Hull confirm that Bragg presented the staff of the GE Research Laboratory with a colloquium on his findings on X-ray studies of crystal structure.(18) Langmuir's notebook doesn't mention the colloquium. But he almost certainly attended, and took important ideas away from it.

Conclusion: Outsider to Insider

In 1912, Langmuir was an unknown outsider. By 1921, when he gave a famous paper on "Chemical Reactions at Surfaces" to the Faraday Society, he was the leading insider on the subjects of adsorption, surface chemistry, and heterogenous catalysis.

He'd done it by combining with his own talents the advantages to be gained by both insider and outsider. As a student insider he'd been exposed to physical chemistry by one of the creators of the field, Walther Nernst. He'd gained an acquaintance with the literature in the field, including the observations of catalysis at platinum and other metal surfaces, and the attempts of such chemists as Bodenstein and Fink to explain those observations.

As an industrial outsider, he'd been fortunate to go to work at a laboratory that was, in 1910, a unique combination of industrial urgency and professional science. General Electric's interest in better light bulbs provided him with a problem. The freedom of choice of methods established by laboratory director Willis R. Whitney allowed him to attack that problem in his own way. The combination of the problem and mode of attack led Langmuir into the realm of kinetic theory. When his interest turned to catalysis, the experience he'd gained from years of low pressure experiments enabled him to reject almost intuitively the Bodenstein-Fink theory.

Finally, to get the key ideas of the checkerboard and chemical forces, Langmuir appears to have drawn once more on his special advantages as both an insider and an outsider. On a November afternoon in 1914, a physicist from Leeds, England, lectured to the employees of an electrical manufacturing company in Schenectady, New York, about some new findings in the areas of X ray spectra and crystal structure. That combination of science and industry epitomizes the opportunity for a creative scientist to combine the viewpoints of the insider and the outsider. No scientist ever took better advantage of that opporunity than Irving Langmuir.

Literature Cited

1. Faraday, M. "Experimental Researches in Electricity" (London: Dent, J.M., 1951), p. 98.
2. Langmuir, I. J. Am. Chem. Soc. 1915, 37, 1139.
3. Roginsky, S.Z. "Introduction to Vol. I," Collected Works of Irving Langmuir; Suits, C.G. and Way, H.I., Eds. (New York, NY, 1962), I, ixviii.
4. Bodenstein, M. and Fink, C.G., Z. Physik. Chem. 1907, 60. 1.
5. Rosenfeld, A. "The Quintessence of Irving Langmuir," Collected Works of Irving Langmuir; Suits, C.G. and Way, H.I., Eds., (New York, NY, 1962), 12, 36-77.
6. Langmuir, I. to Langmuir, S.C., July 16, 1909, Langmuir Collection, Library of Congress.
7. Langmuir, I. to Langmuir, S.C., September 10, 1909, Langmuir Collection, Library of Congress.
8. Langmuir, I. Phys. Rev. 1913, 1, 337.
9. Langmuir, I. Phys. Rev., 1916, 8, 48.
10. Langmuir, I. Jour. Franklin Inst., 1916, 182, 719.
11. Langmuir, I. J. Am. Chem. Soc. 1913, 35, 105.
12. Langmuir, I. and Orange, J.A. Trans. Amer. Inst. of Electrical Engineers 1913, 37, 1913.
13. Langmuir, I., Laboratory Notebook #355, October 6, 1914, 127.
14. Langmuir, I., Laboratory Notebook #581, February 2, 1915, 259.
15. Langmuir, I., Laboratory Notebook #618, April 21, 1915, 30.
16. Langmuir, I. "The Constitution and Fundamental Properties of Solids and Liquids, J. Am. Chem. Soc. 1916, 38, 2221.

17. Guest Book GE Research Lab, on file at GE R&D Center, Schenectady, NY.
18. Hull, A.W. "Autobiography" (undated, ca. 1950), on file at GE R&D Center, Schenectady, NY, and Center for History of Physics, New York, NY.

RECEIVED November 18, 1982

V. N. Ipatieff: As I Knew Him

HERMAN PINES
Northwestern University, The Ipatieff Catalytic Laboratory, Department of Chemistry, Evanston, IL 60201

In the late 1920s, the industrialized nations were greatly concerned about a shortage of petroleum--the rich Texas oil pools and the vast Arabian deposits were unknown. There was therefore much interest in the possible catalytic conversion of coal to motor fuel, and Universal Oil Products Company, then licensor of thermal-cracking processes to the oil industry, had decided to build a research competence in catalysis. The director of research for UOP, Dr. Gustav Egloff, was commissioned to search everywhere for catalytic experts.

Egloff met Ipatieff in Germany and invited him to visit Chicago. Ipatieff arrived for exploratory talks in 1930, and was asked to organize a research laboratory at Universal Oil Products for studying the catalytic conversions of petroleum hydrocarbons. He accepted the offer and returned to Germany to fulfill an industrial consulting commitment and to await a permanent visa to enter the U.S. It was not easy for a Russian citizen to obtain such a visa at that time, and Ipatieff eventually came to the U.S. through the efforts of Ward V. Evans, Chairman of the Department of Chemistry at Northwestern. Ipatieff's academic visa permitted him and his wife to establish their permanent residence here.

Eight years later, in appreciation of Northwestern's friendship and assistance, Ipatieff donated funds for the establishment of the Ipatieff High-Pressure and Catalytic Laboratory at the university, and eventually willed his estate to it.

In the few years between 1930 when Ipatieff first arrrived in the United States, commencing a new career at a time when most men retire, and his 70th birthday party in 1937, arranged by the Chicago Section of ACS, he had published many research papers, was granted 14 U.S. patents, applied for 50 more, written a book, his chemical autobiography, and discovered processes for producing high octane aviation gasoline.

My association with Ipatieff dates from the day of his arrival in this Country and over the years we became close

0097-6156/83/0222-0023$06.00/0

friends. Ipatieff was one of chemistry's great pioneers. His productive work spanned 60 years, from his first research paper in 1892 until his death in 1952 at the age of 85. This was an extraordinary accomplishment for anyone--an almost unimaginable feat for a man trained in his youth for the Russian army, and for one whose life was directly affected by the First World War, the Russian Revolution, the Stalinist police state, and emigration.

Vladimir Nicolaevitch Ipatieff was born into the minor Russian nobility in Moscow on Nov. 21, 1867. His father was an architect.

At his father's wish, Ipatieff entered a military school, where he began as a mediocre student. When he was about 14, his interest was quickened by science, especially mathematics and chemistry. He continued his education at the Mikhail Artillery School in St. Petersburg, where the main fields were mathematics and ballistics. The chemistry curriculum was so ineffective that Ipatieff began a program of self-study. The few Russian chemistry texts became his teachers; he read and reread them and carried out the experiments they described.

In 1887, when he graduated with highest grades as an officer in the Tsar's army, he received money from the government and from his father to purchase a uniform, a saddle, and other equipment. He used part of the money to put together a small laboratory. And so, with limited facilities, he acquired a fundamental knowledge of inorganic chemistry at his leisure.

After two years as a field officer, Ipatieff entered the Artillery Academy in St. Petersburg. The academy offered advanced technical training to officers, but its chemistry laboratory was ill-equipped. Ipatieff requested permission from the authorities to build a small laboratory in his apartment. Permission was needed because in those days of political unrest, home laboratories sometimes produced bombs. In this laboratory he investigated the steel used in making guns, and in 1892 published the first of several hundred research papers. In the same year, he was appointed instructor in chemistry at the academy and he married Barbara Ermakova.

Ipatieff realized that his knowledge of organic chemistry was inadeqaute, and he approached Professor A. E. Favorsky of St. Petersburg University for advice. As a result of Favorsky's recommendation, Ipatieff began to attend Menshutkin's lectures in organic chemistry and at the same time worked under Favorsky's guidance on his dissertation entitled, "The Isomerization of Alkenes and Acetylenes," for which he was awarded the Butlerov Prize. He was promoted soon to the rank of Assistant Professor at the Academy.

In 1896 Ipatieff spent a year in Germany in the laboratory of the great Adolph von Baeyer. In the laboratory he made friends with Moses Gomberg, the discoverer of stable free radicals (University of Michigan) and Richard Wilstater, later recipient of a Nobel Prize. The three of them met in Chicago in 1933 during the ACS meeting (Figure 1). On returning to St. Petersburg, Ipatieff expanded his research started in Germany and presented an advanced dissertation. In 1898 he became the Artillery Academy's first Professor of Chemistry.

The entrance of Ipatieff into catalytic research occurred in 1900. He noticed that when isoamyl alcohol was passed through a heated iron tube, what emerged was not the alcohol but the aldehyde and hydrogen and that when the alcohol was passed through a heated quartz tube, it emerged unchanged:

$$\text{no reaction} \xleftarrow[\text{tube}]{\text{quartz}} CH_3CH(CH_3)CH_2CH_2OH \xrightarrow[\text{tube}]{\text{iron}} CH_3CH(CH_3)CH_2CH{=}O + H_2$$

Ipatieff realized that he had encountered a new phenomenon, contact catalysis. Thus began 50 years of fruitful work in a new field.

An avalanche of discoveries followed--the dehydration of alcohols over alumina, catalytic isomerization of olefins, conversion of ethanol to butadiene. Another major accomplishment was his introduction in 1905 of an autoclave to contain reactions at high pressures and temperatures. His early training as an artillery officer enabled him to develop a tight seal for the autoclave which could withstand high pressures never before attained, thus opening a new era in the investigation of catalytic reactions.

Using his newly discovered high-pressure technique, Ipatieff studied hydrogenation of organic compounds in the liquid phase. The pressure technique enabled Ipatieff for the first time to achieve complete hydrogenation of benzene to cyclohexane and of phenol to cyclohexanol.

Ipatieff used his pressure technique to the study of polymerization of olefins, hydrogenation of sugars, and displacement of metals (forerunner of hydrometallurgy). All this he accomplished with untrained assistants since there were no students specializing in chemistry at the Academy of Artillery.

He was honored repeatedly for his scientific accomplishments, and in 1916 he became a member of the Russian Academy of Sciences and a commander of the French Legion of Honor.

Military promotions accumulated. By 1910, he was a major general, by 1914, a lieutenant general (Figure 2). Although these ranks entitled him to be addressed as "Your Excellency," he insisted that his laboratory colleagues refer to him by his first name and patronymic. (Later, in the U.S., he preferred to be called "Professor").

Figure 1.
V. N. Ipatieff (left), R. Willstatter (center) and M. Gomberg at an ACS meeting in Chicago, IL, 1933.

Figure 2.
Ipatieff as Lt. General of the Russian Army, 1916.

World War I and Revolution

On the outbreak of the First World War, Tsar Nicholas II placed Ipatieff in charge of the nationwide chemical industry to serve Russia's wartime needs and to prepare for the eventual return to peace. In effect, Ipatieff presided over the birth of the modern chemical industry in his country.

Early in 1917, the Russian government ceased functioning. One evening in February, crossing the frozen Neva River on his way home from his army office, Ipatieff encountered a mob of rioters. He later described the event: "I owe my life to the lucky chance of meeting a soldier whom I knew .... It was outright suicide for a general to appear in uniform before the angry mob, and at least 10 generals and many officers were killed that day. Without Romashev's help, I never would have reached home."

The first years of the Revolution were difficult. Lack of facilities, food, and fuel made for inactivity. Aware of Ipatieff's wartime contribution to the development of the Russian chemical industry, Lenin made him a member of the Presidium of the Supreme Council of the National Economy and chairman of the chemical administration. Ipatieff thus became the only nonmember of the Communist Party in the Soviet higher government.

Although his governmental duties occupied much of Ipatieff's time, he still was able to continue his scientific work and to study the destructive hydrogenation of hydrocarbons.

On Lenin's death in 1924, Stalin came to power. Ipatieff was removed from high office, but he organized another laboratory. He devoted his time to research and during 1927 and 1928 Ipatieff published many more research papers, and the honors kept coming. Then, a close associate and friend, E. I. Spitalsky, was arrested by the dreaded state police in 1929, and five of his former students and co-workers had been shot without a trial. Arrests of many other associates followed. He was secretly warned that he was to be arrested. In June of that year he and his wife left Russia for a scientific meeting in Germany. At the Polish border he turned to her and said, "Take a good look at your country, Barbara, as we are leaving it for good."

Emigration

Once settled, Ipatieff divided his activities between Universal Oil Products and Northwestern's Chemistry Department, where he spent two days a week training graduate students, directing research, and carrying on experiments with his own hands (Figure 3). These first months in the new land were grueling. He was in his sixties. He did not know English. His assistants did not know Russian, German, or French. Through sheer perseverance and long hours, his spoken English improved steadily, though he never became fluent. When he was to speak

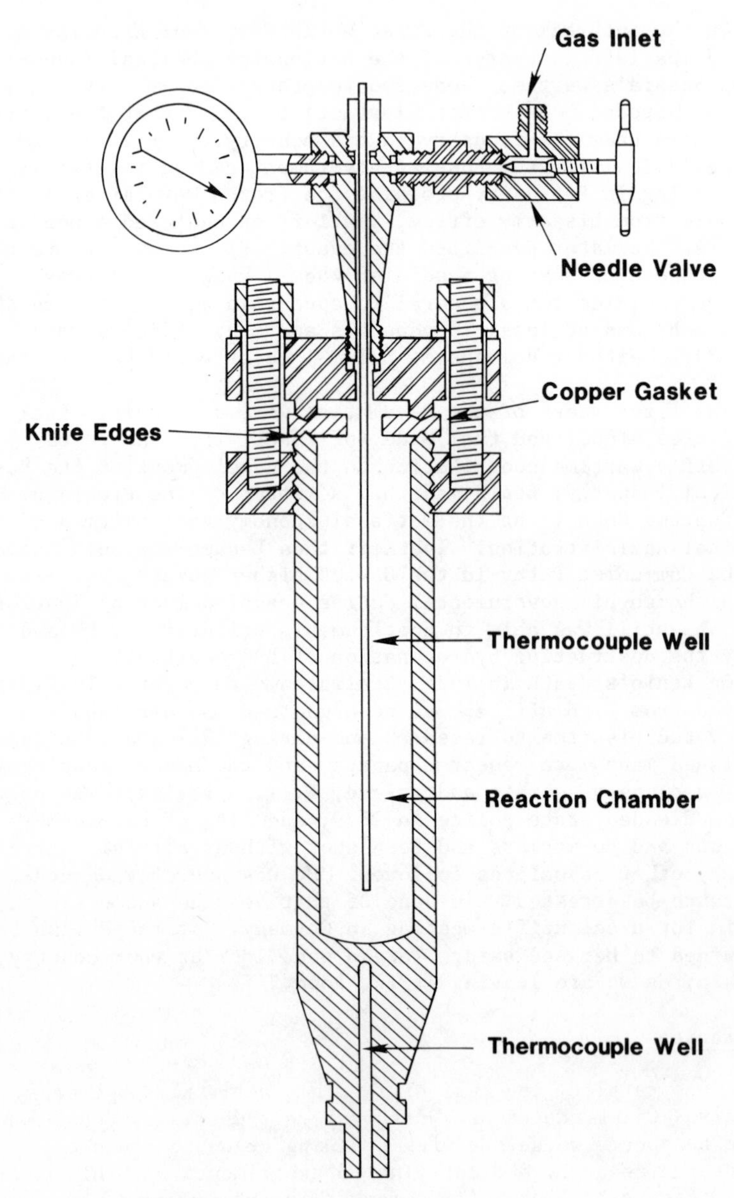

Figure 3.
Ipatieff Type Rotating Autoclave.

publicly, he wrote out his lectures in Russian, had them translated into English, and spent hours reading them to his tutor.

At Universal Oil Products, his research group comprised 10 people. In a short time, his inspiring leadership, practical experience, and canny foresight in assigning problems yielded many discoveries.

His first major discovery in the U.S. was the development of silicophosphoric acid (solid phosphoric acid) as a catalyst for hydrocarbon conversion.

$$H_3PO_4 + \text{Kieselguhr} \xrightarrow{\Delta} \text{"Solid Phosphoric Acid"}$$

SPA

Propylene was polymerized by means of this catalyst to trimers to be blended with gasoline, and to tetramers to be used in the manufacture of detergents.

$$\text{C-C=C} \xrightarrow{\text{SPA}} (C_3H_6)_4$$

Silicophosphoric acid was also used in copolymerization of butenes, which, followed by hydrogenation, yielded high-octane components for gasoline.

$$\text{C-C(C)=C} + \text{C-C=C-C} \xrightarrow{\text{SPA}} \xrightarrow[\text{Ni}]{H_2} \text{C-C(C)(C)-C(C)-C-C}$$

Polymerization reaction was the first commercial catalytic process introduced to the petroleum industry. Ipatieff can thus be considered the father of petrochemistry.

In the Second World War, this catalyst was also used in the manufacture of cumene, a crucial component of high-grade fuel for bombers.

$$C_6H_6 + \text{C=C-C} \xrightarrow{\text{SPA}} C_6H_5\text{-CH}(CH_3)_2$$

The discovery of conjunct polymerization, whereby hydrogen transfer occurs between molecules of olefins, leading to the formation of saturated hydrocarbons and polyenes, became the forerunner of the discovery of alkylation and isomerization of paraffins. During World War II these processes were used for the manufacture of high octane aviation gasoline to be used in pursuit planes, and presently products from the alkylation reaction assumed new importance as a major component of unleaded gasoline (Figure 4).

Figure 4.
V. N. Ipatieff and Herman Pines, 1948.

In the short period between 1935 and 1939 Ipatieff co-authored over 50 research papers, and applied for dozens of patents. He was elected to the National Academy of Sciences and received the coveted Willard Gibbs medal, and his achievements did not go unnoticed in his native land. In 1937, when he refused invitations of Stalin to return to Russia, he was expelled from the Russian Academy of Sciences and deprived of his Soviet citizenship. His son, a professor of chemistry at the Moscow University, was made to denounce his father in a public meeting, and Ipatieff's name was no longer cited in Russian chemical literature.

At Northwestern, Ipatieff was an inspiring teacher (Figure 5). His first graduate student, Vladimir Haensel, invented the UOP platforming process.

Figure 5.
V. N. Ipatieff (1951). Taken by Chicago Sun-Times Photographer and Entitled "The Thinker".

Ipatieff was a man of charm and kindness, of modest desires, and simple habits. He organized his work carefully and was annoyed by interferences in his schedule--and was most upset by his rare illnesses. He hated to waste time, and prepared his lectures and speeches weeks in advance. He was gifted with a fine sense of humor. When his old Swedish friend, Nobel, sent a congratulatory telegram for his 70th birthday, he remarked, "From Nobel I get praises but not prizes." He enjoyed social occasions. He was a generous man, setting up a loan fund to help impoverished graduate students. A number of his countrymen, facing the hardships of immigration in old age, likely owed their very survival to him. These were matters, though, that Ipatieff preferred not to discuss.

Cut off from his homeland and his children (with whom he was forbidden to communicate), he suffered greatly. He followed the events of the Second World War, especially Germany's invasion of Russia, with profound concern. Two of his grandsons perished in that war.

On Nov. 29, 1952, while awaiting my arrival to discuss some laboratory problems, he died suddenly. Ipatieff left a heritage of not only many innovations in petroleum refining and petrochemistry, but also a heritage of many students and associates who regarded working with Ipatieff as a rare privilege. His wife followed him 10 days later.

Fifteen years thereafter, the Ipatieff High-Pressure and Catalytic Laboratory of Northwestern University played host to a Centennial Celebration of Vladimir Nicolaevith Ipatieff. Catalytic experts from all over the world gathered for a week-long symposium in tribute to the father of modern petrochemistry.

RECEIVED November 29, 1982

4

Sir Hugh Taylor: The Man

JOHN E. BENSON [1]
Dickinson College, Carlisle, PA 17103

M. BOUDART [2]
Stanford University, Department of Chemical Engineering, Stanford, CA 94305

For this paper we will present Hugh Taylor as we knew him, as a teacher, counsellor, and friend for almost 30 years. We will try to show his kindness and concern for his students, his personality, his love of science and teaching, and his love of life.

In this introduction we would like to summarize the life of Hugh Taylor from his birth in 1890 in England until his death in Princeton in 1974. A lot of the later anecdotes and details on his accomplishments are drawn from our own contacts with him in the late 1940's until his death. The definitive assessment of Hugh Taylor's scientific career can be found in an excellent memoir by Professor C. Kemball, F.R.S., (1), who also collected biographical information which we have borrowed in this paper. We are also indebted to Miss Lucy D'Arcy, Sir Hugh's devoted and efficient secretary of many years (2).

Hugh Taylor was born in St. Helens, Lancashire, to a happy and devout Roman Catholic family of four brothers and three sisters. After his D.Sc. degree under Professors F.G. Donnan and Henry Bassett, Jr. (nine published papers!), he chose a "wanderjahr" in Stockholm with Arrhenius and later in Hanover with Bodenstein.

In 1914 he was appointed instructor at Princeton for what he called "a brief stay". He never left the United States, except during World War II, when he returned to England to help in the war effort. Intensely British, he could not bring himself to give up his citizenship of birth, so he remained a loyal subject of England, and, at the same time, a devoted but adopted resident of America.

An introduction to Hugh Taylor's long career may be briefly summarized by four tables. The first shows the extraordinary stability of his professional and private life, the second

[1] Retired
[2] To whom queries should be addressed

0097-6156/83/0222-0033$06.00/0

Figure 1. Sir Hugh Taylor, 1890 - 1974 (photograph by Elizabeth Menzies, Princeton, NJ).

the influence of his teaching in a selected list of his students who made their own careers in chemistry and chemical engineering, the third his interest in the application of his science through patents, and the last, the quality and quantity of his scientific writing through books, handbooks, and translations.

TABLE I
HUGH S. TAYLOR

Dates		Years
'90-'74	British Subject, Roman Catholic	84
'14-'74	Resident, Princeton, New Jersey	60
'14-'58	Faculty, Princeton University	44
'19-'58	Married to Elizabeth Sawyer	39
'26-'51	Chairman, Chemistry Department	25
'45-'58	Dean, Graduate School	13
'55-'69	Editor-in-Chief, American Scientist	14
'58-'69	President, Woodrow Wilson National Fellowship Foundation	11

TABLE II
HUGH S. TAYLOR

A selected list of his graduate students, postdoctoral students, and associates in the Department of Chemistry at Princeton University

Akira Amano
Ralph A. Beebe
Hans A. Benesi
John E. Benson
Charles E. Birchenall
Michel Boudart
Robert L. Burwell, Jr.
Alessandro Cimino
Henry Eyring
Everett Gorin
George G. Joris
Joseph C. Jungers
Charles Kemball
George B. Kistiakowsky
Takao Kwan
Walter J. Moore
Kiyoshi Morikawa
Giuseppe Parravano
Thor H. Rhodin
Charles Rosenblum
Hussein Sadek
Felix Sebba
Pierce W. Selwood
Robert Spence
Frank S. Stone
Kenzi Tamaru
Ellison H. Taylor
Charles S. Tuesday
John Turkevich

Figure 2. Some of the people mentioned in Table II are shown in this photograph taken at Northwestern University on the occasion of the Ipatieff Centennial in 1967. From left to right: Robert L. Burwell, Jr., Alessandro Cimino, D. J. Salley, Frank S. Stone, Charles Kemball, Giuseppe Parravano, Sir Hugh Taylor, John Turkevich, Sir Eric Rideal, Michel Boudart, Kiyoshi Morikawa, John E. Benson, (unknown), Kenzi Tamaru.

TABLE III
HUGH TAYLOR'S PATENTS

Title	Assigned to	U.S. Patent No.
Deuterium oxide	USA, Atomic Energy Commission	2,690,380
Isotope-exchange process for concentrating deuterium	USA, Atomic Energy	2,690,381
Conversion of olefins to dienes	M.W. Kellogg Co.	2,401,802
Aromatics from aliphatic hydrocarbons	Process Management Co.	2,357,271
Model atomic units for building three-dimensional models of molecules	Research Corporation	2,308,402
Aromatizing conversions of aliphatic hydrocarbons, as in the production of toluene from heptane or in treating paraffinic gasoline to improve its anti-knock qualities	Process Management Co.	2,336,900
Cyclizing aliphatic hydrocarbons, as in producing aromatic compounds from heptane	Process Management Co.	2,227,606
Catalytic alkylation of aromatic hydrocarbons	M.W. Kellogg Co.	2,168,590
Nickel catalyst substantially free from sulfur	E.I. duPont de Nemours & Co.	1,998,470
Polymerizing ethylene, vinyl compounds, or other substances	E.I. duPont de Nemours & Co.	1,746,168
Hydrogen	--	1,411,760

TABLE IV
BOOKS BY HUGH TAYLOR

Fuel Production and Utilisation

London (Balliere, Tindall and Cox), 1919
New York (D. Van Nostrand Co.), 1920

Catalysis in Theory and Practice (with E.K. Rideal)

London (Macmillan), 1919
New York and London (Macmillan), 1926 - Second Edition

Industrial Hydrogen

New York (Chemical Catalogue Co.), 1921

Elementary Physical Chemistry

New York (D. Van Nostrand Co.), 1927

Elementary Physical Chemistry (with H.A. Taylor)

New York (D. Van Nostrand Co.), 1937 - Second Edition
New York (D. Van Nostrand Co.), 1942 - Third Edition

Catalysis (with R. Spence)

Translation from the German of G.-M. Schwab's book
New York (D. Van Nostrand Co.), 1937

There are other salient facts that do not appear in these lists. Thus, we have not included Hugh Taylor's knighthoods by Queen Elizabeth II and Pope Pius XII and his presidency of the Faraday Society. Nor have we mentioned his great contribution to the Manhattan Project. But it is not our intention within the scope of this paper to list every scientific accomplishment or even every facet of Hugh Taylor's complex personality. After a brief review of his early research, we will try to relay to the reader more personal reminiscences of Sir Hugh.

Early Research

Hugh Taylor's research was eclectic, covering electrochemistry, chain reactions, catalysis, adsorption, photochemistry, tracer work, atomic energy, the properties of polymers, even the dyeing of ivory. Perhaps the two most important concepts in his scientific life were published in his paper in 1925 on active centers (3), and his later paper on activated adsorption (4). Of considerable influence also was his book with E. K. Rideal (5), "Catalysis in Theory and Practice", the first comprehensive book on catalysis, published in 1919 (see Figure 3 and Table IV).

CATALYSIS
IN THEORY AND PRACTICE

BY

ERIC K. RIDEAL
M.B.E., M.A. (CANTAB.), PH.D. (BONN)
AUTHOR OF "INDUSTRIAL ELECTROMETALLURGY"

AND

HUGH S. TAYLOR
D.SC. (LIVERPOOL), ASSISTANT PROFESSOR OF PHYSICAL CHEMISTRY, PRINCETON UNIV., U.S.A.

Figure 3. Frontispiece of "Catalysis in Theory and Practice", by Hugh S. Taylor and Eric K. Rideal, published in 1919.

Personality

Hugh Taylor was a complex man. He has been called (by some who did not know him well) ruthless, but this is not entirely correct. He was an active Roman Catholic with a deep, enduring faith and steadfastness. Even in the dark days of his persecution by a priest at Princeton, he held the conviction that the Princeton chapel should be consecrated so it could offer Mass. And ultimately this goal was reached.

In the scientific side of his life, he was a man of similar conviction and competition. Many years after his 1925 paper on active centers, he was heard to say "In 1925, by a deductive leap in the dark, I suggested that catalyst surfaces could not be homogeneous". After his paper on activated adsorption, he wrote: "This extraordinarily simple generalization seems to have aroused a quite unreasonable amount of opposition".

What may have given people the idea that Hugh Taylor was ruthless was that he knew, better than most people, how to use his time. He never walked: he ran! When he rounded a corner, he was moving at a 45° angle. He did the work of three men, and always allowed exactly the right amount of time for each job to be done. In his work, he was hard-driving, insensitive, and blunt. But he was also, and at the same time, sensitive and generous in his dealings with people. When Lady Taylor died, he gave generously to a fund in her name to help needy wives of graduate students. He helped younger associates in many ways, even financially, but in such a discreet manner that this aspect of Hugh Taylor's personality was hardly known, except by the inner circle of his friends.

The More Personal Aspects

From what we have said so far, it should be clear that Hugh Taylor was indeed a very complex man. In this section we explore more of his complex qualities and personality with memories of events and circumstances during the years we knew him.

While his list of patents attests to his being quite conscious of the material value of his ideas, he never kept a penny of royalties from his patents, so far as we can determine. In fact, while gathering information for this paper, we learned that both his daughter in Princeton, who has his papers, and his secretary, Miss D'Arcy, thought he had no patents. Only by a computer search did we learn of them. He was ever careful not to compromise scholarly work with the temptations of money, and was very careful not to indulge in process research at the university.

While working on the Manhattan Project, he traveled each day into New York City, at the same time carrying a full-time teaching schedule at Princeton -- a crushing workload. He used to tell a story about those years with a bit of wry humor. After years of hard work, he had a store of light water enriched, as we remember, with about 25% of heavy water which he "loaned" to the government. At the end of the war, he was paid back in kind: 100% heavy water diluted back to 25% heavy water.

In the classroom he was superb. He had a phenomenal memory of facts, figures, colors, people, and events. Hence, his lectures combined carefully developed concepts with examples drawn from his own work and that of others. Lectures were on three consecutive days. Assignments were rarely made, although the students could guess from what had been presented previously what was coming next. On the first day of each week Hugh Taylor would lay the background, on the second he would develop the ideas, and on the third day he would tie everything together. All was done in elegant chalk writing, starting on the upper left board and ending at the bottom right board. It was a careful, but seemingly effortless, orchestration of science.

In fact, he could be austere. He couldn't see the point of first-class flights; he used to say he was not about to pay $75 for a Martini on flights to San Francisco. He drove the same car (at high speeds, by the way) for years, and he always washed his car in front of Wyman House, the official residence of Princeton's Dean of the Graduate School. He ate simply, and during his years at Wyman House, he took great pride and pleasure in cultivating a vegetable garden on the grounds of the Graduate College.

Hugh Taylor had a high sense of consideration toward other people. On a visit to Dickinson College to receive the last of his 28 honorary degrees (see Figure 4), he offered to walk the mile to church for Mass, which he attended every day if possible, so as not to inconvenience his host at that early hour. When he visited overnight, he was always very careful to point out that he would do as his host or hostess wished, and certainly did not want to be waited on or deferred to.

One of Hugh Taylor's greatest gifts was his executive organization: the ability to allot to each job the requisite amount of time, and the capacity to reach quickly a carefully considered opinion. As he freely admitted, some of his decisions were wrong. One of his rare failures was an attempt, as Dean of the Graduate School, to push through the idea of a three-year Ph.D. program. This concept was probably an extension of his own hard-driving personality. Many faculty members opposed the idea, however, as being too hurried, particularly in some areas of the humanities, which Hugh Taylor treated, to the surprise of many, as equal to the sciences. As Dean, he played no favorites among the disciplines, despite his own scientific orientation. He simply assumed that everyone, whatever the school or department, would be as hard-working as he. His relations with Princeton's President Dodds were

Figure 4. Dickinson College, June 1966: Presentation of Henry Eyring by Sir Hugh Taylor for the Priestley Award.

excellent; as he once remarked to one of us, "We often disagreed, but could quickly agree on a compromise". The key word was quickly.

This trait of being in constant motion made him seem ruthless and insensitive. In fact, during what he called "the golden years" (1929-1939), when he was at the peak of his scientific output, he was not much liked by his graduate students. The following two stories are indicative of Hugh Taylor's interaction with the graduate students.

As Dean of the Graduate School, Hugh Taylor had a chance to show another side of his complex nature, his love of tradition. Wyman House connected to the glorious dining hall of the residential Graduate College, Procter Hall, through a private door camouflaged by a series of panels. Each Wednesday he would come to dinner to say grace (in Latin, of course) and to chat with the students.

To Hugh Taylor's right was a heavy pedestal on which was a bronze bust of an early Master of the Graduate College (Augustus Trowbridge) and a bronze plaque telling of the Master's life and early death. Before dinner, the students would turn the bust on its swivel base, so the bust and plaque faced each other. Partly through dinner, Sir Hugh would realize something was amiss, but not knowing that the bust was on a swivel, he would lift the entire pedestal and turn it back to its proper orientation.

Some time later, the students in residence at the Graduate College sent a delegation to Hugh Taylor to complain about the obsolete obligation to attend dinner in Procter Hall wearing an academic gown. The delegation explained that a student vote had expressed the will of the majority in this matter. The story is that Dean Taylor reacted quite negatively and abruptly, stating that there were a few issues in life that were not decided by a majority vote, and that wearing gowns for dinner was one of those issues.

In his summer visits to Stanford in the 1970's, Hugh Taylor revealed another side of his character. He would arrive tired and very discouraged about his age and the state of his health. We would set up a carefully spaced series of conferences with the graduate and postdoctoral students, and suddenly Hugh Taylor would begin to sparkle and become again the man we remembered from the 1950's. He would listen intently to the brief summaries of the research, and then launch into a string of insights, suggestions, and references to past research. These conferences left the students exhilarated and Hugh Taylor smiling and content. It became clear to us that, after his retirement from the Woodrow Wilson National Fellowship Foundation, he missed his former outlets for his enormous energy, in particular, contact with students.

Over the years Hugh Taylor developed many friendships. In closing, we shall summarize one of the most enduring and perhaps the most unusual friendships of his long life. In 1931, Hugh Taylor brought Henry Eyring, a Mormon, to Princeton. Henry Eyring stayed until 1946, when he returned to Utah, declining to take over Hugh Taylor's endowed chair, which Taylor offered to him in a desperate effort to keep him in Princeton. During his Princeton years, Henry Eyring developed the absolute rate theory, the long-sought-for link between kinetics and thermodynamics. Naturally, Hugh Taylor had a lively interest in this, and he and Henry Eyring remained close professionally. At the same time, their opposing views on religion led to many spirited, but amicable, discussions on that subject. Their mutual admiration lasted to the end of their lives.

In 1974 Henry Eyring was awarded Dickinson College's Priestley Award. When asked who he wished to present him, Eyring suggested Hugh Taylor, a previous recipient of the award. Although Hugh Taylor was by then blind in one eye, deaf in one ear and very frail, he agreed to do the presentation and made the trip to and from Carlisle unattended and unaided. He died three weeks later at the age of 84.

AVE ATQUE VALE

Literature Cited

1. Kemball, C., Biographical Memoirs of Fellows of the Royal Society, 1975, 21, 517.
2. Private communications.
3. Taylor, H.S., Proc. R. Soc., London, 1925, A108, 105.
4. Taylor, H.S., J. Amer. Chem. Soc. 1931, 53, 518.
5. Taylor, H.S. and Rideal, E.K., "Catalysis in Theory and Practice", Macmillan, New York and London, 1919.

RECEIVED January 28, 1983

5

Paul H. Emmett: Six Decades of Contributions to Catalysis

ROBERT L. GARTEN
Catalytica Associates, Inc., Santa Clara, CA 95051

The distinguished career of Professor Paul H. Emmett has spanned six decades, beginning with his Ph.D. research under A.F. Benton at the California Institute of Technology in 1922. His pioneering contributions to the field of catalysis have provided the foundation for much of the present-day work in the field. Among his most notable contributions is the BET method for determining the surface area of solids, done in collaboration with Stephen Brunauer and Edward Teller. Surface area measurement by the BET method is probably the most widely used characterization method in catalysis today.

In addition to the BET equation, Paul Emmett made enduring contributions to the experimental determination of gas-solid equilibria and the understanding of ammonia synthesis over iron-based catalysts. He also pioneered the development of selective chemisorption methods to estimate the surface composition of multicomponent catalysts and the use of tracer methods to explore the mechanism of Fischer-Tropsch synthesis and catalytic cracking.

This paper presents a perspective on Paul Emmett--the man and his contributions over six decades of research in catalysis.

In his autobiographical remarks at the 1974 conference in his honor in Gstaad, Switzerland(1), Professor Paul Emmett stated: "any accounting of the career of a scientist should relate circumstances that led to his choosing science as a career, the people and events that influenced his work and then, a chronology of his working career and some recounting of a few of the things that may have been accomplished." In recounting

0097–6156/83/0222–0045$06.00/0

the remarkable career of Paul Emmett, I will accept this advice, with some extension to include a perspective for the man, his philosophy of research, and the impact he has had on catalytic science.

Paul Emmett's first serious encounter with science did not occur until his senior year at Washington High School in Portland, Oregon, where he was born in 1900. In the middle of his senior year a faculty advisor, a Miss Workman, suggested that, based on his grades in mathematics, he sign up for physics and chemistry courses for the second semester. The inspirational teaching of Dr. William (Willie) Green in chemistry during that time was to have a profound influence on Paul Emmett, initiating a brilliant and outstanding career in science that was based on pioneering contributions to the field of catalysis. Attending Washington High School a year ahead of Paul Emmett was another student who would go on to make outstanding achievements in the field of chemistry, including winning the Nobel prize. This student was Linus Pauling. The bond of friendship and mutual respect between Paul Emmett and Linus Pauling would continue throughout their illustrious careers.

Following high school, both Paul Emmett and Linus Pauling stayed close to home, attending Oregon Agricultural College, which later became Oregon State University. Paul chose the chemical engineering curriculum because chemistry was offered only in the agricultural department at that time. During his freshman year, Paul was significantly influenced by another inspiring teacher, Professor John F.G. Hicks, who taught freshman chemistry with a strong emphasis on qualitative analysis. One of Professors Hicks' favorite expressions was "In the book it is so--but why?" Professor Hicks emphasized a critical approach to learning. This influence, coupled with three years of varsity debating, provided the foundation for Paul Emmett's approach to research in catalysis.

Following graduation from college, Paul chose to attend a new school headed by a group of outstanding scientists including Professors A.A. Noyes in chemistry, R.A. Millikan in physics, and R.C. Tollman in physical chemistry. The school was the California Institute of Technology, which at the time was only two years old. Influential in Paul's choice of Cal Tech was the decision of his friend, Linus Pauling, to attend the same school. One of the conditions for acceptance at Cal Tech was that they work out all the problems in the first nine chapters of the book <u>Chemical Principles</u> by Noyes and Sherill. They spent the entire summer prior to beginning studies at Cal Tech completing these problems, which in effect was a course in advanced physical chemistry.

At Cal Tech, Paul Emmett began research with Dr. Arthur F. Benton, who had just recently completed his Ph.D. with Professor Hugh S. Taylor at Princeton University. Dr. Benton had a two-year appointment at Cal Tech on a National Research Fellowship. He was intensely interested in adsorption and catalysis and was very skilled in laboratory techniques.

Through Dr. Benton, Paul became knowledgeable in experimental design and measurement techniques to obtain accurate and reliable experimental results. The techniques acquired from Dr. Benton would have a significant influence on Paul's later work which contemporaries would admire and respect as meticulous experimentation of the highest standards. Paul's thesis work with Benton concerned investigations of autocatalysis in the reduction of metal oxides with hydrogen(2,3).

During his graduate work at Cal Tech, Paul made an interesting discovery about his own physiology. He became somewhat dismayed to find that during the afternoon seminars, which featured such speakers and Einstein and Ehrenfest, he frequently was very sleepy and had difficulty staying awake. A Dr. Geiling, who at the time was carrying out X-ray diffraction studies on insulin at Cal Tech, advised him to try taking a short nap in the middle of the day as a possible remedy. Paul soon discovered that this was an effective solution. To this day he is able to set his internal clock for a ten-minute nap at almost any location, and awaken refreshed for an afternoon of work.

After two years, Dr. Benton left Cal Tech and Paul looked for an opportunity to broaden his background during the final year of his graduate studies. He decided to work with his fellow student, Linus Pauling, on the crystal structure of barium sulfate and published a paper on this work(4).

Figure 1 is a photograph of the faculty and graduate students at Cal Tech in 1924--an outstanding assembly of past and present talent in the field of chemistry.

Following the completion of his Ph.D. work at Cal Tech, Paul Emmett spent a year teaching at his alma mater, Oregon Agricultural College, which had then become Oregon State University. He then joined the Fixed Nitrogen Laboratory in Washington, D.C. for what was to become a very productive and significant eleven-year period of research. He was a skilled experimentalist entering the field of catalysis at a time when most of the fundamental work was still to be done.

The major achievements accomplished during this period, from 1926 to 1937, are listed below:

- BET method for surface area determination
- Thermal diffusion effects in equilibrium measurements
- Selective chemisorption methods applied to Iron ammonia catalysts
- Kinetics and mechanism of ammonia synthesis.

Figure 1. Faculty and Graduate Students at the California Institute of Technology in 1924. Paul Emmett, front row, fourth from the right; Linus Pauling, front row, first from the left.

Since the work during this period is described in more detail in Professor Emmett's paper in this volume, I will only briefly highlight some of his accomplishments at the Fixed Nitrogen Laboratory.

In attempting to understand why some ammonia synthesis catalysts would be good catalysts and others not so good, Paul Emmett reasoned that information on the surface area of the catalytic materials would be necessary. The idea for using low-temperature gas adsorption to measure surface area actually originated with Dr. Benton. He had obtained some nitrogen adsorption isotherms at liquid-nitrogen temperature, and the resulting curves contained kinks at 13.5 and 45 cm pressure, which he suggested might be associated with the completion of a monolayer and a second layer of adsorbed nitrogen. Following the lead provided by Dr. Benton, Paul Emmett and his colleagues, particularly Dr. Stephen Brunauer, began an extensive series of investigations of a number of adsorbates on a variety of materials to develop a surface area method. They soon found that the kinks observed by Benton were removed when proper consideration was given to deviations from ideal gas behavior. The low-temperature adsorption isotherms for a variety of gases and materials consisted of smooth curves which bend over into a straight line between about 0.1 and 0.5 relative pressure. For convenience of identification, the various portions of the isotherms are labeled ABCDE. The knee of the curve, labeled point B, appeared to be the best choice for the completion of the first monolayer coverage needed to calculate the surface area of the material. The consistency of the point B method in surface area determination for a variety of gases and solids lent credence to the method(5).

Stephen Brunauer became interested in the possibility of developing a theory for the low-temperature adsorption curves. Brunauer suggested that he talk with a physicist at nearby George Washington University. The physicist was Dr. Edward Teller who had emigrated from Hungary in 1933. Brunauer and Teller's common Hungarian ancestry may have been important in initiating and facilitating this interaction. In a short time, the familiar BET equation:

$$\frac{P/Po}{V(1 - P/Po)} = \frac{1}{VmC} + \frac{(C - 1)P/Po}{VmC}$$

was developed, which adequately describes the adsorption isotherms and became the basis for the now-accepted classic method of measuring surface areas(6).

A contribution which Professor Emmett identifies as being one of the most satisfying of his career resulted from work at the Fixed Nitrogen Laboratory. It concerned thermal-diffusion effects in equilibrium measurements. The problem that confronted Paul Emmett in 1930 is shown in Figure 2. Indirect measurements of the water-gas shift equilibrium by combining results of equilibrium measurements from the Fe-H_2-H_2O system with those from the Fe-CO-CO_2 system were about 40% smaller than values obtained from the direct measurement of the water-gas shift equilibrium constant. Paul Emmett identified the source of this discrepancy as an error caused by the influence of thermal diffusion in equilibrium measurements in the Fe-H_2-H_2O system. In general, previous work on this equilibrium had been carried out in a static system following the design of Deville(7), as shown in Figure 3. The partial pressures of water and hydrogen over a solid were determined by establishing the water vapor pressure via a thermostatted reservoir and then measuring the total pressure of the system at equilibrium. In an elegantly designed and executed series of experiments reported in 1933(8), Paul Emmett and his assistant, J.F. Shultz, showed that in the static system, the presence of large temperature gradients induces thermal diffusion. This thermal diffusion tends to concentrate the heavy molecules, namely water, in the cold zones of the apparatus and the lighter molecules, namely hydrogen, in the hot zones of the apparatus. The concentration differences in the various zones of the tube were shown to account for the 40% discrepancy in the measurement of the equilibrium constant for the Fe-H_2-H_2O system. This work is an outstanding example of Paul Emmett's logical, thorough, and critical approach to a problem, combined with meticulous design of experiments to obtain accurate results.

A third important contribution of Emmett's during the period at the Fixed-Nitrogen Laboratory was the development of selective chemisorption methods(9,10). During the extensive work on low-temperature physical adsorption methods for the determination of surface areas, one adsorbate, carbon monoxide, gave unusual results with pure iron catalysts, in that calculated surface areas were generally about twice as great as those obtained for nitrogen adsorption. Further experimentation showed that the excess carbon monoxide was chemically adsorbed on the iron surface, and hence a method evolved for assessing the amount of exposed iron surface relative to the total surface area of the sample. Further work with CO_2 established that this adsorbate could be used to estimate the surface area of promoters such as potassium oxide and aluminum oxide, which are common in the iron ammonia-synthesis catalyst. A combination of these methods on a given iron catalyst allowed, for the first time, the relative fractions of the surface covered by promoter and by metallic iron to be estimated. These same techniques are still widely used in modern catalysis research. The use of CO chemisorption has been extended to the measurement of metal surface areas in many other systems, including the majority of the Group VIII metals.

Finally, this productive period at the Fixed Nitrogen Research Laboratory resulted in many important contributions to

INDIRECT MEASUREMENT

$$\begin{array}{l} Fe + H_2O = FeO + H_2 \\ FeO + CO = Fe + CO_2 \\ \hline H_2O + CO = H_2 + CO_2 \end{array}$$

DIRECT MEASUREMENT

$$H_2O + CO = H_2O + CO_2$$

Discrepancy (1930) = 40%

Figure 2. Thermal Diffusion Effects in Equilibrium Measurements.

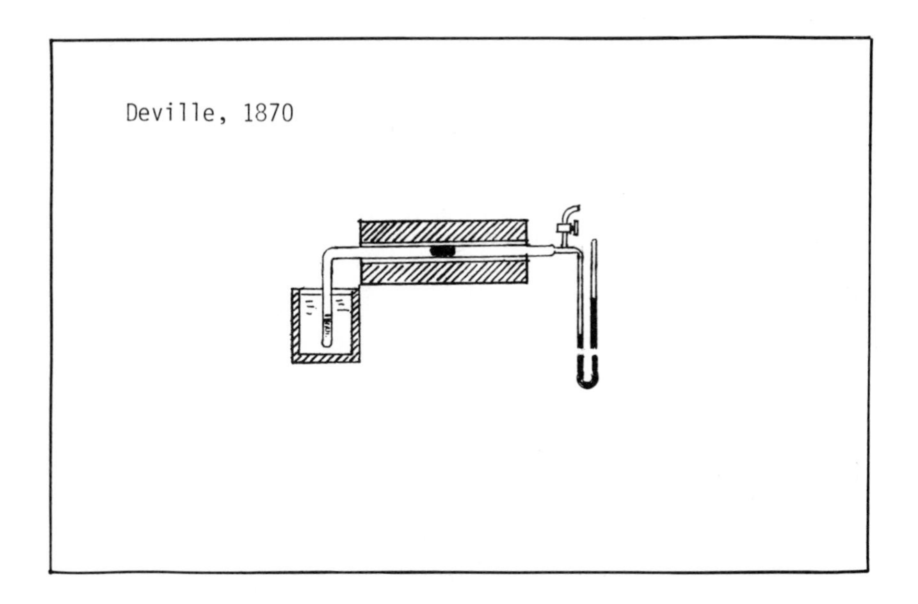

Figure 3. Apparatus for the Measurement of H_2O/H_2 Vapor Pressure Ratio over Solids.

the understanding of the kinetics and mechanism of ammonia synthesis over iron catalysts. As mentioned previously, these contributions are documented and discussed in a separate paper by Professor Emmett.

It is of interest to note that Professor Emmett had his first graduate student during the period at the Fixed Nitrogen Research Laboratory. This was Rowland Hansford, who was attending George Washington University, where Professor Emmett was teaching a course in catalysis. Rowland Hansford wrote a thesis on the Activity of Iron Catalysts for the Hydrogenation of Ethylene(11), and went on to a distinguished career in industrial research with Union Oil.

In 1937, Paul Emmett accepted the chairmanship of the newly formed Chemical Engineering Department at the Johns Hopkins University in Baltimore, Maryland. Over the next seven years, he built up an accredited department and continued his research on the development of surface-area methods and hydrogenation over metals and alloys. It was during this time that he became close friends with Professor E. Emmet Reid, who had retired from the Chemistry Department in 1935. With Professor Reid, he published the book Catalysis, Then and Now(12), which included a translation by Reid of Sabatier's Organic Catalysis and an updating by Professor Emmett of recent developments in catalysis. Professor Reid was a remarkable man who did much of his publication after retirement and who continued to be active until he passed away just before of his 102nd birthday.

In 1943 and 1944, Paul Emmett was associated with the Manhattan Project. When this program ended, he left Johns Hopkins for a new position with the Mellon Institute for Industrial Research on a Multiple Petroleum Fellowship financed by Gulf Oil Company. This began another productive period of important scientific contributions, which continued for the next eleven years. Some of the major areas of contribution during this time are listed below:

- Mechanism of Fischer-Tropsch synthesis
- Development of C^{14} tracer techniques applied to catalysis
- Catalytic cracking - mechanism, nature of catalysts.

The early work at Mellon concentrated on the Fischer-Tropsch synthesis reaction and was stimulated by a national concern for the dwindling supplies of petroleum, a concern which has stimulated the resurgence of interest in the Fischer-Tropsch synthesis in recent years. In carrying out this work, Professor Emmett introduced a new and important technique for the study of the mechanism of catalytic reactions. This was the use of C^{14}

tracers to investigate the intermediacy of various moieties in the overall Fischer-Tropsch reaction(13). One should keep in mind the difficulties of doing such experimentation in 1945. Carbon-14 compounds could not simply be purchased from a catalog, but had to be prepared in the laboratory. In addition, gas chromatography was not available for the separation and identification of reaction products. Reaction products had to be collected by low-temperature trapping, separated by fractionation and, in the case of the C^{14} measurements, converted to CO_2 by combustion and then counted with a Geiger counter. The C^{14} tracer work is yet another example of the experimental skill and excellence which characterized the work of Paul Emmett throughout his career. His careful and thorough experimentation was admired by many of his contemporaries. In particular, Professor Hugh Taylor of Princeton University regarded Paul's work as the standard of excellence at that time(14).

Figure 4 shows a photograph of Paul Emmett taken in the laboratory of the Mellon Institute. The co-worker in the picture is Dr. Joe Kummer, now at the Ford Motor Company. This picture appeared in the New York Times in the late 1940's with the caption "Gulf Oil Co. Makes Radioactive Gasoline." This, of course, referred to the C^{14}-tracer work. One can imagine what a newspaper reporter would do with such a caption today! Paul relates the story that the apparatus behind them in the photograph was designed and built by Joe Kummer, who is 6'6" tall. The next student who occupied this laboratory was 5'7" and consequently found himself spending a great deal of time on a stepstool in order to use the apparatus. That student is now well-known in the field of catalysis--Prof. W.K. Hall of the University of Wisconsin, Milwaukee.

During the latter part of his stay at the Mellon Institute, Professor Emmett became interested in catalytic cracking further extending the use of the C^{14}-tracer technique to study the chemisorption of hydrocarbons on cracking catalysts. The publication of his now famous volumes on "Catalysis" was also initiated and continued over the next five years.

In 1955 Paul Emmett was invited once again to join the Johns Hopkins University, this time as the W.R. Grace Professor of Chemistry. In the ensuing years between 1955 and 1971, Professor Emmett, his students, and associates continued work on the Fischer-Tropsch synthesis reaction and catalytic cracking with C^{14}-tracer techniques. Catalytic cracking is, of course, a very complex process involving a variety of chemical reactions. By adding a selected number of C^{14}-labeled compounds to a stream of cetane being cracked over silica-alumina catalysts, much insight was gained concerning the various reactions of primary products which are produced in catalytic cracking(15-18).

In 1971 Professor Emmett retired from the Johns Hopkins University. He did not, however, retire from his active

Figure 4. Photograph of Paul Emmett and J.T. Kummer in the Catalysis Laboratory of the Mellon Institute in 1948.

participation in catalysis. He returned to his home state and joined the faculty of Portland State University as Professor Emeritus of Chemistry. There he has continued to write papers, give lectures, consult, and continue research. His most recent publication(19) describes studies on the effect of oxidation on the pore structure of bituminous coal, and utilizes the contribution for which he will always be remembered, the BET surface-area method.

Over the now six decades, and entering the seventh, of active work in catalysis, Paul Emmett has been an inspiration to the many students and associates listed in Tables I and II. Many of these individuals have gone on to make their own mark in catalysis or other fields.

Table I
Graduate Students of Paul H. Emmett, 1937-1980

Rowland Hansford	Russel Hay
John Gray	Luther Browning
Ben Harris	W. Van Hook
N. Skau	Joe Hightower
J.T. Kummer	Jane Phillips
Dave Lipnick	Gervin Harkins
Keith Hall	Robert Garten
Robert Zabor	Burton Bartley
Hugh Tobin	Kunichi Matsushita
R. Haldeman	John Bordley
Mr. Starnes	Tom Quaie
Don McIver	Salvador Lyon

Table II
Research Associates of Paul H. Emmett, 1937-1970

Dr. R.J. Kokes	Dr. Harry Leftin	Dr. Y. Huang
Dr. Stephen Brunauer	Dr. Robert Kunin	Dr. G. Blyholder
Floyd Shultz	Dr. Lowell Wayne	Dr. G. Keulks
Jew Yam Yee	Dr. Leslie Joyner	Dr. Y. Kubokawa
Katherine Love	Dr. Harry Podgurski	Dr. Toyoshima
Dr. Roy Harkness	Dr. Louis deMourge	Dr. N. Takezawa
Dr. Thomas De Witt	Dr. Geoffrey Pass	Dr. P. Fejes
Dr. Martin Cines	Dr. J. Wishlade	Dr. G. Kreminic
Dr. Robert Anderson	Dr. Campbell	Dr. Burtron Davis
Dr. Herbert Pohl	Dr. Gharpurey	Dr. A. Solbakken
Dr. Ray Davis	Dr. Stoddard	Mrs. A. Solbakken
Dr. Joe Kummer	Dr. Hank Yao	Dr. M. Klotzkin

Paul Emmett's philosophy and approach to research are best summarized in his own words from an interview with the Chairman of this Symposium, Burtron Davis, in 1978(20). In indicating what he thought had been most important in shaping his attitude toward research, Paul said, "The important thing is to try to develop a critical attitude towards problems, observations, and statements. The methods to be used in helping students to achieve sound reasoning are beyond my humble ability to advise. In my own case, I feel that the qualitative analysis course as taught by my Professor Hicks and four years of college debating have been outstanding aids in my search to attain a sound reasoning capability that may be briefly expressed by the word 'logic'." Many of us have seen that remarkable logic in action, fortified by an incredible memory which recalls details from papers Professor Emmett either read or published years ago.

Concerning his own personal goals in research, Professor Emmett stated in the same interview, "I think that the kind of research that appeals to many people is to be able to follow ideas that they have in carrying out research, regardless of whether they are going to be applicable or not. This is a personal goal that I have always cherished. I have, accordingly, directed my career in such a way as to remain capable of doing the type of work that I wanted to do, following the leads that I ran across in performing the experiments that appealed to me, mostly in the catalytic research field. As a matter of fact, today I think it is very unfortunate that it is so hard to get support for basic pure research, because it is out of basic research that the ideas come that tomorrow will be practical and applicable."

With regard to the people with whom he has worked over the years and who have made significant impact in their own right in the field of catalysis, he said, "I was inclined to think the best of people and to be optimistic about them and in most cases this has worked out very effectively."

Figure 5 shows a photograph of Professor Paul H. Emmett in his laboratory at Portland State University. Behind him is the familiar BET adsorption apparatus, similar in many respects to the system which he used many years ago. He continues to follow the advice given to him many years ago by his good friend Professor E. Emmet Reid of the Johns Hopkins University. That advice was "The way to be active is to be active." Paul Emmett is certainly the personification of that statement. Paul Emmett, a warm and sensitive human being, a gentleman above all, and an outstanding scientist whose contributions have left their indelible mark on the field of catalysis, we salute you.

Figure 5. Professor Paul H. Emmett in his laboratory at Portland State University, Portland, Oregon.

Acknowledgments

I wish to express my appreciation to Professor Paul Emmett for helpful discussions and for providing the photographs presented in the paper. I am also indebted to Prof. Michel Boudart, Dr. Burtron Davis, and the many other friends of Paul Emmett for discussions and contributions.

Literature Cited

1. Emmett, P. H. "The Physical Basis for Heterogeneous Catalysis"; Draughs, E. and Jaffee, R. I., Ed.; Plenum Press: 1975; p. xi.

2. Benton, A. F.; Emmett, P. H. J. Am. Chem. Soc. 1924, 46, 2728.

3. Benton, A. F.; Emmett, P. H. J. Am. Chem. Soc. 1926, 48, 632.

4. Pauling, L.; Emmett, P. H. J. Am. Chem. Soc. 1925, 47, 1027.

5. Brunauer, S.; Emmett, P. H. J. Am. Chem. Soc. 1937, 59, 2682.

6. Brunauer, S.; Emmett, P. H.; Teller, E. J. Am. Chem. Soc. 1938, 60, 309.

7. Deville, S. C. Compt. rend. 1870, 70 1105.

8. Emmett, P. H.; Shultz, J. F. J. Am. Chem. Soc. 1933, 55, 1376.

9. Emmett, P. H.; Brunauer, S. J. Am. Chem. Soc. 1937, 59, 310.

10. Emmett, P. H.; Brunauer, S. J. Am. Chem. Soc. 1937, 59, 1553.

11. Hansford, R.; Emmett, P. H. J. Am. Chem. Soc. 1938, 60, 1185.

12. Emmett, P. H.; Sabatier, P.; Reid, E. E. "Catalysis Then and Now" Franklin Publishing Co., 1965.

13. Kummer, J. T.; DeWitt, T.; Emmett, P. H. J. Am. Chem. Soc. 1948, 70, 3632.

14. Boudart, M.; personal communication.

15. Van Hook, W. A.; Emmett, P. H. J. Am. Chem. Soc. 1962, 84, 4410.

16. Van Hook, W. A.; Emmett, P. H. J. Am. Chem. Soc. 1962, 84, 4421.

17. Van Hook, W. A.; Emmett, P. H. J. Am. Chem. Soc. 1963, 85, 697.

18. Hightower, J. W.; Emmett, P. H. J. Am. Chem. Soc. 1965, 87, 939.

19. Leon, S.; Klotzkin, M.; Gard, G.; Emmett, P. H. Fuel 1981, 60, 673.

20. Davis, B. H. J. Chem. Educ. 1978, 55, 248.

RECEIVED November 17, 1982

6

The Contributions of Eugene J. Houdry to the Development of Catalytic Cracking

ALEX G. OBLAD

University of Utah, College of Mines, Department of Fuels Engineering, Salt Lake City, UT 84112

This paper is dedicated to Eugene J. Houdry, who conceived and developed the most important petroleum refining process in the history of the industry. The refining process is catalytic cracking. Today the U.S. capacity for catalytic cracking approaches 5.5 million bbls/day. It was the leading process in volume of petroleum processed per day until sulfur limitations brought on by catalytic reforming and environmental considerations led to large expansions in catalytic hydrorefining-hydrotreating capacity in the late sixties and early seventies. Catalytic cracking is still the leading process for making high octane components for today's gasolines.

The discovery and development of catalytic cracking as a process came about as a result of fulfilling a need. The petroleum industry of the U.S.A. began as a supplier of lubricants and fuels for lighting purposes based on native crude petroleum. With the invention of the internal combustion engine the need for fuels for this device became apparent and, as the automobile caught on with the common man, larger amounts and better fuels began to be produced by the industry as a result of research by trained scientists and engineers.

Fuel quality has always been a limitation on engine performance and efficiency. The developers of the internal combustion engine very early learned of the phenomenon of "knocking" and its limitation on performance. Fuel quality with respect to antiknock properties has always been, and is even today, an overriding consideration in the production of gasolines. The development of thermal cracking of heavier petroleum fractions to produce gasolines began in earnest in 1912 with the success of the Burton process. This process converted heavier petroleum fractions into gasolines of improved burning quality and as a result, the petroleum industry surged to meet the demands of the growing automobile population. Thermal cracking became the main means of making high octane gasoline. With the discovery and commercial production of lead tetraethyl as a knock suppressing agent the quality of gasoline advanced further. Even with this combination the fuel

0097–6156/83/0222–0061$06.00/0

quality, particularly for aviation use, was still a limitation in the early 1930's because the onward progress in the understanding of the workings of the internal combustion engine had advanced engine designs to a point where even higher quality fuels were necessary. Some new inexpensive way of producing gasoline of even higher quality from the abundant heavier fractions of petroleum was required. The new technology developed by Mr. Houdry met that need. How this came about when it did is the next part of this story.

Early Life of Eugene J. Houdry

Eugene J. Houdry was born in Dormont, France, near Paris, on April 18, 1892, to a family owning and operating a successful steel fabricating business. He was encouraged to take part in the family business and to prepare himself for this he enrolled in École des Arts and Métiers, an engineering school at Chalons-sur-Marne. He graduated in 1911 with a degree in mechanical engineering and was honored by the French government with a gold medal for being the first in his class in his scholastic record. Not only was he an honor student, but he was also captain and halfback on his school's soccer team which won the French national championship in 1910.

Following his graduation, Houdry joined his father's company as chief of engineering. This was not for long as Europe was on the way to World War I and young Houdry was called to military service. He was in training in field artillery when the war started in 1914. He soon became a lieutenant in the French Tank Corps and took part in the first battle of WWI in which tanks were used as a military assault weapon. Later on he was seriously wounded in action at the battle of Juvincourt, April 16, 1917. For "having organized the repair of the disabled vehicles on the battlefield under heavy fire in spite of extremely difficult field conditions" he was awarded the Croix de Guerre and made a chevalier in the Legion of Honor. With the end of the war in November of 1918 Houdry returned to the family business to continue his career in engineering.

Career Definition Eugene J. Houdry had developed a keen interest in the motor car. Many young men develop an early interest in automobiles, mainly for what the car can do for them, but this was not the entire reason for Houdry's interest. He was also interested in the automobile as a machine, particularly in the engine that powered it and the fueling of the engine. Very early in his experience with his favorite racing car, a Bugatti, and through his interest in automobile racing, he learned of the limitations on performance of an engine resulting from fuel quality, particularly its knocking tendency. This phenomenon was first recognized by W.A. Otto in his early work on the internal combustion engine back in 1878 and it remained a prominent barrier through the first one hundred years of the motor car. Racing cars at the time of Houdry's youth was a sporting event but it was, and still is, a means of developmental testing and advancement of the technology of the automobile in Europe and the U.S.A.

Along with his work in the structural steel business, he became a director in a number of companies in France, one of which manufactured automobile parts, and through the latter he became associated with technical people in France who were interested in automotive engineering.

In 1922 he visited the U.S. for the first time with the express purpose of taking in the famous Memorial Day 500 mile race at Indianapolis and visiting the Ford Motor Company in Detroit. Talks with his engineering peers in the United States added to his determination to investigate methods for better fuel production that would in turn lead to advances in engine design, higher performance and greater efficiency.

Returning to France, he soon began a new career. His wartime experiences had taught him the vital need for new sources of fuels for France. In addition, there was great interest in a number of European countries, lacking domestic sources of petroleum, to become more self-sufficient in a source of fuels based on indigenous materials such as lignite, shale and coal. These interests stemmed primarily from military and economic (balance of payments) considerations. Houdry was very much aware of these desires in France and, knowing that France had large proven reserves of lignite, he decided to do something about the production of synthetic fuels from this resource.

Status of Fuel Technology in 1922

To put the Houdry development in perspective, at the start of his work, the petroleum industry was expanding rapidly, mainly to supply gasoline for the automobile, as has been pointed out. Distillation and thermal cracking of the heavier fractions were the main routes for producing gasoline. Thermal cracking had been commercial beginning in 1912. The application of catalysis for the production of fuels was in its infancy. Some attempts were made prior to 1923 to commercialize catalytic developments such as aluminum chloride cracking in the U.S.A. but this and several other cracking processes using solid catalysts containing metals did not catch on.

After World War I there was a substantial effort on the part of the automobile manufacturers, the petroleum refining industry, and, to some extent, the chemical companies to investigate the properties of fuels such as chemical composition of gasolines, volatility, knocking tendency, storage problems, etc. The effort in the U.S.A. was especially notable. The phenomenon of knocking and its limitations on engine output and performance (compression ratio) and the relationship of chemical structures of fuels and knocking tendency and the effect on knocking by non-hydrocarbon additives such as alcohols, ethers, nitrogen compounds, metal carbonyls and metal alkyls were under investigation. The effect of lead alkyls was discovered at this time but the addition of this material to gasoline on a commercial scale did not arrive until 1926. The octane rating scale now in use was not developed until the late 1920's. During the 1920's and 1930's an R & D effort on the part of the U.S. and world refinery industry was begun to develop processes for producing higher quality fuels from crude oil. This did not really get going until the late 1930's and it was greatly accelerated by the Houdry development of catalytic cracking and the vast requirements brought about by World War II.

Beginning of a Career - First Steps

At the end of 1922, Houdry became aware of a scheme to produce gasoline from lignite developed by a biologist working in Nice, France. Very soon thereafter he visited the biologist's laboratory to observe the actual operation of the process. He was impressed with what he had learned in Nice for upon his return to Paris, Houdry met with his friends and supporters and together they formed a company to pursue the evaluation of the process and to make a decision regarding their participation in the development of the lignite process. To compensate for his inexperience in the field of chemistry, he brought together a group of experts, including university professors, to help him evaluate the potential of the process.

The group visited Nice in February 1923 to observe the process in operation. The process consisted of what we would call today a mild steam pyrolysis of lignite followed by various non-catalytic and catalytic treatments of the vapors produced to remove H_2S and to hydrotreat the products to form a stable gasoline. Gasoline was produced but in small quantity and no material balances were obtained. Neither the Houdry group nor the Nice group understood what was happening in the process. They did understand that quantitative data was not obtained, that the catalyst life of the metals used in hydrotreating was too short and that the purification steps of the product were inadequate.

In spite of these shortcomings, Houdry and his group made the decision to continue the research. Critical to this decision were the high quality of the gasoline produced (as compared with available fuels) and the fact that low temperatures and atmospheric pressure were employed in contrast to the high pressures and temperature being used elsewhere in producing synthetic fuels. Looked upon with favor by the group was the feeling that the process would be an industrial fulfillment of the works of Sabatier if successful. Substantiation of their own feelings about the process was obtained when they learned that an Italian group planned to move ahead with a larger unit at Milan.

The Houdry group designed and built a larger unit at Beauchamp near Paris which came on stream in May 1923. To their intense dismay, no gasoline product was obtained nor was any obtained in the Milan installation which came on stream later. In transferring from the laboratory to larger units the results were not the same. After much work and many conferences it was concluded that their working hypotheses of the chemistry of the process had to be changed. The process was modified at Houdry's Beauchamp laboratory to improve the yield of distillate tar which was then "hydrotreated" as before. After intense effort and around-the-clock operation, they began to obtain consistently high quality gasoline in yields of about 4-5 wt.% from lignite. With these results, which were confirmed to the Houdry backers by an impartial expert who supervised numerous runs, a decision was made in 1924 to continue support of Houdry's work. This was formalized by the incorporation of La Society Anonyme Francaise pour La Fabrication des Essenes et Petroles, which later became the Houdry Process Company of France.

Houdry had survived his initiation into catalytic work and process development. During the course of the work that led to their success Houdry had begun to develop an appreciation and understanding of catalysis and the action of the latter on hydrocarbons. Likewise, he learned some of the secrets of scale up of equipment and above all the necessity of small scale research and development to pursue the development of catalytic processes and catalysts. Gradually results became better and more predictable. The program was altered to include research as well as development scale studies. Over the next several years, involving much work which was guided by empirical and intuitive approaches, Houdry concluded that the lignite derived tars had cracked to gasoline and were upgraded by the catalytic steps employed downstream from the lignite pyrolysis. It was also determined that the catalysts were being fouled during the reaction and that regeneration was necessary and could be accomplished by air burning.

The research effort pursued several objectives: (1) studies of the lignite pyrolysis, (2) research on catalyst improvement for cracking and upgrading of the volatile tars (sulfur, oxygen and nitrogen removal), (3) catalyst regeneration studies including how to handle the substantial heat developed.

Progress was such that in 1927 erection of a demonstration plant was begun to process 60 tons of lignite day. Engineering design of this plant was done entirely by the Houdry group which also supervised the construction. The plant included provisions for recovery and drying of the lignite, pyrolysis, tar condensation and distillation, catalytic upgrading of the tars, purification of products and recovery of sulfur. Catalysts for the operation were made in a separate plant--15,000 liters of catalyst/month and 20,000 liters/month of absorbents.

Operation of the plant was started in June 1929, but was only partially successful in that yields were only about 70% of those expected. When funding finally ran out and the French government refused further grants, the plant was shut down in June of 1930. Even though this plant was not as successful as hoped, it demonstrated that lignite could be processed to yield light petroleum products which had excellent antiknock quality as gasoline.

From this experience with lignite Houdry accumulated much know-how on catalysis. He learned that cracking of the tars was a necessary step to produce light distillates and that this could be done best by catalysts containing silica-alumina with low metals content. He also learned to appreciate and understand catalyst activity, life, regeneration, heat effects, heat transfer, recycle operation and catalyst preparation in laboratory and semi-commercial scales. Thus, while the lignite experience was expensive and very disappointing and discouraging, it was a tremendous learning experience which prepared Houdry for the very successful future adventure in catalytic cracking development.

It is quite surprising that a person of Mr. Houdry's background became interested in a field that was thought to be mainly chemical. The field of fuel production from petroleum and even synthetically from lignite and coal, which was of interest in Europe, particularly in Germany, during and following World War I, was largely developed by trained chemists as was the practice of catalysis, which was a rapidly growing field at the time. There were important contributions to the industry by mechanical engineers, such as the development of high pressure equipment used in the synthesis of ammonia and the Bergius process for coal liquefaction. Bosch, a mechanical engineer and Haber, a chemist, were responsible as a team for the synthetic ammonia and they later shared the Nobel Prize in Chemistry for their joint development.

In an important way it turned out to be fortunate that Mr. Houdry had training and talent in mechanical engineering. The mechanical problems encountered and solved in developing catalytic cracking into a commercially feasible process were formidable and proved to be as important as the chemical problems.

The Development of Catalytic Cracking

Work of the 1927-1930 Period As the lignite work progressed, fostering a much better understanding of the role of the employed catalysts, Houdry became convinced that he should try cracking petroleum with his promising lignite tar upgrading catalysts. This occurred during early 1927. He knew that with the use of catalysts he could bring about cracking reactions which, prior to his experience with lignite tars, were known only to occur non-catalytically at high temperatures and pressures. He was also aware from his own work and that of others earlier, that his catalysts rapidly lost activity while in contact with heavier hydrocarbons but he had developed a way around this by treatment of the coked catalyst with air at elevated temperatures. Such a procedure restored the catalyst to previous activity but handling the heat was a problem. From his experiments with reactor design, he was confident he could solve this. Since he was heavily engaged in the lignite work at that time, he did not act immediately to experiment with petroleum. It must be remembered that he had at his command a laboratory and personnel to carry out his requests.

At the appropriate time he began experiments on cracking using a Venezuelan gas oil and employing his lignite distillate, upgrading catalysts which were mainly metals supported on porous inert materials such as pumice and kaolin. His first attempts at cracking petroleum were with a kaolin supported nickel in a simple apparatus consisting of a vaporizer-preheater, reactor and condenser-receiver. He had turned to kaolin as a support material when he determined that free silica in other supports (kieselguhr) mineralized the nickel and caused a loss of activity. No regeneration was tried in these earlier attempts. The results were poor, giving some yield of liquid but much gas and coke causing the catalyst to quickly lose activity. The apparatus was modified to provide catalyst regeneration which was frequent on a time scale. An empirical search for better catalysts making less coke was soon underway. He tried many metals which were believed to be required as well as various other porous supports including fullers earth and other available clays. The results varied considerably but always too much coke and gas were made, catalyst activity decreased rapidly, the liquid products were prone to form gum and regeneration was difficult, slow and frequent.

Houdry had to get over the long-believed concept that metals were required for catalysis to occur. He finally became convinced to try the support without the metal. When he did so the results were much better. He was now on the right course and on the way to a breakthrough.

Trying out various clays, including acid activated clays (used as adsorbents to purify lubricating oils), he soon achieved very promising results. With a sample of clay from San Diego which had been acid activated by Pechelbronn Oil Refining Company before delivery to Houdry, results were obtained for which he had long been seeking: high liquid yields, low gas and coke and facile regeneration. The San Diego clay pelleted readily with only small amounts of binder. Receiving this clay proved to be a stroke of great luck.

With this new catalyst he studied catalytic cracking of petroleum oils in depth along with his lignite tars. The effects of the operating conditions were determined. Development work was also done on the catalytic cracking apparatus. Two reactors of 20 and 40 liter capacity with oil feed rate of 2.5 kg/hr. and a larger unit of 160 kg/hr. capacity were built and used for the main research and development effort. The apparatus in each case consisted of a vaporizer directly connected to a cylindrical catalytic reactor which was provided with a series of special nozzles for dispersing the regeneration of air evenly within the surrounding catalyst bed, condensing and distillation means and product refining sections to remove sulfur and hydrogenate unsaturates. These latter sections were a holdover from the lignite work. At this time the units were operated over cycles as long as 8 hours before regeneration and this made regeneration very slow. Better means for handling the heat of regeneration were needed, so in July 1930 Houdry settled on a fixed-bed heat exchanger type of reactor. To cool the catalyst bed during regeneration, water was circulated through the tubes. This was so successful that a long list of gas oils were tested along with lignite tars from France, Germany and Hungary. With good testing equipment available, catalyst development work led to optimization of catalyst properties. Clays having compositions of close to 80% silica-20% alumina with low impurity levels, particularly iron, and good porosity were found to be optimum. Clays from many sources were examined. Catalysts having these compositions and formed as 3/6 mm pellets became the standard and were used in all later development work and commercial units. The gasolines were extensively evaluated in the laboratory and on the road. Road tests proved the gasoline to have good anti-knock properties. Octane rating could not be done because the test was new in 1930, and procedures and apparatus for the test were not yet developed. It was quite some time before it was known how good the gasoline really was.

By the summer of 1930 the process was ready for testing on a pilot plant scale. Houdry and his group were always open with results and as early as November 1928 the large petroleum refining companies of the world began investigating the new cracking process. Among these companies were Anglo-Iranian Oil Company (now British Petroleum), N.V.De Bataafsche Petroleum Matschappij (part of Royal Dutch and Shell Oil), several French oil refining companies, including Pechelbronn Oil Refining Company which had supplied the San Diego clay to Houdry earlier, and Vacuum Oil Company of U.S.A. Demonstration tests were made for these interested parties. Vacuum Oil Company was very impressed with the demonstration and after examining the patent position, made an offer to continue the development in the U.S.A. The deal required Houdry and several assistants to travel to Paulsboro, New Jersey during November 1930, set up the 200 liter Beauchamp unit and demonstrate the process in a continuous run of 15 days and nights. If it was successful, Vacuum would proceed with a pilot plant scale unit. The offer was readily accepted by the Houdry group since by this time they had spent over 46 million francs and they knew that the next steps to commercialization would be much more costly. Houdry took great pains to obtain patents on his inventions. By 1930 he had over 50 patents and by the time of commercialization of the process he had registered over 100 patents. Houdry and his group did not publish their results in the conventional sense until the late 1930's.

The apparatus and equipment that the Houdry team used throughout the entire program, from earliest experiments to the large scale pilot units, were well designed and constructed, indicating much planning and a high level of mechanical engineering capability. This was Houdry's field, but he was not as much at home in chemistry, catalysis and chemical engineering. He had to learn these first hand by himself as there was not much information in the literature pertaining to what he was doing. Mr. Houdry, in selecting his associates, was careful to include personnel with excellent backgrounds in chemistry. He and his associates did study the available literature on catalysis and were very much aware of the work of Bergius, Fischer, Sabatier and others in the field. Houdry relied a great deal on his intuitive powers, which were formidable.

Development of the Process in the U.S.A.

The Period November 1930-June 1933 The fifteen-day test in the Paulsboro laboratories of Vacuum Oil Company was successful. The information provided by the extensive testing available at Vacuum, new to Houdry, was the impressive stability of the catalytic gasoline compared with that of thermal gasoline. For the first time a detailed analysis of the gasoline was made and it showed the presence of isoparaffins which accounted, in part, for the anti-knock quality.

The next step was design and construction of a pilot unit of 60 bbls/day capacity which came on stream in May 1931. Results were as expected which added momentum to the project. On July 25, 1931, the Houdry Process Corporation was founded and on July 31, 1931, Vacuum Oil Company was merged with Socony to become Socony-Vacuum Oil Company. Support for development was continued by the new company and by 1933 a 200 bbl/day fixed bed pilot plant was operating.

In spite of the encouraging results which were accumulating, a decision was made by Socony-Vacuum in the spring of 1933 to discontinue support for the development. The Great Depression was on in the U.S. and funds were tight so Houdry was encouraged to seek support from other companies. He accordingly made contact with five U.S. companies during the period May 1 to June 15, 1933. An agreement was made with Sun Oil Company to continue the development. With this agreement Houdry was certain that he was in the last stage of catalytic cracking development and that it would finally go commercial.

Sun Oil Company was an independent oil company with headquarters in Philadelphia and a sizeable refinery across the Delaware River a little north of the Paulsboro, New Jersey refinery of Socony-Vacuum where Houdry's development activities had been centered. The story goes that Houdry met with Arthur E. Pew, son of one of the Founders of the Sun Oil Company and Vice-President of Refining, and quickly the two made a deal for Sun to continue the development of the Houdry Process. The arrangement involved his moving across the river to the Marcus Hook refinery of Sun where there were extensive laboratory facilities. Sun's interest in the Houdry process probably stemmed from its reputation as a maverick in the industry. At that time and for some time into the future Sun did not go along with the industry-wide practice of using tetra ethyl lead to improve the octane quality of its gasolines. They continued to feature clear, lead-free gasoline. Gasoline quality was increasing at the time, and Sun could foresee difficulty in remaining competitive over the long term. The Houdry process, if successful, would not only be a means for Sun to remain competitive with a clear gasoline but would put them out in front of the competition. This foresight proved valuable for Sun. They acquired half of the holdings of Houdry and his associates with the approval of Socony-Vacuum which maintained its previous holdings. The ownership was now approximately one third each for Houdry and his associates, Socony-Vacuum and Sun Oil.

Commercial Development - The Final Stage

June 1933-March 1937 Mr. Houdry and his close associates, who had come with him from France, were invited by Sun Oil to move to Pennsylvania and set up shop at the Marcus Hook Refinery. Following this move, work was carried forward on both sides of the Delaware River. In 1933 a 175-200 bbl/day pilot plant was running.(1) Meanwhile research and development work on the fixed bed concept and suitable catalysts based on bentonite-type clays were intensified at Marcus Hook. Intensive work on the mechanical equipment for the fixed bed apparatus was carried out. The internals in the fixed bed required for introducing oil vapors, steam and regeneration gases into the bed, removing the cracked vapors and regeneration products from the bed and providing for heat removal in regeneration and heat addition during endothermic cracking were pressing problems needing solutions before commercial design could begin.

Earlier than 1933, contact clays had been developed for the purification of oils, fats and waxes. Modification of such processes for treating thermally cracked gasolines were extensively used in the petroleum industry. It was appropriate for Houdry Process Corp. to turn to the manufacturers of these clay-treating materials for help in developing commercial cracking catalysts. They sought cooperation from the Filtrol Corporation to achieve the goal of a successful operating catalyst. A wide variety of clays were investigated and from these, the best were selected for activation by acid treatment, washing, pelleting and heat treating to yield an acceptable catalyst. The catalyst had to meet activity, selectivity and life specifications, be mechanically strong and able to survive in the environment of use. Catalysts meeting these requirements were produced which were based on bentonitic clays. A typical commercial catalyst contained silica (SiO_2) and alumina (Al_2O_3) in a weight ratio of 3.0:4.5, was low in sodium (0.5 wt.% as Na_2O), iron (2.0% as Fe_2O_3), calcium (1.8% as CaO) and magnesium (3.8% as MgO).

The mechanical engineering design of the fixed bed reactor went through a number of versions. The first version to reach demonstration size stage employed combustion gas recycle with controlled oxygen content (5%) to remove coke and provide a means for removing the heat of combustion of coke. Alternately, in a fixed cycle, vaporized oil feed and enriched flue gas flowed into the bed through many orifices in the internal piping. Direct transfer of heat between the gaseous feed stream and the catalyst was thus made possible. More efficient heat transfer, using water circulation within heat exchanger pipes within the bed during the regeneration cycle, was next developed. The water-containing tubes were shielded from the catalyst so as to prevent catalyst cooling below combustion temperature by providing jackets which were perforated to allow access for the regeneration gases.

This second design was the basis for the 2000 bbl/day semicommerical unit built at Socony-Vacuum's Paulsboro refinery. This unit went on stream in April 1936 and attained full operation in June of that year. This was three years after the move of Houdry Process Corp. to Marcus, Hook, Pennsylvania and indicates not only a tremendous research and development accomplishment, but a great cooperative effort on the part of Houdry and his team, Sun Oil and Socony-Vacuum.

With the performance of the 2000 bbl/day unit, decisions were made on the part of Sun Oil and Socony-Vacuum to proceed to full scale commercialization. The operation was controlled manually; the cycles of cracking, purging and regeneration were long.

The first large scale commercial fixed-bed Houdry unit was at Marcus Hook. This unit came on stream in 1937 charging 15,000 bbls/day of a heavy gas oil. This version of the process was fully automated, operating on short cycles. The unit utilized a cycle timer for automatic control, for a first, and also used for the first time a gas driven turbo-compressor-electric generator to provide regeneration air and electricity for process use. Large valves for controlling flows of the process streams and operating at high temperature were put on stream for the first time.

By the time of the annual meeting of the American Petroleum Institute in November of 1938, three units were operating (2,3), a 2000 bbl/day unit at Paulsboro, NJ (Socony-Vacuum), a 3000 bbl/day in a Socony-Vacuum unit in Europe and a 15,000 bbl/day unit at the Marcus Hook refinery of Sun Oil Co. The latter unit went on stream in March of 1937. Ten units were under construction (two units at Sun Oil refineries and eight units at Socony-Vacuum refineries). By this time a molten salt liquid was being used in the place of the previously used flue gas and water as a heat exchange medium. The Socony-Vacuum unit at Beaumont, Texas was the first to employ the molten salt, a eutectic mixture of potassium nitrate and sodium nitrite, melting at 284°F.(4) The molten salt was circulated through the bed in special tubes placed within the catalyst bed to absorb the heat produced during the regeneration cycle. The hot salt was then exchanged with incoming oil feed and with water to make high pressure steam.

The announcement of the commercialization of the Houdry cracking process was made at the annual meeting of the American Petroleum Institute held in Chicago in November 1938. The paper was titled "Catalytic Processing by the Houdry Process," authored by Eugene J. Houdry of Houdry Process Corp., Wilmington, DE, Wilbur F. Burt of Socony-Vacuum Oil Co., New York City, NY, Arthur E. Pew, Jr. of Sun Oil Co., Philadelphia, PA, and W.A. Peters, Jr. of E.B. Badger & Sons Co., Boston, MA. The paper was presented by Arthur E. Pew and its impact was sensational.

As a young scientist just beginning his career, the author of this historical review was present at the API meeting and heard the presentation and the discussion that followed. It was very obvious that the audience was greatly impressed and concerned about the impact of the new process. The editorial which accompanied the publication of the full paper in the November 30, 1938, issued of the National Petroleum News had this to say about the Houdry cracking process:

> "Catalysis is playing a progressively more vital role in refining operations; catalytic treatment of lubricating oils, catalytic polymerization of gases, hydrogenation of unsaturates, all have become commercial operations in one or a large number of plants. The most interesting recent development is the Houdry Process, reputedly a catalytic cracking process, several huge units of which have been built or are under construction.
>
> Papers presented at the Institute meeting on this intriguing development discuss the major principles employed in cracking heavier oils to gasoline at relatively low temperatures and pressures to produce motor fuel blending naphtha of very high octane number. There are those who believe that modern refining will follow the lead of those few catalytic developments, and that catalysis will become the major principle in refining processing of the future. There is a considerable amount of evidence to indicate that this opinion may be essentially correct, though prophecy at this point is too rash. . . ." (2)

The authors of the API paper stated in summary that

> "they believe that in these processes (variations of catalytic cracking) the industry has at its command

a new and profitable means to achieve a better economic balance between gasoline, furnace oil, fuel oil, and crude requirements; to meet the need for higher octane gasoline, and at the same time to adjust refinery operations to the seasonal variations in demand for the principal products, gasoline and heating oils. It appears obvious that the application of this process will assist the industry in meeting the ever increasing demand for distillate products and at the same time in conserving our crude oil reserves. The ability of the process to refine economically any type of crude would appear to make available for use many crudes not now suitable for conventional refinery operations and thus, again to increase our reserves."(3)

The announcement stated that the Houdry process for catalytic cracking, treating, etc. were covered by 96 U.S. patents and many pending patents and that the processes were available for licensing from the Houdry Process Corp. It is estimated that the development costs of catalytic cracking through the first commercial unit were at least $15,000,000 (in 1939 dollars). This was a very large amount considering the times.

Progress in the use of the process was rapid. By 1939 Socony-Vacuum and Sun Oil had 15 Houdry units operating or under construction for a total of 212,000 bbls/day. By 1944 24 units were operating or under construction with over 330,000 bbls/day capacities. These units made a tremendous contribution to the war effort of the allies during World War II. For the first part of the war the Houdry Process provided most of the high octane gasoline for aviation and other use. The Houdry Process was the first process to produce a cracked gasoline meeting U.S. Army and Navy specifications for aviation fuel blends. Prior to the advent of the Houdry Process the high percentages of unsaturated hydrocarbons in thermally cracked gasolines had limited aviation fuel bases to straight run gasolines. Thus, the production in quantity of 100 octane and higher aviation gasoline through the use of Houdry catalytic gasolines greatly aided the development of high performance aircraft for civilian and military use.(5) The fluid bed process did not come on stream commercially until May 1942. The TCC moving bed process went commercial in 1943. At the end of World War II the installed capacity for catalytic cracking was over 900,000 bbls/day.

The development of catalytic cracking and its subsequent commercial usage before and after the war contributed greatly not only to the winning of World War II but to the development of the automotive industry after the war by making premium quality gasoline available in large quantities at reasonable cost. This, in turn, had a tremendous influence on transportation, which is continuing to the present day.

In July 1940, Houdry Process brought on stream the first

synthetic silica-alumina cracking catalyst. The synthetic catalyst eliminated the variability in properties that was experienced in developing catalysts from clay. In general, at a given conversion level, clay catalysts produce more coke, less gas, less C_4 hydrocarbons and more gasoline of lower octane number than synthetic silica-alumina. Of great importance during the war years was the better quality aviation gasoline produced by synthetic silica-alumina catalyst. Synthetic catalysts soon became the dominant catalysts in catalytic cracking.

In 1943 Houdry Process Corp. announced the Houdry adiabatic process for catalytic cracking which eliminated the molten salt heat transfer system. This process did not go commercial for catalytic cracking, but it was successfully used for butane dehydrogenation to produce butenes for synthetic rubber production during World War II.

Many people assisted Mr. Houdry in his quest to develop a successful catalytic cracking process. Some of these are:

Houdry Processing Corp.
R. C. Lassiat
J. B. Maerker
T. H. Milliken
G. A. Mills
W. A. Joseph
T. B. Prickett
J. W. Harrison
W. F. Faragher
G. R. Bond, Jr.
A. G. Peterkin
W. J. Cross
H. A. Shabaker

Sun Oil Co.
A. E. Pew, Jr.
C. H. Thayer

Socony-Vacuum Oil Co.
F. H. Sheets
W. T. Simpson
G. S. Dunham
C. S. Teitsworth

It should be pointed out that Mr. Houdry and the Houdry Process group's work on catalytic cracking were greatly aided by outstanding engineering talent from both the Sun Oil Co. and Socony-Vacuum Oil Co. This was particularly true after 1933 when Houdry moved to Marcus Hook. It was at this time that decisions were made to move as rapidly as possible to commercialization and engineers experienced in large-scale refinery design, construction and operation were brought into the final development of the process. Their contributions were critical for the success of the process.

Mr. Houdry received many awards for his development of catalytic cracking. Among these were the Modern Pioneer Award of the National Association of Manufacturers in 1940 and the E. V. Murphree Award in Industrial Chemistry, American Chemical Society in 1962. He received honorary Doctor of Science degrees from Pennsylvania Military College in 1940 and from Grove City College in 1943. He was founder of the France Forever Group and assisted in the foundation of the International Congress on Catalysis and its first meeting in Philadelphia, Pennsylvania in 1956. He died in June 1962 at age 70.

Houdry Innovations

1. First successful catalytic cracking process.
2. First large scale catalytic process to practice air regeneration.
3. First catalytic process to employ automatic control of the cycles--the development of the cycle timer.
4. First use of large, high temperature operating valves.
5. First large scale use of gas turbine driven compressor.
6. First commercial production of cracking catalyst from naturally occurring clays.
7. First commercial production and use of synthetic silica-alumina catalyst.
8. First commercial large scale use of a catalyst for oxidation of carbon monoxide.

Acknowledgments

The information used in preparing this paper came from references 6-9, the Houdry News, my personal files as well as my own remembrances of personal contacts with Mr. Houdry and his many associates who were still at Houdry Process Corp. when I arrived there in 1947. I am grateful to G. A. Mills and W. J. Cross for the information they supplied me.

Literature Cited

1. National Petroleum News, 1933, Vol. 25, No. 52, p. 27.
2. National Petroleum News, 1938, Vol. 30, p. R569.
3. Houdry, E.J.; Burt, W.F.; Pew, A.E.; Peters, W.A. Proc. A.P.I. 19 III, 1938, pp. 133-148.
4. VanVoorhis, M.G., National Petroleum News, Aug. 23, 1939, Vol. 31, p. R346.
5. VanVoorhis, M.G., National Petroleum News, Oct. 30, 1940, Vol. 32, p. R386.
6. Lassiat, R.C.; Thayer, C.H., Oil & Gas Journal, Aug. 3, 1946, Vol. 45, p. 84.
7. Ardern, D.B.; Dart, J.C.; Lassiat, R.C., Progress in Petroleum Technology, 1951, American Chem. Soc.
8. Lassiat, R.C.; Shimp, H.G.; Peterkin, G.A., The Science of Petroleum, Vol. 5, pp. 224, 1953, Oxford Univ. Press, London.
9. Houdry, E.; Joseph, A., Historique du Cracking Catalytique dans l'Industrie du Petrole, Bulletin de l'A.F.T.P., No. 117, May 31, 1956.

RECEIVED November 29, 1982

7

Herman Pines and Organic Heterogeneous Catalysis

NORMAN E. HOFFMAN
Marquette University, Todd Wehr Chemistry Building, Milwaukee, WI 53233

Herman Pines for over fifty years has made significant contributions to heterogeneous catalysis. He has emphasized hydrocarbon reactions. Among the areas he has researched are acid catalysis, base catalysis, aluminas, aromatization and dehydrogenation catalysts and metal hydrogenation catalysts. His students and colleagues over the years recall his industrial and his academic career.

It may be best to begin with a very brief sketch of Herman Pines' career in organic heterogeneous catalysis and then present a more detailed review afterward. He began his career in this field over fifty years ago when he joined Universal Oil Products Company. In the early 1940's, while still working full time at Universal Oil Products, he joined the Northwestern University faculty as a part time member, assistant professor and assistant director of the Ipatieff High Pressure and Catalytic Laboratory. Professor Ipatieff and he worked together at Universal Oil Products, and the Professor, as Ipatieff was called by his coworkers and students, had already become a Northwestern faculty member. Pines considered Ipatieff his mentor.

In the early 1950's Herman Pines resigned from his full time position of Coordinator of Exploratory Research at Universal Oil Products to become a full time member of Northwestern's faculty. He was apointed Vladimir Ipatieff Research Professor and Director of the Ipatieff High Pressure and Catalytic Laboratory. In 1970, he became Vladimir Ipatieff Emeritus Professor.

Thus, Herman Pines brought to an academic career a successful background in industry and was able to help his students understand the industrial significance of the chemistry they were learning in the classroom and investigating in the research laboratory. However, I do not want to leave the impression that practical industrial research was being done in the Ipatieff Laboratory. On the contrary, the research was basic, but often it had industrial implications. As his former students Joseph Ar-

0097-6156/83/0222-0077$06.00/0

rigo, Robert Kozlowski and Stanley Brown say, Herman Pines prepared his students well for an industrial career while collaborating with them on basic research problems.

I would like to return to the early days at Universal Oil Products. Herman Pines was earning his doctorate part time evenings and weekends at the University of Chicago in his first few years of working at Universal Oil Products. I often wondered how he was able to carry that kind of workload. Among the characteristics that his former students like Ralph Olberg, Jean Germain, and Alexander Edeleanu recall today were his energy and dedication to chemistry.

Surely these must have been part of the answer. Herman Pines was organized too. Joseph Chenicek, who was a full time graduate student at the University of Chicago with Pines and who later also joined Universal Oil Products, recalls that Pines would plan his experiments in advance and, therefore, know the chemicals and supplies he needed from the stockroom. He would ask Chenicek to obtain them during the day so that Pines was able to work at night and on weekends. In 1935 Herman Pines received his doctorate presenting the graduate faculty with the dissertation entitled "A Study of the Electronegativities of Organic Radicals."

But prior to writing his dissertation he published his first paper (1). It was a paper on the dehydration of butanol. Herman Pines' earliest discovery in acid catalysis was what he and Professor Ipatieff called conjunct polymerization (2). When low molecular weight olefins, for example, a butylene, are dissolved

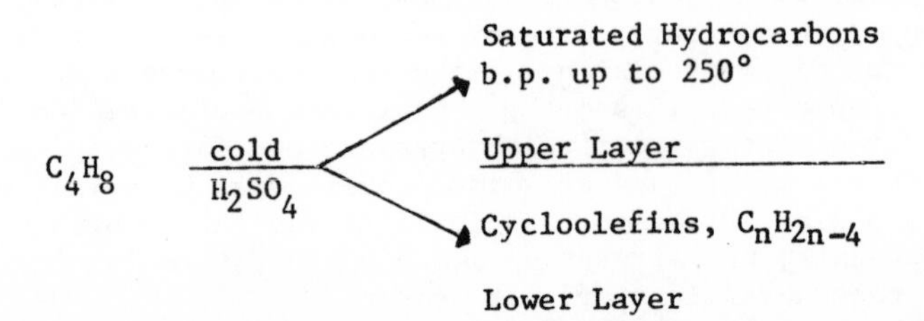

in cold sulfuric acid, they produce on standing paraffins and cycloparaffins by hydrogen transfer, and highly unsaturated hydrocarbons.

This work was typical of the kind of research Herman Pines did at Universal Oil Products in that he worked with pure hydrocarbons to study a reaction. His colleagues at Universal Oil Products, Herman Block, Julian Mavity and Ralph Thompson remember that this was always his approach. And that was in contrast to the use of petroleum fractions, which was typical at that time in a petroleum research laboratory.

But Herman Bloch notes that Pines was able to elucidate the basic underlying chemistry of hydrocarbons through working with pure compounds and many commercial processes stemmed directly or indirectly from his work. And one of these reactions was the alkylation of isoparaffins and cycloparaffins with olefins (3). In 1932 Pines and Ipatieff discovered the reaction and it continues today, perhaps even more so, to be a truly significant petroleum refining reaction. An example is the alkylation of isobutane with ethylene to give 2,3-dimethylbutane.

$$CH_3-\underset{CH_3}{\overset{CH_3}{\underset{|}{\overset{|}{C}}}}-H + CH_2{=}CH_2 \xrightarrow[HCl]{AlCl_3} CH_3-\underset{CH_3}{\underset{|}{CH}}-\underset{CH_3}{\underset{|}{CH}}-CH_3$$

It was this kind of reaction that convinced Pines that "paraffin" was a completely erroneous textbook designation. To this day he continues this theme that alkanes and cycloalkanes are not without chemical affinity. They react readily under the proper conditions.

Another discovery was soon to follow for Pines and Ipatieff, the isomerization of saturated hydrocarbons. Louis Schmerling, a colleague at Universal Oil Products, recalls the day Pines awaited the Professor's arrival to tell him of their analytical results. They had obtained much isobutane from *n*-butane. Schmerline remembers that Pines' enthusiasm was so great it was as though he was saying, "Eureka, we have it!" This enthusiasm was typical of his approach to research. His former students Herbert Appel and Donald LaZerte called it contagious. It is one of their strongest memories of their student days with Herman Pines.

In their isomerization work Pines and Ipatieff found it difficult to reproduce their results. Herman Pines set out to determine what the problem was. He and his assistant Richard Wackher did careful experiments using high vacuum lines for purification and isolation of the reaction from the atmosphere. They discovered pure n-butane was unreactive. Traces of an olefin were sufficient to cause reaction (4). Later Pines and Wackher showed that a source of cations was necessary for the isomerization and oxygen, water or ultraviolet radiation could produce cations (5, 6).

Pines, Block and Schmerling proposed the following mechanism (7).

$$RCH{=}CH_2 + HCl + AlCl_3 \longrightarrow R\overset{+}{C}HCH_3 + AlCl_4^-$$

$$R\overset{+}{C}HCH_3 + CH_3CH_2CH_2CH_3 \longrightarrow CH_3\overset{+}{C}HCH_2CH_3 + RCH_2CH_3$$

$$CH_3\overset{+}{C}HCH_2CH_3 \longrightarrow CH_3\underset{CH_3}{\underset{|}{C}}H\overset{+}{C}H_2 \longrightarrow CH_3\underset{CH_3}{\underset{|}{\overset{+}{C}}}CH_3$$

$$CH_3\overset{+}{C}(CH_3)CH_3 + CH_2CH_2CH_2CH_3 \longrightarrow CH_3CH(CH_3)CH_3 + CH_3\overset{+}{C}HCH_2CH_3$$

Here an olefin is the cation source. Formation of the cation is the chain initiation step. The chain is propagated by carbocation abstraction of an hydride from n-butane. Skeletal rearrangement occurs to form the more stable tertiary carbocation.

I'd like to comment on the experimental methodology used to study the reaction. Herman Bloch points out that Pines always was careful and used the best available techniques to do his work. The use of high vacuum technique here illustrates this point. The isomerization of cycloalkanes was fully studied by several students collaborating with Pines (8-17). Again they used high vacuum technique. And toward the end of this research carbon 14 labeling was done. Again, this illustrated Bloch's point: use of the best techniques.

Herman Pines joined Professor Ipatieff at Northwestern University at the time the Ipatieff Laboratory was being relocated from the basement of University Hall to the Technological Institute. Leon Gershbein, a student of Ipatieff and Pines at that time, recalls that Pines' thoughts and plans were constantly with the new Ipatieff Laboratory. He was successful in making it one of a kind among university chemistry laboratories throughout the world.

In the mid 1940's, Pines and Ipatieff began publishing what was to become a long series of papers in the studies of terpenes. In addition to offering interesting structures and optical activity, Louis Schmerling believes the terpenes allowed Pines and Ipatieff to study reactions at Northwestern that were of less interest to management at Universal Oil Products. In 1956, Herman Pines received the Fritzsche Award of the American Chemical Society for his research in catalytic and thermal reactions of terpenes.

Herman Pines' first graduate student, Ralph Olberg, did his research in the terpenes and recalls that Pines was always a highly motivating influence on him. One of the things that Herman Pines did that was motivating was to visit us in the laboratory frequently and ask us about our results and talk about future experiments. We realized he was very interested in our work. Many of us in fact sometimes got a little concerned about it. Pines, in his zest for chemical results would have one experiment after another to add to what we'd done. And, when, after three, four, or five years of work, we thought we would begin writing our dissertation, he was ready to add more experiments. Donald LaZerte recalls that, when he was ready to write, Pines was planning to add a whole new problem to his work. Somehow we managed to complete our work and satisfy Pines' scientific curiosity.

In the late 1940's Herman Pines started teaching a course in catalytic organic chemistry. The course emphasized hydrocarbon reactions, but it also covered other reactions such as hydrogenation of functional groups, dehydration of alcohols, and dehydrogenation of primary and secondary alcohols. The course was designed for senior undergraduates and graduate students in chemistry and chemical engineering. Pines would lecture on Wednesdays for two hours because he came to Northwestern only two days of the week, Saturdays and Wednesdays.

I recall from talking to other students in the course at that time that their first impressions were much like mine. Here I was two minutes into the first lecture that was going to last two hours and Pines was speaking in what he still calls the international language of science, "broken English". What had I gotten myself into? But by the time half an hour had passed, I understood everything he was saying. It was for most of us new and adventurous - a subject not found in textbooks. Pines was an excellent teacher because of his interest in and knowledge of the subject. He was energetic and enthusiastic in his presentation and we learned a great deal.

There was a teaching laboratory associated with the course. It was in a room located next to the research laboratory. I'm not sure at the time that we all appreciated the experience we were getting in that teaching laboratory. It was the first teaching organic laboratory that most of us had where we were given a wrench and told to go to work. We used batch apparatus, including different kinds of autoclaves and flow equipment at atmospheric and also moderate pressure. We learned how to experiment with pressured liquified gases. It was something unique that Northwestern offered to chemistry students.

There is a story about the design of this teaching laboratory that shows Herman Pines occasionally made mistakes too. George Czajkowski, an assistant to Pines and the Professor, recalls that he and Pines designed a system for cleaning batchtype autoclaves in the teaching laboratory. One could blow water, steam, or air through a tube into the inverted autoclave by choice of valves. Well, after their system was installed, Czajkowski found one morning that water had run out of the air lines into test equipment in the Mechanical Engineering Department the night before, and by morning, water had been coming out of air lines throughout the building. Obviously, the autoclave cleaning design was not without flaws.

Stanley Brown remembers how he chose Herman Pines as his research advisor. During their interview, Pines talked to him for an hour about the wonders of catalytic chemistry, escorted him through the Ipatieff Laboratory, introduced him to all the workers, and loaded him down with dozens of reprints. It was this kind of approach that enlarged the Ipatieff Laboratory group. By the late forties Pines and the Professor had attracted a good sized group of graduate students and postdocs. I think we were pleased that we were large enough to have others to talk to about common research interests and to conduct group seminars but not so large that we couldn't frequently see our research advisor. Leon Gershbein and Jean Germain remember that Pines was never too busy to drop whatever he was doing to help a student.

Herman Pines tried to have each student build some piece of equipment to add to the research laboratory, for example, an ozonizer or a boiling point apparatus. We had excellent equipment, and we were able to get a great deal done. The laboratory got the latest equipment appropriate for the type of research in the Ipatieff Laboratory. For example, the Ipatieff Laboratory had a gas chromatograph before most academic laboratories did.

But the salesman who thought he could convince Herman Pines to buy equipment wasn't successful usually. I recall two cases, a distillation column salesman who thought he could sell Pines a spinning band column and a salesman trying to sell an autoclave. We got the spinning band column and the autoclave, but Pines convinced the manufacturers to give the equipment to the laboratory because they could say their equipment was being used in the Ipatieff Laboratory. Also, publications showing our use of the equipment in the "Experimental" sections would help advertise it.

In the last year of the Professor's life, Pines and he began doing research in base catalyzed hydrocarbon reactions at Northwestern. After the Professor's death and after Pines left Universal Oil Products, he devoted increasing effort to these kinds of reactions. Among them was the isomerization of olefins (18, 19).

$$Na^+B^- + RCH_2CH{=}CH_2 \longrightarrow BH + (R\text{-}\bar{C}HCH{=}CH_2 \longleftrightarrow RCH{=}CH\bar{C}H_2)\ Na^+$$

$$RCH{=}CH\bar{C}H_2\ Na^+ + RCH_2CH{=}CH_2 \longrightarrow RCH{=}CHCH_3 + (RCH\text{---}\bar{C}H\text{---}CH_2)\ Na^+$$

A strong base such as sodium-benzylsodium abstracts a proton to produce an allylic carbanion. A chain propagation follows through the carbanion abstracting a proton to produce the allylic intermediate and an isomerized olefin.

Limonene isomerizes but it also dehydrogenates to p-cymene (20).

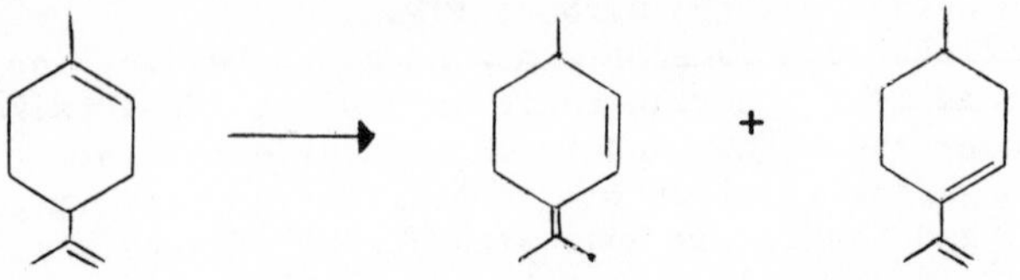

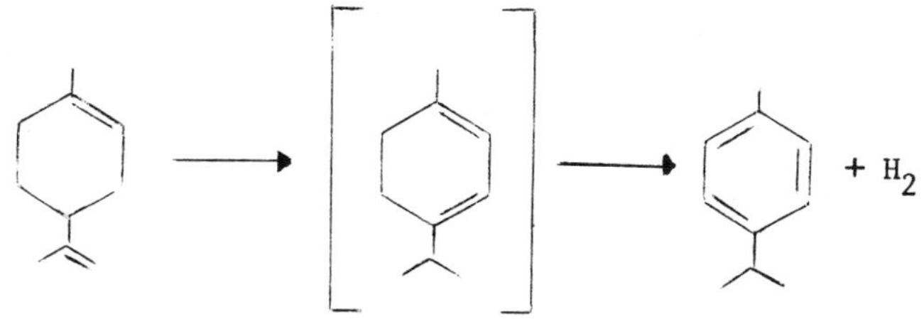

$$\longrightarrow \left[\ \right] \longrightarrow + H_2$$

A geminal dimethyl group does not prevent aromatization (21).

CH_3 CH_3 CH_2 $\longrightarrow$ CH_3 CH_3 $+ CH_4$

Alkylbenzenes are alkylated on benzylic carbons by olefins (22).

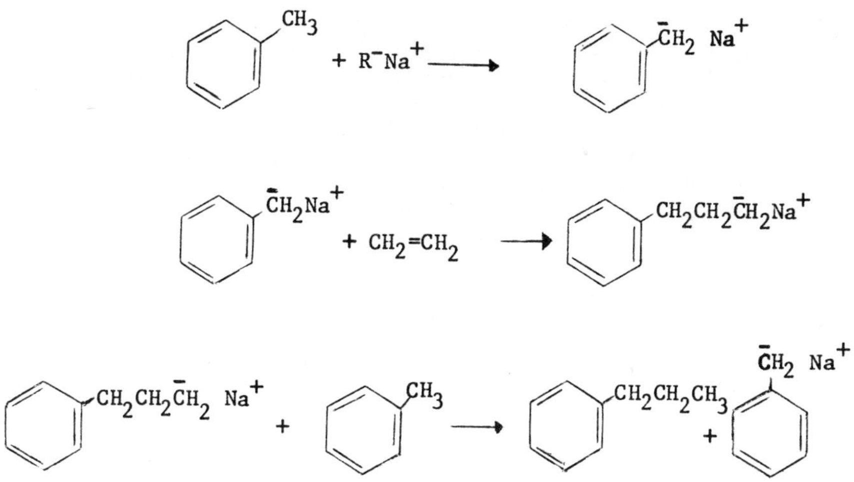

$$C_6H_5CH_3 + R^-Na^+ \longrightarrow C_6H_5\bar{C}H_2\ Na^+$$

$$C_6H_5\bar{C}H_2Na^+ + CH_2{=}CH_2 \longrightarrow C_6H_5CH_2CH_2\bar{C}H_2Na^+$$

$$C_6H_5CH_2CH_2\bar{C}H_2\ Na^+ + C_6H_5CH_3 \longrightarrow C_6H_5CH_2CH_2CH_3 + C_6H_5\bar{C}H_2\ Na^+$$

Alkenes dimerize and oligomerize in the presence of strong bases (23).

$$CH_3CH{=}CH_2 + B^- \longrightarrow BH + \bar{C}H_2CH{=}CH_2$$

$$\bar{C}H_2CH{=}CH_2 + CH_3CH{=}CH_2 \longrightarrow CH_2{=}CHCH_2\underset{\displaystyle CH_3}{\underset{|}{C}}H\bar{C}H_2$$

$$CH_2{=}CHCH_2\underset{\displaystyle CH_3}{\underset{|}{C}}H\bar{C}H_2 + CH_3CH{=}CH_2 \longrightarrow CH_2{=}CHCH_2\underset{\displaystyle CH_3}{\underset{|}{C}}HCH_3 + \bar{C}H_2CH{=}CH_2$$

In the mechanism of dimerization, an allylic carbanion is formed and adds to propylene so as to give the more stable primary ion. Thus, the dimer is branched.

For almost every reaction presented thus far - acid and base catalyzed - a mechanism was shown. The understanding of a reaction in terms of its mechanism has always been a primary goal in Pines' research. For several of his students like Eugene Aristoff and Bozidar Stipanovic, his interest in mechanisms is one of their strongest memories of his research approach. Herbert Appel recalls that Pines taught him never to be satisfied with a mechanism until it explains all the products.

Workers in the Ipatieff Laboratory were always aware of the differences in activity, particularly acid activity, of the silica and alumina they used. In the late fifties and early sixties, Pines and his students studied the intrinsic acidity and catalytic activity of aluminas as a function of their preparation method (24). They investigated the dehydration of alcohols (25-31) and many excellent papers were contributed in the alumina area and in the area of alumina supported dehydrogenation and aromatization catalysts (32, 33, 34).

In the early fifties, Pines and Ipatieff studied a very useful synthetic reaction, reductive dehydroxymethylation of primary alcohols (35, 36, 37).

$$2H_2 + RCH_2OH \xrightarrow{Ni} RH + CH_4 + H_2O$$

$$RCH_2OH \rightleftarrows RCHO + H_2$$

$$RCHO \longrightarrow RH + CO$$

$$CO + 3H_2 \longrightarrow CH_4 + H_2O$$

Pines, in pursuing this reaction, found that running the reaction in the presence of thiophene or methyl sulfide gave only reductive dehydroxylation of a primary alcohol (34).

$$H_2 + C_6H_5CH_2CH_2CH_2OH \xrightarrow[R_2S]{Ni} C_6H_5CH_2CH_2CH_3 + H_2O$$

R_2S = thiophene or methyl sulfide

When an alcohol prone to rearrange in the presence of acids was used, rearrangement occurred in the presence of sulfur but not in its absence. 3,3-Dimethyl-1-butanol gives neopentane when sulfur is absent but substantial amounts of 2,3-dimethylbutane when sulfur is present.

$$C\text{-}C(C)(C)\text{-}C \xleftarrow[H_2]{Ni} C\text{-}C(C)(C)\text{-}C\text{-}C\text{-}OH \xrightarrow[H_2]{Ni,\ R_2S} C\text{-}C(C)(C)\text{-}C\text{-}C + C\text{-}C(C)\text{-}C(C)\text{-}C$$

The latter compound was believed to be formed by dehydration and acid catalyzed rearrangement prior to hydrogenation.

A further extension of the nickel studies was the observation that nickel containing nickel oxide could be used as a catalyst for ether formation (39-49). Herman Pines viewed this as a concerted dehydration.

$$R\text{-}CH_2(OH \rightarrow Ni^{\delta+}) \leftarrow H\text{-}O\text{-}CH_2\text{-}R\ (H \leftarrow O^{\delta-}) \longrightarrow RCH_2OCH_2R + H_2O$$

The reaction was investigated using other metals, platinum, palladium, iridium, and rhodium (50, 51).

A sketch of Herman Pines would not be complete without reviewing some of his awards. After the A. C. S. Fritzsche Award, Herman Pines received the Midwest Award of the St. Louis Section of the American Chemical Society in 1963. In 1981 he received the American Chemical Society Award in Petroleum Chemistry. In 1981 he also received the Eugene J. Houdry Award in Applied Catalysis of the Catalysis Society. At its recent meeting in Las Vegas, he received the 1982 Chemical Pioneer Award of the American Institute of Chemists. Space limitations prevent elaboration of all of his honors.

But I do have enough space to give you perceptions of him by his students and colleagues in addition to those already mentioned. Herman Pines took a personal interest in the progress and well being of each of his students. For example, more than one of them needed more financial help at times than was available through their graduate student stipends. Herman Pines found sources who were willing to pay for preparative organic work, and the student overcame his financial difficulty by doing some organic preps.

He counseled students when they were finishing their graduate or postdoctoral work and seeking an employer. He helped his students in academic careers to get their names in print by coauthoring books and chapters in books with them.

He was always soft spoken. We never saw him angry. He was genuinely frank and never maliciously took apart a colleague. We all enjoyed his good sense of humor.

Herman Pines is still contributing to the chemical literature; I'm sure he will publish many more papers.

Literature Cited

1. Pines, H. J. Am. Chem. Soc. 1933, 55, 3892
2. Ipatieff, V. N.; Pines, H. J. Org. Chem. 1936, 1, 464.
3. Ipatieff, V. N.; Grosse, V.; Pines, H.; Komarewsky, V. I. J. Am. Chem. Soc. 1936, 58, 913.
4. Pines, H.; Wackher, R. C. J. Am. Chem. Soc. 1946, 68, 595.
5. Pines, H.; Wackher, R. C. J. Am. Chem. Soc. 1946, 68, 599.
6. Wackher, R. C.; Pines, H. J. Am. Chem. Soc. 1946, 68, 1642.
7. Bloch, H. S.; Pines, H.; Schmerling, L. J. Am. Chem. Soc. 1946, 68, 153.
8. Pines, H.; Abraham, B. M.; Ipatieff, V. N. J. Am. Chem. Soc. 1948, 70, 1742.
9. Pines, H.; Aristoff, E.; Ipatieff, V. N. J. Am. Chem. Soc. 1949, 71, 749.
10. Pines, H.; Aristoff, E.; Ipatieff, V. N. J. Am. Chem. Soc. 1950, 72, 4055.
11. Pines, H.; Aristoff, E.; Ipatieff, V. N. J. Am. Chem. Soc. 1950, 72, 4304.
12. Pines, H.; Pavlik, F. J.; Ipatieff, V. N. J. Am. Chem. Soc. 1951, 73, 5738.
13. Pines, H.; Pavlik, F. J.; Ipatieff, V. N. J. Am. Chem. Soc. 1952, 74, 5544.
14. Pines, H.; Huntsman, W. D.; Ipatieff, V. N. J. Am. Chem. Soc. 1953, 75, 2315.
15. Pines, H.; Aristoff, E.; Ipatieff, V. N. J. Am. Chem. Soc. 1953, 75, 4775.
16. Pines, H.; Myerholtz, R. F., Jr.; Neumann, H. M. J. Am. Chem. Soc. 1955, 77, 3399.
17. Pines, H.; Myerholtz, R. W., Jr. J. Am. Chem. Soc. 1955, 77, 5392.
18. Pines, H.; Vesely, J. A.; Ipatieff, V. N. J. Am. Chem. Soc. 1955, 77, 347.
19. Pines, H.; Haag, W. O. J. Org. Chem. 1958, 23, 328.
20. Pines, H.; Eschinazi, H. E. J. Am. Chem. Soc. 1955, 77, 6314.
21. Pines, H.; Eschinazi, H. E. J. Am. Chem. Soc. 1956, 78, 5950.
22. Pines, H.; Vesely, J. A.; Ipatieff, V. N. J. Am. Chem. Soc. 1955, 77, 554.
23. Mark, V.; Pines, H. J. Am. Chem. Soc. 1956, 78, 5946.

24. Pines, H.; Haag, W. O. J. Am. Chem. Soc. 1960, 82, 2471.
25. Pines, H.; Haag, W. O. J. Am. Chem. Soc. 1961, 83, 2847.
26. Pines, H.; Manassen, J. Adv. Catal. 1966, 16, 49.
27. Blanc, E. J.; Pines, H. J. Org. Chem. 1968, 33, 2035.
28. Pines, H.; Pillai, C. N. J. Am. Chem. Soc. 1961, 83, 3270.
29. Schappell, F. G.; Pines, H. J. Org. Chem. 1966, 31, 1735.
30. Watanabe, K.; Pillai, C. N.; Pines, H. J. Am. Chem. Soc. 1962, 84, 3932.
31. Pines, H.; Brown, S. M. J. Catal. 1971, 20.
32. Pines, H.; Chen, C. T. J. Am. Chem. Soc. 1960, 82, 3562.
33. Pines, H.; Csicsery, S. M. J. Am. Chem. Soc. 1962, 84, 292.
34. Pines, H.; Goetschel, C. T. J. Catal. 1966, 6, 371.
35. Ipatieff, V. N.; Czajkowski, G. J.; Pines, H. J. Am. Chem. Soc. 1951, 73, 4098.
36. Pines, H.; Rodenberg. H. G.; Ipatieff, V. N. J. Am. Chem. Soc. 1953, 75, 6065.
37. Pines, H.; Rodenberg, H. G.; Ipatieff, V. N. J. Am. Chem. Soc. 1954, 76, 771.
38. Pines, H.; Shamaiengar, M.; Postl, W. J. J. Am. Chem. Soc. 1955, 77, 5099.
39. Pines, H.; Steingaszner, P. J. Catal. 1968, 10, 60.
40. Pines, H. and Kobylinski, T. P., J. Catal. 1970, 17, 375.
41. Kobylinski, T. P.; Pines, H.; J. Catal. 1970, 17, 384.
42. Pines, H.; Kobylinski, T. P. J. Catal. 1970, 17, 394.
43. Hensel, J.; Pines, H. J. Catal. 1972, 24, 197.
44. Pines, H.; Hensel, J.; Simonik, J. J. Catal. 1972, 24, 206.
45. Simonik, J.; Pines, H. J. Catal. 1972, 24, 211.
46. Pines, H.; Simonik, J. J. Catal. 1972, 24, 220.
47. Licht, E.; Schachter, Y.; Pines, H. J. Catal. 1973, 31, 110.
48. Licht, E.; Schachter, Y.; Pines, H.; J. Catal. 1974, 34, 338.
49. Licht, E.; Schachter, Y.; Pines, H. J. Catal. 1975, 38, 423.
50. Licht, E.; Schachter, Y.; Pines, H. J. Catal. 1978, 55, 191.
51. Licht, E.; Schachter, Y.; Pines, H. J. Catal. 1980, 61, 109.

RECEIVED November 10, 1982

8

Early Catalysis Research with Para-Hydrogen and Heavy Hydrogen

ADALBERT FARKAS
Boca Raton, FL 33431

The Kaiser Wilhelm Institute

In the fall of 1928, I came to Berlin to start my Ph.D. thesis under Dr. Karl-Friedrich Bonhoeffer in the Kaiser Wilhelm Institute for Physical Chemistry and Electrochemistry. I was accepted as a doctoral student at the recommendation of my brother Ladislas who was then private assistant to Fritz Haber, the director of the Institute and a coworker of Dr. Bonhoeffer.

Haber's Institute was located in the prosperous Berlin suburb of Dahlem in a campus-like setting together with other K. W. institutes devoted to biochemistry, inorganic chemistry, silicate and fiber research. It was one of the most famous and most generously endowed research centers in the world. The Institute had modern equipment, workshops, a library, access to a luxurious clubhouse and even two tennis courts. It served as the focal point for seminars and meetings on physical chemistry in which Fritz Haber, the winner of the 1918 Nobel prize was the undisputed leader.

For me this environment was an astounding change from the pre-World War I facilities of the Vienna Technical University, its antiquated lecture halls and laboratories.

Haber had assembled in his Institute a brilliant group of scientists whose research interest centered around the then frontiers of physical chemistry: photochemistry, spectroscopy, the role of atoms and radicals in chemical reactions and colloid chemistry. The department heads included R. Ladenburg, H. Freundlich, M. Polanyi, K.-F. Bonhoeffer, and later Paul Harteck.

0097-6156/83/0222-0089$08.25/0

Fritz Haber (Figure 1) was a legend in his time. His synthetic ammonia process brought him world wide fame. Everybody addressed him as "Herr Geheimrat" (Mr. Privy Councillor), a high honorary title given to prominent officials. Haber had an encyclopedic knowledge and was an accomplished story teller. In scientific meetings, Haber was a master in summarizing and analyzing divergent views and in clarifying complicated relations.

Bonhoeffer was a scion of an old cultured family of professors, ministers and doctors. He was a delightful person with a boyish charm and a brilliant scientist.

I was originally given the task of looking for the hydrides of gold and silver and determining the dissociation energy of these hypothetical compounds.

Ortho- and Parahydrogen

In the meantime Bonhoeffer and Harteck were looking for the two mysterious modifications of hydrogen, the existence of which was predicted by the quantum theory on the basis of a curious decline of the specific heat of hydrogen at low temperatures and of the intensity variations of a series of lines in the molecular spectrum of hydrogen.

The two modifications were supposed to show a difference in the orientation of the nuclear spins of the two atoms in the hydrogen molecule and in the rotational quantum numbers. One modification, the para-form, was thought to have antiparallel spin orientation and to rotate with even quantum numbers, while the other, the ortho-form would have parallel spin orientation and odd rotational quantum numbers.

The theory postulated that ordinary hydrogen was a mixture of para- and ortho-hydrogen in the proportion 1:3 at room temperature and that this proportion would shift to favor the para modification at lower temperature if one could somehow facilitate the "forbidden" transitions between even and odd rotational states.

Since the conversion of ortho-hydrogen to parahydrogen causes an increase in the specific heat and heat conductivity of the hydrogen, Bonhoeffer and Harteck used a thermal conductivity to detect the conversion. In February 1929 they were able to report (1) that the expected conversion at $-193^{o}C$ was very slow at low pressures but reached equilibrium at 350 atmospheres in about a week. They also found that the conversion was very noticeable in one day old liquid hydrogen and that activated charcoal was a

Figure 1.
Dr. Fritz Haber, director of the Kaiser Wilhelm Institute for Physical Chemistry and Electrochemistry.

very effective catalyst for converting ordinary hydrogen into a 1:1 ortho-para-mixture at -193°C and essentially pure para-hydrogen at -253°C.

Bonhoeffer and Harteck showed that para-hydrogen was a very stable gas at room temperature and that it was converted into ordinary hydrogen on exposure to high temperature, high pressure or to an electric discharge and by passage over platinized asbestos.

This sensational discovery was first published in Naturwissenschaften in February 1929 (1). The editor of the chemistry section of this journal was A. Eucken, a prominent professor of physical chemistry at Gottingen University who happened to be searching for the two hydrogen modifications at about the same time. Alerted by the Bonhoeffer-Harteck communication passing over his desk, Eucken hastily wrote up his own results and managed to have his note appear on the same page of Naturwissenschaften.

Soon thereafter Bonhoeffer and Harteck published two papers (2-3) on the preparation, reactions and properties of para-hydrogen.

The news of the existence of the two hydrogen modifications electrified the scientific world. Bonhoeffer and Harteck received innumerable invitations to present papers on para-hydrogen in seminars and meetings. Later in 1929, Bonhoeffer went on a triumphal lecture tour of American Universities.

By the summer of 1929 I had obtained positive spectroscopic and gravimetric results to prove the existence of the hydrides of gold and silver and to estimate their heats of dissociation (4). I asked Bonhoeffer whether I could work on some aspect of the ortho-para-hydrogen conversion. He suggested that I try to find out first the mechanism of the thermal para-hydrogen conversion and then that of the catalytic conversion.

The Thermal Para-Hydrogen Conversion

My study of the thermal para-hydrogen conversion carried out in stationary and flow systems at 600-750°C in quartz reactors gave clear-cut results. The reaction was homogeneous and had an activation energy of 58.7 kcals/mole. It followed first order kinetics as far as the disappearance of the excess para-H_2 concentration was concerned, and one and one half order in its pressure dependence.

These results indicated a hydrogen atom-catalyzed conversion reaction in which properly oriented H-atoms formed by thermal dissociation of molecular hydrogen interacted with H_2 molecules according to (5,6):

$$H + p\text{-}H_2 \underset{2}{\overset{k_1}{\rightleftharpoons}} o\text{-}H_2 + H$$

$$\downarrow + [\uparrow\uparrow] \rightleftharpoons [\downarrow\uparrow] + \uparrow$$

where $k_1/k_2 = 3$.

I mentioned this mechanism while it was still in the tentative state to a young American National Research fellow named Henry Eyring who was there working with Professor Michael Polanyi on the theory of atomic reactions. Henry was very enthusiastic about my experimental results and their interpretation. He thought that my work concerned the interaction of three hydrogen atoms - one of the simplest chemical reactions - and thus would be amenable for theoretical treatment (7).

The Catalytic Para-Hydrogen Conversion

In 1930, Dr. Bonhoeffer at 31 was appointed professor of physical chemistry at the University of Frankfurt am Main and invited me to come along as his assistant.

We decided to study first the catalytic conversion of para-hydrogen on electrically heated platinum, nickel and tungsten wires and then the formation of para-hydrogen at low temperatures under pressure and in the presence of charcoal.

This work was started in the old physical chemistry laboratory and then continued in a brand-new modern institute built according to Bonhoeffer's specifications.

Two Types of Catalytic Mechanisms

The results of these studies carried out in 1930 and 1931 led to the following conclusions (8,9):

1. The temperature and pressure dependence of the various conversion reactions suggest that there are essentially two types of catalytic ortho-para-hydrogen conversion reactions

2. The catalytic conversions occurring at elevated temperatures on metals involve atomic or

dissociative adsorption in which the bond between the two atoms in the hydrogen molecule is loosened to such an extent that the individual atoms are bonded more strongly to the metal than to each other. Thus the atoms lose their original spin orientation and yield, on recombination, ortho- and para-molecules in a proportion corresponding to the temperature of the catalyst

3. The atomic adsorption on metals is not caused by van-der Waals forces, but depends on forces similar to those postulated by Slater for explaining metallic cohesion

4. The high temperature catalytic para-H_2 conversion has a positive temperature coefficient, follows a reaction order less than one and is affected by poisons

5. The para-H_2 formation on charcoal at low temperatures involves molecules held by van der Waals forces, depends little on temperature and follows first order kinetics.

Towards the end of 1931, Dr. Bonhoeffer and I wrote a paper (10) on this topic for the Faraday Society General Discussion on "Adsorption of Gases on Solids" scheduled for January 1932 in Oxford.

The Faraday Society Discussion of 1932

To my pleasant surprise, Dr. Bonhoeffer decided that I present our paper at Oxford. My trip to England, disregarding the extremely rough channel crossing, turned out to be a tremendous success. It also proved of great impact to me in the following year.

As an overseas visitor, I was given the full VIP treatment as soon as the boat train arrived in London. The President of the Faraday Society, Dr. Robert L. Mond (the grandson of Ludwig Mond of the Mond Process fame) entertained the visitors at dinner at the then brand-new Dorchester Hotel in London the night before the Discussion. Professor F. G. Donnan of University College and Professor E. K. Rideal of Cambridge University acted as cohosts. At the dinner I met Professor H. S. Taylor of Princeton and two old acquaintances from the K. W. Institute, Professors H. Freundlich and M. Polanyi. These three and Professor Rideal were the four principal speakers at the Symposium.

There was a tragicomical incident at the dinner that went unnoticed by all but me. After duly toasting the King, our hosts toasted the heads of state of the visitors. So in turn we toasted President Roosevelt, Fieldmarshal Hindenburg and finally, in my honor, Admiral Horthy, the autocratic regent of Hungary. Thus it came about that I had to drink a toast to the health of a despicable anti-semite whose infamous regime invented the "numerus clausus" more than a decade before Hitler came to power.

After spending the night at the Dorchester, we were taken by train - first class, of course - to Oxford the next day.

The papers had been distributed in advance in preprint form and were considered as having been read. Each author was given only a few minutes for the presentation of the highlights of his paper.

Our paper (5), presented in the section entitled "Energetics and Kinetics of Adsorption" emphasized the thesis that high-temperature para-H_2 conversion on metals is caused by dissociative adsorption of hydrogen while the formation of para-H_2 on charcoal at low temperatures involved adsorbed molecules.

Our paper was followed by a paper by Taylor and Sherman (11) which sought to establish a connection between ortho-para-H_2 conversion and activated adsorption - a temperature dependent sorption process observed on a variety of metals and oxides. Professor Taylor ascribed the low temperature para-H formation on charcoal also to activated adsorption and claimed that the reaction was bimolecular.

However this explanation of the low temperature conversion became rather unconvincing in the light of a communication of Professors Polanyi and Cremer (12,13) which showed that there is a homogeneous para-H_2 formation in solid hydrogen at 12° K not unlike the conversion observed in liquid H_2 by Bonhoeffer and Harteck. Polanyi and Cremer pointed out that in the solid state the conversion very likely occurs as a consequence of van der Waals forces and not via atomic exchange and that a similar mechanism might be operative in the adsorption layer on such catalysts as charcoal.

In the Discussion I referred to a paper presented by F. E. T. Kingman (14) describing activated adsorption of H_2 on charcoal occurring above 400° C and requiring an activation energy of 30 kcals/mole. Because of this observation and the fact that charcoal was an active catalyst for para-H_2 conversion at -253° C I thought it very unlikely that activated adsorption had any role in this case.

Professor Rideal was apparently very impressed by our work and asked me whether I would like to write a book on ortho- and para- hydrogen for the Cambridge Series of Physical Chemistry of which he was general editor.

Of course, I felt very flattered and agreed to do so without hesitation. The contract was signed the same year but was subsequently revised to cover heavy hydrogen as well.

An Experimental Flaw

Shortly after I returned to Frankfurt I found out that there was a flaw in my paper on the catalytic para-H_2 conversion on tungsten and nickel wires.

The flaw was caused by a curious physical effect. The uniform temperature distribution along electrically heated wires can become unstable if the temperature coefficient of the electric resistance of the wire exceeds the temperature coefficient of the heat loss. In such a case the temperature distribution can become non-uniform showing hot spots, and the temperature of the wire can not be calculated from the resistance.

In my experiments this effect caused certain discontinuities in the rate measurements which I ascribed to the formation and decomposision of surface hydride layers.

Needless to say, the error was duly corrected (15). To avoid this complication we abandoned further catalytic experiments with wires and continued our studies on evaporated metal films, metal tubes and on charcoal.

Paramagnetic Para-Hydrogen Conversion

In connection with the elucidation of the catalytic conversion on charcoal we gained an illuminating insight from a striking discovery made by my brother Ladislas and H. Sachsse (16-18) at the K.W. Institute while investigating detonating gas mixtures containing para-H_2.

They found that in the presence of O_2, parahydrogen was rapidly converted to normal hydrogen. Closer study showed that the conversion was due to the paramagnetism of the oxygen molecule and that other paramagnetic molcules and ions had similar effects.

In this case the para-ortho conversion was a consequence of the large perturbation caused by the very strong inhomogoneous magnetic field of these paramagnetic molecules and ions which lifted to some extent the prohibition of the even-odd rotational transitions.

Since the oxygen induced conversion showed little dependence on temperature as did the conversion on charcoal we suggested that the latter conversion is also caused by magnetic interaction. This concept was confirmed by the striking accelerating effect of small amounts of added oxygen on the catalytic activity of charcoal (19).

Taylor and Diamond (20) provided further evidence for this mechanism by demonstrating the efficiency of paramagnetic oxides as low temperature conversion catalysts.

In the beginning of 1933 we published a paper entitled "The Heterogeneous Catalysis of the Para-Hydrogen Conversion" in which we summarized all our results obtained on metals, charcoals and salts (21). Our conclusion was that there are two mechanisms for the conversion. At low temperatures the conversion is monomolecular, occurs in the van der Waals adsorption layer and is probably caused by magnetic effects. At higher temperatures the conversion involves atomic adsorption.

The latter mechanism became known as the Bonhoeffer-Farkas mechanism and seemed to enjoy general acceptance for a few years.

Exit From Frankfurt

In an effort to extend the catalytic studies to lower pressures I developed a modified thermal conductivity method which required only samples of $2\text{-}3 \times 10^{-3}$ cm^3 of hydrogen for the para-hydrogen analysis (22).

The method worked perfectly in our laboratory and also at the K.W. Institute in my brother's hands. But my plans for more extensive work were rudely wiped out when I lost my position because of the advent of the Nazi regime.

When Dr. Bonhoeffer officially had to give me notice, I almost felt more sorry for him than for myself because I knew how much pain the Nazi takeover caused him. He was an outspoken critic of Nazism and, surprisingly, he survived the regime unscathed.

His brother Dietrich, also an opponent of the Nazis, did not fare so well. Dietrich was a theologian of renown as the author of widely acclaimed books and articles on ecumenism. During World War II he was a member of military intelligence but worked secretly for the resistance movement. Dietrich was arrested in 1943 and was executed two years later when some seized documents connected him with the assassination attempt on Hitler.

Rideal's Laboratory in Cambridge

The summer of 1933 found me in beautiful Cambridge where Professor Rideal extended most generous hospitality in his laboratory and in his lovely home. Eric K. Rideal was the coauthor with Hugh S. Taylor of a pioneering text entitled "Theory and Practice of Catalysis". He was a gentleman and a scholar in the truest sense of the phrase. Brilliant, witty and easy going, Rideal made you comfortable the moment he started to talk to you.

Rideal's right-hand man at the laboratory was Dr. Owen H. Wansbrough-Jones, an old acquaintance of mine from the K.W. Institute where he worked with my brother a few years before. He was most helpful in looking after me while I got settled and later in smoothing out the rough spots in my English manuscript.

Rideal had then a relatively cramped laboratory for his Department of Colloid Science on School House Lane. He told me I could start my work in a corner of his big office which was in the attic. I suggested that I set up my microconductivity method and check whether it would work for the analysis of heavy hydrogen which then started to become available.

Within a few weeks I installed my thermal conductivity apparatus after many hours of glass blowing while standing on one foot and pumping the bellows with the other for my blowtorch since there was no compressed air in the laboratory.

In the meantime I found out that Paul Harteck was working next door in Lord Rutherford's Cavendish Laboratory on the recovery of heavy water from water containing 0.3% deuterium supplied by the Ohio Chemical Company. The concentration was done by electrolysis, and we joined forces on this project.

Soon thereafter my brother arrived from Berlin, and we moved downstairs into a spacious laboratory that became available. We incorporated the provisional equipment from the attic into a more elaborate vacuum line which allowed the production of heavy hydrogen from heavy water, its analysis and the study of its reactions.

The microconductivity method worked splendidly for deuterium analysis - better than we ever hoped.

Its economical sample requirement was a great advantage in view of the scarcity and cost of deuterium.

Toward the end of 1933 Professor Haber came to visit Lord Rutherford. Paul Harteck brought him over to our laboratory and we had a brief reunion. But by now Haber was not the imposing Geheimrat anymore. He was a sick, broken man who had lost his position, his home, his fatherland and most of his income. He died a few months later in Basel.

In the room next to ours, Dr. J. K. Roberts was studying the adsorption of hydrogen on very pure tungsten wires using original techniques developed by him. Little did we know that a few years later he would question the validity of the Bonhoeffer - Farkas mechanism of the catalytic para-hydrogen conversion.

The first report on our heavy hydrogen experiments appeared in Nature on December 9, 1933 (23). We used heavy water donated by Paul Harteck and demonstrated the occurrence of the thermal and catalytic reaction $H_2 + D_2 = 2HD$, the exchange reaction $D_2 + H_2O = HD + HDO$, and the partial separation of hydrogen-deuterium mixtures by diffusion through palladium (24).

The Royal Society Discussion on Heavy Hydrogen

Professor Rideal presented a more complete report on this work (25) in a Royal Society Discussion presided over by Lord Rutherford and held in London early in 1934 (26).

Paul Harteck (27) described the electrolytic concentration of heavy water and demonstrated the difference in the vapor pressures of light and heavy water. Professor Polanyi, by then in Manchester, reported on the exchange reaction involving gaseous heavy hydrogen and liquid water in the presence of platinum black and explained it by ionization. Professor Bernal discussed the physical properties of heavy water and pointed out its strikingly higher viscosity and lower surface tension compared to the properties of light water.

This Discussion had also some amusing aspects. Professor F. Soddy, the winner of the Nobel Prize in 1921 for his work on the radioactive elements, took exception to calling heavy hydrogen an isotope of hydrogen. He reminded the audience that he coined the name "isotope" in 1911 to designate variants of elements that differ in atomic weight but are chemically inseparable. He insisted that heavy hydrogen can not be called an "isotope" since it can be separated from light hydrogen.

Lord Rutherford did not like "deuterium", the name given by Urey to heavy hydrogen. Disregarding the traditionally acknowledged privilege of the discoverer to name his discovery, he proposed the name "diplogen", composed of the root of "diplos", the Greek word for double and the ending "gen" of hydrogen. This etymological monstrosity was used in deference to Lord Rutherford during the Discussion and for a short time thereafter.

Experiments with Deuterium

Thanks to the efficiency of the microthermal conductivity method, we published a series of papers concerned with the relative reactivity of H_2 and D_2 with C_2H_4 (28) and Cl_2 (29), the separation of the hydrogen isotopes by chemical reactions (30) and electrolysis (31), and the equilibrium $H_2O + D_2 = HDO + HD$ (32).

We found that the addition of D_2 to ethylene in the presence of catalytically active nickel wire is accompanied by the exchange reaction

$$C_2H_4 + D_2 \rightarrow C_2H_3D + HD$$

above 60°C. We ascribed this reaction to the dissociative adsorption of ethylene. This mechanism differed from that proposed by Horuiti and Polanyi (33,34) for the D_2-benzene exchange according to which hydrogenation and exchange are connected by a half-hydrogenated intermediate formed by

$$D + C_6H_6 \rightarrow C_6H_6D \rightarrow C_6H_5D + H$$

Decomposition of this intermediate leads to exchange while further addition of hydrogen to it results in hydrogenation.

Jointly with Harteck (35) we demonstrated the existence of the ortho-para modifications of deuterium using active charcoal as the catalyst at 20°K. Because of a breakdown in the hydrogen liquefier at Cambridge, we had to fetch the liquid hydrogen necessary for our experiments from Oxford. The ride back was a hair-raising experience. My brother had to hold a five-liter Dewar vessel in his lap in the rumble seat while the liquid hydrogen sloshed around wildly.

In the case of deuterium, the normal mixture 2/3 ortho-D_2 and 1/3 para-D_2 and the stable form at 20° K is ortho-D_2. The reconversion of ortho-D into normal D_2 in the presence of oxygen was about ten times as slow as the corresponding conversion of para-H_2, indicating the magnetic moment of the deuteron nucleus is about a fourth of that of the proton.

Chemical Separation of Deuterium

As we noted that in most hydrogen-consuming or -producing reactions there is a separation of the isotopes. We became intrigued with the possibility of finding a chemical separation method which would not consume enormous quantities of electricity and could make deuterium and heavy water inexpensive.

We found that the equilibrium constants of the exchange (32)

$$H_2O + HD \rightleftharpoons HDO + H_2$$

are substantially different at low and high temperatures (3 at 3°C and 2 at 100°C) and conceived a dual temperature catalytic exchange process by which the deuterium in ordinary water could be readily concentrated.

Professor Rideal submitted our idea to the I.C. Industries who sponsored our work at that time. In due course their chief engineer, Mr. Appleby, came to see us in Cambridge. We discussed the proposal in detail. Mr. Appleby agreed that the suggestion sounded workable but wanted to know whether there was a large scale use for heavy water that would justify commercial production. We had to admit that we knew of none, whereupon the matter was dropped.

There was a large scale application for heavy water in the offing but nobody knew it yet in 1934. It was the use of heavy water as a moderator for fast neutrons in a nuclear chain reaction.

A few years later in 1940, we learned that there was great interest in the U.S.A. in heavy water for a very important secret process.

When I arrived in the U.S.A. in 1941 I submitted a memorandum to Dr. Leo Szilard on the various methods for concentrating heavy water or deuterium that we knew.

Dr. Szilard acknowledged my memorandum and informed me that he forwarded same to Professor Urey. Shortly thereafter I received an acknowledgement from

Professor Urey dated January 19, 1942, which was stamped "Secret" and had a warning that the letter contained information affecting the national defense of the U.S. within the meaning of the espionage act (Figure 2).

About thirty years later, I came across a memorandum dated October 26, 1942, in which the I.C.I. submitted our Cambridge idea for concentrating heavy water to the Atomic Energy Commission of the U.S.A.

As it turned out, the dual temperature H_2O-H_2 exchange process was never used because the exchange was too slow, and a much better exchange reaction was found in the system H_2O-H_2S which does not require a catalyst.

Interestingly the Atomic Energy Commission of Canada has been working on the H_2O-HD exchange reaction up until recently even though they are operating enormous heavy water plants based on the H_2O-H_2S exchange.

Two Invitations

Towards the end of 1934 my brother was invited by Professor Weizmann to join the faculty of the Hebrew University in Jerusalem as professor of physical chemistry. He decided to accept and to leave the following year.

When Professor Harold Urey was awarded the 1934 Nobel Prize for chemistry, he visited us in Cambridge on his way to Stockholm. He seemed very interested in our work and invited me to come to the U.S.A. for a lecture tour in April 1935.

In anticipation of our impending separation my brother and I continued our planned work at a furious pace. We compared the rates of the thermal reconversion of parahydrogen and orthodeuterium with that of the $H_2 + D_2 = 2HD$ reaction (36) and showed that the effect on the rates of the differences in the zero-point energies of the hydrogen molecules H_2, D_2 and HD are almost completely compensated by corresponding differences in the zero-point of the respective triatomic activated complexes H_3, H_2D, HD_2 and D_3.

We determined the ratio of the magnetic moments of the proton and the deuteron with improved accuracy (37). We demonstrated that even thoroughly outgassed materials such as glass, quartz and metal wires retain a certain amount of hydrogen, the presence of which becomes evident by exchange with deuterium at elevated temperature (38).

SECRET

Columbia University in the City of New York

DEPARTMENT OF CHEMISTRY

January 19, 1942

Dr. A. Farkas
Hotel Ontario
Ontario, Calif.

Dear Dr. Farkas:

Your interesting suggestions in regard to the production of heavy water have been turned over to me by Dr. Szilard. I am very glad indeed to have them, and will transmit them to the proper authorities.

I hope that you are being successful with your business affairs in California, and that we shall see you back in New York before long.

With best regards.

Very sincerely,

Harold C. Urey

Harold C. Urey

THIS DOCUMENT CONTAINS INFORMATION AFFECTING THE NATIONAL DEFENSE OF THE U. S. WITHIN THE MEANING OF THE ESPIONAGE ACT, 50 U. S. C., 31 AND 32. ITS TRANSMISSION OR THE REVELATION OF ITS CONTENTS IN ANY MANNER TO AN UNAUTHORIZED PERSON IS PROHIBITED BY LAW.

Figure 2.
Professor Urey's acknowledgment of suggestion for production of heavy water.

Early in April 1935, I sailed from Liverpool on the Aquitania to New York. Before the ship docked, I received word that Professor Urey would meet me at the pier. He was accompanied by L.I. Rabi, professor of physics at Columbia. After I passed customs, these two gentlemen insisted that they carry my luggage to Dr. Urey's car parked nearby. As I was not even 29 then, I was very embarrassed at having a Nobel Prize winner handle my suitcases. Actually it turned out to be worse than that: Dr. Rabi was awarded the Nobel Prize in physics in 1944 so that I had the services of two Nobel Prize men!

Professor Urey arranged lecture engagements for me at Princeton, Columbia, Brown, Cornell and MIT-Harvard (Figure 3). I also spent some time sightseeing and visited Dr. Langmuir in his home in Schenectady. I attended the A.C.S. National Meeting then held in New York which to my great surprise included a dance.

At the beginning of 1935, my book on "Orthohydrogen, Parahydrogen and Heavy Hydrogen" was published by the Cambridge University Press (39), and I became the proudest man in Cambridge when all the reviews turned out to be very flattering.

The Russians must have liked the book, too, because without much delay, without permission or royalty payment, they published a Russian translation of it (Figure 4).

On My Own in Cambridge

After my brother left for Jerusalem I continued the work we had planned jointly which included the study of catalytic deuterium exchange in the systems D_2-NH_3 (40) and D_2-H_2O (41) conjunction with using the para-H conversion as a means for monitoring the activity and surface coverage of the catalyst. In both systems the para-H_2 conversion measurements showed strong adsorption of ammonia and water respectively.

The catalytic H_2O-D_2 exchange study led to a simple method for the analysis of the deuterium content of heavy water involving the catalytic equilibriation of water vapor with hydrogen on a platinum wire spiral and the measuring the deuterium content of the gas (42). This method was presented at the 1936 Royal Society Conversazione - an annual exhibition of scientific discoveries and techniques (Figure 5).

The para-H_2 conversion method was also used in elucidating the mechanism of the diffusion of

HARVARD UNIVERSITY
MASSACHUSETTS INSTITUTE OF TECHNOLOGY

DR. A. FARKAS

OF CAMBRIDGE, ENGLAND

WILL LECTURE ON

The Elementary Reactions of
Light and Heavy Hydrogen

THURSDAY, APRIL 18, 1935, AT 4 P.M.

Mallinckrodt MB8
12 Oxford Street, Cambridge

Figure 3.
Notice for Harvard University - M.I.T. lecture during a U.S. visit in 1935.

А. ФАРКАС

ОРТОВОДОРОД,
ПАРАВОДОРОД
И
ТЯЖЕЛЫЙ ВОДОРОД

Перевод с английского
Р. Х. БУРШТЕЙН и С. Д. ЛЕВИНОЙ

ОНТИ 1936
ГЛАВНАЯ РЕДАКЦИЯ ХИМИЧЕСКОЙ ЛИТЕРАТУРЫ • МОСКВА

ORTHOHYDROGEN, PARAHYDROGEN
AND
HEAVY HYDROGEN

by

ADALBERT FARKAS
Dr.phil.nat. (Frankfurt), Dr.-Ing. (Vienna).

CAMBRIDGE
AT THE UNIVERSITY PRESS
1935

Figure 4.
Title page of the English and Russian version of the book "Orthohydrogen, Parahydrogen and Heavy Hydrogen." The Russian translation was published shortly after the English edition without permission or royalty payment.

THE

ROYAL SOCIETY

SIR WILLIAM BRAGG, O.M.,

PRESIDENT

CONVERSAZIONE

BURLINGTON HOUSE

May 28th 1936

22. *Dr. A. Farkas, in collaboration with Professor L. Farkas*

Micro-Methods for the Analysis of Heavy Hydrogen and Heavy Water.

The micro-heat-conductivity method for the analysis of heavy hydrogen is based upon the variation of the specific heat of hydrogen with temperature. It allows us not only to determine the percentage deuterium in a few cubic millimetres of gas within a few minutes but also to estimate the concentrations of the molecular species H_2, HD and D_2 and to discriminate between ortho and para modifications of deuterium or hydrogen. Applying the same method for the analysis of heavy water not directly the deuterium content of the water is estimated but that of hydrogen brought in equilibrium with the water to be analysed. The equilibrium $HDO + H_2 = H_2O + HD$ is established in the vapour phase on a hot platinum filament in a few minutes.

Figure 5.
Title page of the catalogue of the 1936 Royal Society exhibit and the description of a micro-method for the analysis of heavy hydrogen and heavy water.

hydrogen through palladium (43) and of the electrolytic separation of the hydrogen isotopes.

In the fall of 1936 I left Cambridge and joined my brother in Jerusalem as lecturer in physical chemistry (Figure 6).

The Jerusalem Years

By then my brother had established a modern research institute in the physics building on the Mount Scopus campus of the Hebrew University. We had a small library, an excellent glassblower and access to liquid air and to industrial gases. However, we had no natural gas. Instead, we used a non-explosive air-gasoline mixture piped into the laboratory. And of course we had compressed air too!

In the years 1937-1939 we studied the interaction of deuterium with benzene (44), acetone (45), acetylene (46), alcohols (47), propane (48), *n*-butane (49), *n*-hexane and cyclohexane (50,51) using the para-hydrogen conversion whenever applicable. We also re-invesigated the ethylene-deuterium reaction (52).

From the results we concluded that the dissociative mechanism originally suggested for the ethylene-deuterium exchange is valid and is responsible for the exchange involving other unsaturated, saturated and aromatic hydrocarbons and oxygenated compounds.

On the basis of the para-hydrogen conversion data obtained in connection with ethylene hydrogenation experiments, we postulated (53) that under mild conditions on active catalysts, the two atoms of a given hydrogen molecule were added simultaneously to the olefine, according to this concept the addition occurs in the cis-position yielding cis-olefins from substituted acetylenes, meso-compounds from substituted cis-olefins and a racemic mixture from substituted trans-olefins. We also speculated that catalytic hydrogenation under severe conditions or addition of nascent hydrogen should give the thermodynamically stable isomer.

A search of the literature showed that a large majority of the examples confirmed these concepts.

Subsequently, I wrote a paper entitled "The Activation of Hydrogen in Catalytic Reactions of Hydrocarbons" (54) for presentation at the 1939 Faraday Society Discussion on hydrocarbon chemistry. This article proposed that dissociative adsorption of hydrogen and of hydrocarbon molecules on metal catalysts can explain a wide variety of reactions in

Figure 6.
My brother Ladislas and I in the Jerusalem laboratory in 1937.

addition to deuterium exchange. By giving examples from the literature I showed the applicability of this concept to the formation of stereo-chemical isomers in hydrogenation, double bond isomerization, disproportionation, decomposition and polymerization.

At that time, it seemed to us that the publication of this paper (I had not attended the Faraday Meeting) set off a veritable storm of criticism of the Bonhoeffer-Farkas mechanism of the para-hydrogen conversion and the dissociative mechanism of the deuterium exchange.

To top it all, our most severe critics were Rideal, Polanyi and their respective pupils!

The Bonhoeffer-Farkas Mechanism: True or False?

In a room next to ours in Rideal's laboratory, J. K. Roberts developed two elegant original methods for studying chemisorption of hydrogen on tungsten (55,56).

Roberts determined the heat of hydrogen on thoroughly outgassed fine tungsten wires from the change of the wire temperature on exposure to small amounts of hydrogen. He also measured the fraction of the surface of the wire covered by hydrogen by determining the accommodation coefficient (α) of neon on the wire defined by

$$\alpha = \frac{T_2' - T_1}{T_2 - T_1}$$

where T_1 and T_2' the temperatures of the neon atoms striking and leaving the wire respectively, and T_2 is the temperature of the wire.

The accommodation coefficient is a measure for the efficiency of the heat exchange between the neon molecules and the wire and depends on the nature of the wire surface. It is low for the bare wire, increases with the extent of the hydrogen layer and reaches a maximum value on complete coverage.

From his adsorption results Roberts drew the following conclusions:

1. There is very rapid chemisorption of hydrogen on W even at $-185^{\circ}C$, occurring with an activation energy not exceeding a few hundred calories.

2. The hydrogen layer formed at a pressure of 3×10^{-4} mm mercury covered essentially the whole surface of the wire.

3. The hydrogen layer was very stable and started to evaporate only at 400°C.

Then Roberts reasoned that since para-hydrogen conversion can be observed on W at -100° C such a reaction can not be explained by the Bonhoeffer-Farkas mechanism which involves rapid adsorption-desorption processes.

Based on Roberts' results Rideal then suggested (57) that the para-hydrogen conversion proceeds via a reaction between impinging H_2 molecules and chemisorbed H-atoms according to

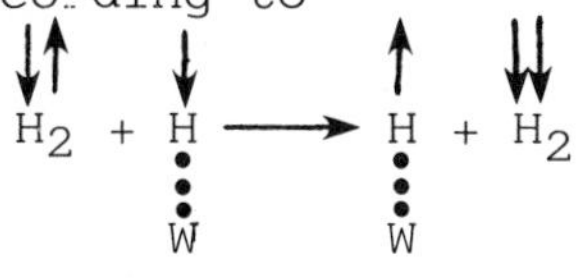

This mechanism was subsequently modified by Eley (58,59) to include an atomic interchange between chemisorbed H atoms and H_2 molecules adsorbed in a van der Waals layer.

This mechanism became known as the Rideal-Eley mechanism. It was used by Rideal's school for explaining the para-hydrogen conversion and related reactions connected with deuterium exchange notwithstanding a serious flaw. The Rideal-Eley mechanism would require a reaction order of one or larger, while actually the order of the para-H_2 conversion is much less than unity.

It took over ten years to show that Roberts' experimental results were faulty, the Rideal-Eley mechanism was unnecessary and that Bonhoeffer-Farkas mechanism was correct. And this was done by none other than Professor Rideal (60-62) and his student H. M. W. Trapnell (63,64-67).

Their careful adsorption measurements proved that the chemisorption layer on W was far from complete at the low pressures used by Roberts and that the heat of reversible chemisorption fell from 15 kcal/mole at 80% coverage to 3 kcal/mole at 100% coverage.

Rideal and Trapnell proved that there is reversible chemisorption of H_2 on W down to -185°C, and that the accommodation coefficient method used by Roberts was not sensitive enough to detect small changes in the coverage.

They also showed that on chemisorption of incremental amounts of H_2 the heat of chemisorption dropped over the whole area of chemisorption and that reversible desorption with the lower heat of adsorption could take place from any part of the surface.

Associative Mechanism vs. Dissociative Mechanism

Originally Horiuti, Ogden and Polanyi (68) considered both the dissociative and associative mechanism as possible explanations of the isotopic hydrogen exchange between liquid benzene and heavy hydrogen in the presence of platinum black. Subsequently, Horiuti and Polanyi (69,33,34) decided in favor of the associative mechanism on the basis of the comparison of exchange rates of the reactant pairs liquid C_6H_6-D_2O, C_2H_4-D_2O and C_2H_4-C_6H_6 with each other and with the exchange rates of the pairs liquid C_6H_6-D_2 and C_2H_4-D_2.

From the very slow exchange rates in the absence of gaseous hydrogen and the particularly slow exchange rates in the case of the C_2H_4-C_6D_6 pair, Horiuti and Polanyi concluded that the dissociative adsorption of the hydrocarbon moiety cannot be the rate-controlling reaction in the exchange.

In 1939 Polanyi and coworkers (70,71) carried out additional exchange studies with groups of three exchange partners A, B and C, by determining the exchange rates for the three pairs A-B, B-C and A-C and for the mixture of all three, A-B-C, in the group.

The groups studied were D_2, D_2O and cyclohexane or isopentane; D_2O , benzene and cyclohexane; D_2O , cyclohexane and isopentane; D_2 , HCl or DCl and cyclohexane; D_2, H_2O or D_2O and benzene.

Polanyi and coworkers also developed kinetic equations for both the associative and dissociative mechanisms for comparison with their experimental results.

Their main arguments against the dissociation mechanism were these:

1. The rate of exchange in liquid benzene increases with the D_2 pressure indicating that the dissociative adsorption of benzene can not be rate controlling.

2. The exchange rate between C_2H_4 and C_6H_6 is slower than the exchange rates between water and either of the two hydrocarbons.

3. Unsaturated hydrocarbons exchange more slowly with D_2O than with D_2.

4. The exchange of cyclohexane with DCl is

slower than with D_2, indicating that the dissociation of cyclohexane is not rate controlling.

5. The D_2 exchange is faster with unsaturated compounds than with saturated compounds.

Twigg and Rideal (72-74) also favored the associated mechanism for these reasons:

1. C_2H_4 and C_2D_4 do not show hydrogen exchange as postulated by the dissociative mechanism.

2. Propene and higher olefins exchange all their hydrogen atoms including those of the methyl groups which would not be expected to be reactive according to the dissociative mechanism.

3. The metal-catalyzed isomerization of butene-1 to butene-2 does not occur in the absence of hydrogen.

However they admitted that the dissociative mechanism must be operative in the exchange of saturated hydrocarbons.

We and others countered these arguments by invoking effects of mutual displacement of exchange partners in the chemisorption layer (75,76), the likelihood of the allyl radical formation from propene (77) and the proof for the occurrence of exchange reaction between deuturated ethylene and butene-1 (78).

Epilogue

In the fall of 1941 I left Jerusalem for the U.S.A. My brother got involved in research relating to problems caused by the war.

By the time T. I. Taylor wrote his classical review "Hydrogen Isotopes in the Study of Hydrogenation and Exchange" for Volume V of Emmett's "Catalysis" published in 1957 (79), he had to cover more than 250 articles and the complexity of the reactions turned out to exceed the expectation by far.

Taylor came to the conclusion that the olefin exchange and related reactions occurred by the associative mechanism. On the other hand in the case of the benzene-deuterium interaction on metal films he quoted the work of Anderson and Kemball (80-82) which indicated the validity of the dissociative mechanism for the exchange.

Apparently the debate on the mechanisms was still open two decades later. In 1976 Goult Ledoux, Masini and Roussy reported (83) that on Ni or Fe at low temperatures butene-1 interacting with D_2 or perdeuteropropene exchanges only its three vinylic hydrogen atoms. They claimed that they "proved unambiguously for the first time the mechanism of exchange suggested more than 40 years ago by Farkas and Farkas."

Literature Cited

1. Bonhoeffer, K. F. and Harteck, P. Naturwiss 1929, 17, 182.
2. Bonhoeffer, K. F. and Hartec, P. Sitzber. preuss. Akad. Wiss. 1929, 103.
3. Bonhoeffer, K. F. and Harteck P. Z. physik. Chem. 1929, B4, 113.
4. Farkas, A. Z. physik. Chem. 1929, B5, 467.
5. Farkas, A. Z. physik. Chem. 1930, B10, 419.
6. Farkas, A. Z. Elektrochem. 1930, 36, 782.
7. Polanyi, M. and Eyring, H. Z. phys. Chem. 1931, B12, 179.
8. Bonhoeffer, K. F. and Farkas, A. Z. physik. Chem. 1931, B12, 231.
9. Farkas, A. Z. physik. Chem. 1931, B14, 371.
10. Bonhoeffer, K. F. and Farkas, A. Trans. Faraday Soc. 1932, 28, 242, 561.
11. Taylor, H. S. and Sherman, A. Trans. Faraday Soc. 1932, 28, 247.
12. Cremer, E. and Polanyi, M. Z. physik. Chem. 1933, B21, 459.
13. Cremer, E. and Polanyi, M. Trans. Faraday Soc. 1932, 28, 435.
14. Kingman, F.E.T. Trans. Faraday Soc. 1932, 28, 269.
15. Farkas, A. and Rowley, H. H. Z. phys. Chem.1933, B22, 335.
16. Farkas, L. and Sachsse, H. Sitzber. preuss.Akad. Wiss. 1933, 268.
17. Farkas, L. and Sachsse, H. Z. physik.Chem. 1933, B23, 1, 19.
18. Farkas, L. and Sachsse, H. Trans. Faraday Soc. 1934, 30, 331.
19. Rummel, K. W. Z. physik. Chem. 1933, A167, 221.
20. Taylor, H.S. and Diamond, H. J. Am. Chem. Soc. 1933, 55, 2613.
21. Bonhoeffer, K. F., Farkas, A. and Rummel, K. W. Z. physik. Chem. 1933, B21, 225.
22. Farkas, A. Z. physik. Chem. 1933, B22, 344.
23. Farkas, A. and Farkas, L. Nature 1933, 132, 894.
24. Farkas, A. and Farkas, L. Proc. Roy. Soc. 1934, A144, 467.
25. Rideal, E. K. Proc. Roy. Soc. 1934, A144, 16.
26. Rutherford, Lord Proc. Roy. Soc. 1934, A144, 1.

27. Harteck, P. Proc. Roy. Soc. 1934, A144, 9.
28. Farkas, A., Farkas, L. and Rideal, E.K. Proc. Roy. Soc. 1934, A146, 630.
29. Farkas, A. and Farkas, L. Naturwiss. 1934, 22, 218.
30. Farkas, A. and Farkas, L. Nature 1934, 133, 934.
31. Farkas, A. and Farkas, L. Proc. Roy. Soc. 1934, A146, 623.
32. Farkas, A. and Farkas, L. J. Chem. Phys. 1934, 2, 468.
33. Horiuti, J. and Polanyi, M. Nature 1933, 132, 931.
34. Horiuti, J. and Polanyi, M. Trans. Faraday Soc. 1934, 30, 1164.
35. Farkas, A. , Farkas, L. and Harteck, P. Proc. Roy. Soc. 1934, A144, 481.
36. Farkas, A. and Farkas, L. Proc. Roy. Soc. 1935, A152, 124.
37. Farkas, A. and Farkas, L. Proc. Roy. Soc. 1935, A152, 152.
38. Farkas, A. and Farkas, L. Trans. Faraday Soc. 1935, 31, 821.
39. Farkas, A. "Ortho-Hydrogen and Heavy Hydrogen,"; Cambridge University Press: Cambridge, England, 1935.
40. Farkas, A. Trans. Faraday Soc. 1936, 32, 416.
41. Farkas, A. Trans. Faraday Soc. 1936, 32, 922.
42. Farkas, A. Trans. Faraday Soc. 1936, 32, 413.
43. Farkas, A. Trans. Faraday Soc. 1936, 32, 1667.
44. Farkas, A. and Farkas, L. Trans. Faraday Soc. 1937, 33, 827.
45. Farkas, A. and Farkas, L. J. Am. Chem. Soc. 1939, 61, 1336.
46. Farkas, A. and Farkas, L. J. Am. Chem. Soc.1939, 61, 3396.
47. Farkas, A. and Farkas, L. Trans. Faraday Soc. 1937, 33, 678.
48. Farkas, A. and Farkas, L. Trans. Faraday Soc. 1940, 36, 522.
49. Farkas, A. Trans. Faraday Soc. 1940, 36, 522.
50. Farkas, A. and Farkas, L. Nature 1939, 143, 244.
51. Farkas, A. and Farkas, L. Trans. Faraday Soc. 1939, 35, 917.
52. Farkas, A. and Farkas, L. J. Am. Chem. Soc.1938, 60, 22.
53. Farkas, A. and Farkas, L. Trans. Faraday Soc. 1937, 33, 837.
54. Farkas, A. Trans. Faraday Soc. 1939, 35, 906.

55. Roberts, J. K. Proc. Roy. Soc. 1935, A152, 445, 464, 477.
56. Roberts, J. K. Trans. Faraday Soc. 1935, 31, 1710.
57. Rideal, E. K. Proc. Camb. Phil. Soc. 1939, 35, 130.
58. Eley, D. D. and Rideal, E. K. Proc. Roy. Soc. 1941, A178, 429.
59. Eley, D. D. Proc. Roy. Soc. 1941, A178, 452.
60. Rideal, E. K. and Trapnell, B.M.W. Discussions Faraday Soc. 1950, 8, 114.
61. Rideal, E. K. and Trapnell, B.M.W. J. chim. Phys. 1950, 47, 120.
62. Rideal, E. K. and Trapnell, B.M.W. Proc. Roy. Soc. 1951, A205, 409.
63. Taylor, H. S. J. Am. Chem. Soc. 1931, 53, 578.
64. Trapnell, B.M.W. "Catalysis," Vol. III, p. 1., Edited by Emmett, P. H.; New York: Reinhold Publishing Corp., 1955.
65. Trapnell, B.M.W. Proc. Roy. Soc. 1951, A206, 39.
66. Trapnell, B.M.W. Trans. Faraday Soc. 1952, 48, 160.
67. Trapnell, B.M.W. Proc. Roy. Soc. 1953,A218, 566.
68. Horiuti, J., Ogden, G. and Polanyi, M. Trans. Faraday Soc. 1934, 30, 663.
69. Horiuti, J. and Polanyi, M. Nature 1933, 132, 819.
70. Greenhalgh, R. K. and Polanyi, M. Trans. Faraday Soc. 1939, 35, 520.
71. Horrex, C., Greenhalgh, R. K. and Polanyi, M. Trans. Far. Soc. 1939, 35, 511.
72. Twigg, G. H. Trans. Faraday Soc. 1939, 35, 934.
73. Twigg, G. H. and Rideal, E. K. Proc. Roy. Soc. 1939, 171A, 55.
74. Twigg, G. H. and Rideal, E. K. Trans. Faraday Soc. 1940, 36, 533.
75. Farkas, A. Trans. Faraday Soc. 1939, 35, 941, 943.
76. Farkas, A. and Farkas, L. J. Am. Chem. Soc. 1942, 64, 1594.
77. Jost, W. Trans. Faraday Soc. 1939, 35, 940.
78. Aman, J., Farkas, L. and Farkas, A. J. Am. Chem. Soc. 1948, 70, 727.
79. Taylor, T. I. "Catalysis," Vol. 5, p. 257, Edited by Emmett, P. H.; New York: Reinhold Publishing Corp.., 1957.
80. Anderson, J. R. and Kemball, C. Proc. Roy. Soc. 1954, A223, 361.
81. Anderson, J. R. and Kemball, C. Trans. Faraday Soc. 1955, 51, 966.

82. Anderson, J. R. and Kemball, C. "Advances in Catalysis," Vol. IX, p. 51, Academic Press: New York, 1957.
83. Goult, F. G., Ledoux, M., Masini, J. and Roussy, G. "Proc. Sixth International Congress on Catalysis," p. 469, The Chemical Society: London, 1977.

RECEIVED January 25, 1983

LATER RESEARCHERS

Allan T. Gwathmey and His Contributions to the Science of Catalysis

HENRY LEIDHEISER, JR.

Lehigh University, Department of Chemistry and Center for Surface and Coatings Research, Bethlehem, PA 18015

It was my privilege to be a student and longtime friend of Allan Talbott Gwathmey who by his personality and experimental observations played an important role in directing the attention of scientists to the role of the atomic geometry of catalysts in controlling the rates of heterogeneous catalytic reactions. I accept with pleasure the invitation to attempt to place his work in perspective. It is my intention to give a picture of the man as well as the technical contributions he made in the field of catalysis. I will draw on two previous publications (1,2) where appropriate.

The Man

Allan Gwathmey was born in Richmond, Virginia on July 29, 1903. He attended preparatory school in Richmond and received his B.S degree from Virginia Military Institute in 1923. Following several years of employment as an engineer, he elected to return to school and he earned the B.S. degree in Electrochemical Engineering from Massachusetts Institute of Technology. After several years of industrial research, Allan Gwathmey entered the Graduate School of the University of Virginia during the severe depression years of the 1930's and earned the Ph.D. degree in Chemistry in 1938. He continued at the University of Virginia as a research associate until his appointment as a member of the Chemistry Department faculty in about 1947, and he remained a member of the faculty until his death in 1963.

His interests at the University of Virginia were campus wide. He lived for many years in the Colonnade Club on the Lawn, that beautiful terraced green lined with grand trees, that connects Cabell Hall with the building that symbolizes the University, the Rotunda. He fought in many ways to retain the dignity and charm of the Colonnade Club to the end that it was the focal point for faculty social life. He served on the University

0097–6156/83/0222–0121$06.00/0

Figure 1. Allan Talbot Gwathmey. Photograph taken in 1955. (Reproduced with permission from Ref. 2.)

Senate and did what he could to encourage deemphasis of intercollegiate athletics. It is interesting to conjecture what his view would be if he lived during the 1981 and 1982 seasons when Ralph Sampson was the outstanding star of intercollegiate basketball. His viewpoints with respect to athletics were different from those of many of his friends, associates, and students, but these views did not inhibit his friendships. He was called on by the President of the University many times to host distinguished visitors because of his knowledge of the University, his charm and his genteel manner. He was a constant thorn in the sides of deans and presidents when they were not aggressively seeking to build excellence into departments. He played important roles in encouraging wealthy alumni to support their university. His intellectual interests were broad. He taught a course in aesthetics in the Department of Philosophy, and he was writing a book, which he never completed, on economics.

Professor Gwathmey treated his graduate students as one big family. They were welcome to confide in him and to seek his advice on technical as well as non-technical matters. His office door was always open and he would interrupt whatever he was doing to talk to students. Since he married late in life and had no children of his own, his students filled a void in his life. When things did not go well in the laboratory or when the demands of the experiment were greater than the available resources, his favorite expression was "The toughening discipline of a complex experiment." Years later when his students communicated among themselves by letters, wire or telephone, it was not unusual to have a postscript which read TTDOACE.

He loved his native state with a passion. The beauty of the Lawn at the University, the Spirit of VMI, and his admiration for Thomas Jefferson were often subjects for comment or discussion among his friends and almost always crept into his public lectures. The loss of William Barton Rogers to Massachusetts, and his founding of MIT in the north rather than at William and Mary, bothered him enormously. He prevailed upon the Virginia Academy of Science to fund an institute devoted to basic research. They did in 1948 and the Commonwealth of Virginia in 1949 provided a grant of $20,000 plus free rental of an old building in a park in Richmond and the Virginia Institute for Scientific Research came into being. Virgil Straughan and I were the first full-time employees, and the hardships endured in pulling an institution up by its bootstraps were only bearable because of the enthusiasm and inspiration of Allan Gwathmey. This institution grew to a staff of approximately 40 and sufficient funds were raised to build a modern laboratory on 20 acres of ground in the beautiful west end of Richmond. The building was dedicated in July 1963 and was named the Allan Talbott Gwathmey Laboratory. He lived to see it completed and occupied. The Institute is now part of the University of Richmond.

The written word and the spoken word were important to Dr. Gwathmey. He would draft a letter, put it aside for several days, rewrite the letter, perhaps revise it several more times before it was considered worthy of mailing. Manuscripts were revised many times before submittal for publication and he insisted upon replication of research results before he would accept them as correct. Social events were planned carefully and meetings with important people, especially on a controversial issue, involved strategy plotting of a high order. He discussed strategy often with close associates and enjoyed the planning and the confrontation with a relish that was contagious.

Allan Gwathmey was not blessed with good health and was a rather frail child. His father died while Allan was young, and this event made him aware of the importance of good health. In the early 1950's it was recognized that he had a severe case of diabetes and he soon became a "brittle diabetic". His pockets always contained sugar, candy or fruit so that he could titrate the injected insulin as his body advised him. Blackouts were not uncommon in the early days of his treatment and one of them resulted in a broken knee. In 1961 he began to experience impairment in the muscles that control breathing, swallowing and talking. Chemotherapy was of some help, but a creeping paralysis began that culminated in his death on May 12, 1963. The last few months of his life were difficult beyond expression, but he never lost his optimism and his interest in the world around him.

The Scientist and His Contributions

Allan Gwathmey began his studies of surface structure as related to the activity of different crystal faces of a single crystal in 1934, at which time he decided to utilize low voltage electron diffraction as a technique in studies leading to the Ph.D. degree. He was captivated by the diffraction experiments reported by Davisson and Germer in 1927 and he recognized the applicability of this phenomenon to a better understanding of the surface properties of metals. In the style of the times, he built the apparatus himself, neutralized the stray fields that might interfere with the trajectory of the diffracted electron beams, and developed the technique for preparing, cutting and surfacing the copper single crystals to be used as samples. He received no technical support from his thesis advisor, A. F. Benton, and moral support only in the form of a pat on the back and a "get to it".

During the course of the diffraction experiments he decided to study the properties of an oxidized copper surface. His experimental sample was shaped in the form of a sphere with a small holding shaft so that he could manipulate the sample in the diffraction chamber and expose different crystal faces to the electron beam. The oxidation of the electropolished sphere resulted in the development of a beautiful interference color pattern which reflected the symmetry of the crystal. This observation was the beginning of his dedication to the application of metallic single crystals to studies of surface phenomena.

Allan Gwathmey began his work in catalysis when the emphasis of much of the work in the field was an attempt to understand the kinetics of the reactions by means of adsorption studies. The development of the Brunauer-Emmett-Teller relationship allowed one to calculate the surface areas of catalysts and to determine the amounts of adsorption and the energetics of the adsorption process. It was a time when typical papers with a catalytic orientation were filled with information on adsorption hysteresis, heats of adsorption, changes in surface area with heat treatment, and pore sizes. People such as Hugh Taylor of Princeton played a dominant role in molding opinion and in setting the standards for research in catalysis and it was Hugh Taylor that Dr. Gwathmey took as his favorite antagonist in meetings such as the Gordon Conference on Catalysis. His objection to the thrust of the research was that it often focused on elaborate measurements on poorly prepared materials. He fought for experiments which were carried out on well characterized materials. Others who also began preaching the same sermon included F. P. Bowden with his beautiful work on the surface energy of mica crystals and E. W. Müller with his pioneering work on field emission and later field ion emission experiments.

The first publication of his work was in the Transactions of the Electrochemical Society (3) in which he described the anisotropic properties of a copper single crystal submitted to various chemical and electrochemical etching treatments. His second paper described the method for preparing, shaping, and surfacing copper single crystals prior to an experiment (4) and rapidly following papers described the beautiful interference patterns obtained when an electropolished copper sphere was oxidized in air (5,6). The first public report, other than local presentations through the Virginia Academy of Science, on catalysis appeared in 1942 in which he described the surface rearrangements that occurred on copper single crystals as a consequence of the hydrogen-oxygen reaction. In his own words (7): "The reaction between hydrogen and oxygen renders the surface sufficiently mobile to develop certain crystal facets. In order to produce the rearrangement, the effective temperature at the surface of the metal when the reaction is taking place must be considerably above the measured temperature of the reaction vessel. Since the reaction between hydrogen and oxygen becomes appreciable at the same temperature at which the development of facet takes place, this experiment suggests that these facets may be related to the positions of enhanced catalytic reactivity for this particular reaction." The work on the hydrogen-oxygen reaction on copper and nickel was further elaborated on by Leidheiser (8), by Cunningham (9), and by Wagner (10).

Although the work on the hydrogen-oxygen reaction received attention, the large difference in the activity of the different crystal faces of nickel for the carbon monoxide decomposition reaction, $2CO = C + CO_2$, was dramatically shown by the different rates of carbon deposition (11). See Figure 2. The ability to display this difference in catalytic activity by means of a photograph aided in convincing the unbeliever of the importance of surface geometry of atoms in catalytic reactions. The superb photographs beginning to emerge from field emission and ion emission microscopy provided complementary support for the concept.

Another catalytic reaction—the deposition of a metal on the surface of a different metal using a reducing agent—was also shown to be anisotropic in nature. The deposition of cobalt from a potassium formate melt, the deposition of nickel from a potassium formate melt, and the deposition of nickel from a hypophosphite solution all occurred on copper single crystals at rates which depended on the crystal face exposed at the surface (12). This work was carried out in Dr. Gwathmey's laboratory but he did not wish to be a coauthor since he had not contributed to the planning or execution of the experiments.

Perhaps the most important contribution that Dr. Gwathmey made was to develop the recognition among those working in materials science that the surface properties of materials were anisotropic and that different crystal faces of the same material had different properties. He demonstrated this phenomenon in catalysis, corrosion, electrochemical properties, electrodeposition, wetting, friction, wear, oxidation, and reaction with hot gases such as bromine, chlorine, and hydrogen sulfide. He developed the recognition that anisotropy of surface behavior was universal.

A chance meeting with A. B. Winterbottom of the University of Trondheim led Gwathmey to apply the technique of ellipsometry to non-destructive studies of the rate of growth of oxide films on specific crystal faces of copper. The heroic exploit of Tronstad, the developer of the technique as applied to metal surface studies, and his death during a raid on a heavy water plant in Norway during WW II, gave the technique an exotic flavor that appealed to Dr. Gwathmey's imagination. He gave the assignment to his student, Fred W. Young, Jr., who along with fellow student John Cathcart brought the technique to the point that it was able to provide precise, quantitative information on the thickness of an oxide film on a single crystal copper substrate. This technique was further developed by graduate student Jerome Kruger, who later applied the technique to many interesting corrosion problems at the National Bureau of Standards. Another student of Prof. Gwathmey's, Kenneth Lawless, spent a year in Trondheim working with Prof. Winterbottom.

Coincident with the research on the anisotropy of the chemical behavior of metal single crystals there was growth in the development of the concept of imperfections, both structural and chemical, in semiconductors and metals. Two leaders in the field, John Mitchell and Nicolas Cabrerra, joined the physics faculty at the University of Virginia. A strong bond of friend-

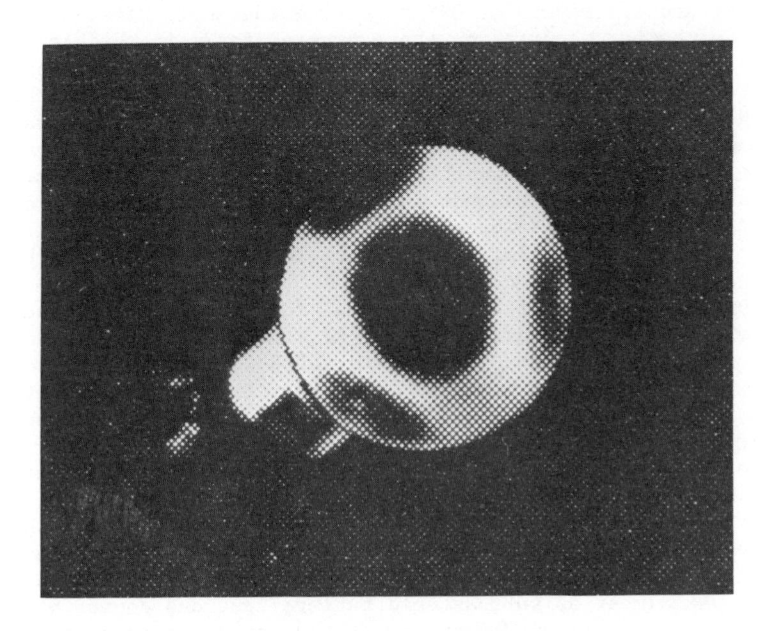

Figure 2. The selective deposition of carbon on the (111) faces of a nickel single crystal during the catalytic decomposition of carbon monoxide at 550 C. (Reproduced from Ref.11. Copyright 1948, American Chemical Society.)

ship built up among these men, especially between Mitchell and Gwathmey. This friendship and a joint empathy for the aesthetic continued until Dr. Gwathmey's death. Dr. Mitchell was a tower of strength for Roberta Gwathmey during the difficult two years preceding his decease. The first significant discussion of the effect of dislocations, lattice defects, and electronic and geometric factors on the catalytic properties of single crystals occurred in a paper with Cunningham (13). This same paper included studies of the hydrogenation of ethylene, the first catalytic reaction that was studied by Gwathmey that did not lead to surface rearrangement.

The concept that crystal orientation and defects jointly were important in controlling surface reactivity was carried forth by Cathcart at the Oak Ridge National Laboratory (ORNL). Cathcart and colleagues at ORNL observed that high rates of oxidation of copper occurred on those crystal faces in which there were several possible epitaxial relationships between the copper substrate and the cuprous oxide oxidation product and that the low rates of oxidation occurred on those faces where there was one dominant orientational relationship. Benard and colleagues in Paris had earlier shown that the oxidation of copper occurs initially at local sites that increase in size by lateral growth. Imperfections tended to be a maximum on those faces with multiple orientational relationships because of lattice mismatch in those regions where merging occurred.

Precise studies of the structural imperfections in copper single crystals were carried out by F. W. Young, Jr. after he received the Ph.D. degree and joined the Solid State Branch of ORNL. Young developed etchants that brought out the dislocations which intersected the surface. He was thus able to determine the number of imperfections that was present as a consequence of the single crystal growth process and to determine those that were introduced by various mechanical treatments.

Many of the metals with which his graduate students worked were very difficult to shape when in the form of a pure single crystal. Metals such as zinc readily split along (0001) planes; indium was almost as soft as cold butter; lead and copper deformed under their own weight if not supported properly; and gold tended to smear rather than cut when handled with a normal cutting tool on a lathe. It was necessary to develop proper cutting speeds, rake angles, and cutting angles for each different metal. Dr. Gwathmey decided to develop his own machine shop and he was fortunate to find a talented machinist in the person of Virgil Straughn. This machine shop grew in competence in handling single crystals of metals such that single crystals of lead and zinc could be machined in the form of spheres where recrystallization was confined to the outer layer in the case of lead and where shearing was prevented in the case of zinc.

Dr. Gwathmey's activities after about 1944 filled the entire lowest floor of Cobb Chemical Laboratory. This floor was known as the sub-basement, even though it was above ground level on one side. The drainage system for the organic chemistry laboratories on the floor above passed through terra cotta pipes below the ceiling level in the sub-basement. The passage of time resulted in sagging of the pipe in some locations and the acids and organic solvents that improperly entered the drainage system caused deterioration of the cemented joints. Students constructed a secondary drainage system with the aid of funnels and tubing to collect the exudate and carry it to the sink. The novelty of the arrangement was always a source of discussion with many of the famous scientists who visited the laboratory. When disagreements broke out between the "organics" and the "physicals", the "organics" on the floor above held the upper hand since they had recourse to lachrimators with which Prof. Lutz's students occasionally worked.

Although Dr. Gwathmey had a good sense of humor there is only one case where I can remember that laughter overwhelmed him. An undergraduate student had begged for summer employment and Dr. Gwathmey took him into the group and suggested that the graduate students might utilize his services in some of the more routine operations that were performed. The young man proved to have little capability for experimental work and especially so for single crystals that had to be handled with gentleness and loving care. As a consequence, the assignments that were given to the undergraduate ended up being mundane and uninteresting. One day he visited Prof. Gwathmey to share his unhappiness. He rested against the open office door during the conversation and when it was completed, he turned to leave the room. Unfortunately, his left rear pocket had become ensnared by the doorknob, his foot prevented the door from following his departure, and the rear of his pants was torn severely. Dr. Gwathmey was very kind and compassionate about his misfortune, but after the student had left the building to change clothing, he was convulsed with laughter that could not be controlled for several minutes. The incident reinforced very convincingly the "all thumbs" label with which the graduate students had identified the undergraduate.

Many scientists came by to visit Prof. Gwathmey and each visitor represented a challenge to convince the visitor of the significance of metal single crystals as research tools and as a cutting edge of materials science. A local carpenter was given the assignment to construct a display case in which the single crystals could be easily viewed. Prof. Gwathmey maintained the case, filled with metal single crystals exhibiting some important principle, in his office. It was a delight to see such well known scientists as J. Beams, W. A. Noyes, Jr., H. H. Storch, N. F. Mott, H. Eyring, and C. J. Davisson leaning over the case and being enthralled by the enthusiasm of their host. Prof. Gwathmey's eyes would become teared when he became emotionally involved in his subject and this act had a favorable effect on the visitor who recognized the sincerity of the speaker.

Literature Cited

1. Leidheiser, H. Jr. Virginia J. Science 1964, 15 (1), 1.
2. Leidheiser, H. Jr. Corrosion 1976, 32, 191.
3. Gwathmey, A. T.; Benton, A. F. Trans. Electrochem. Soc. 1940, 77, 211.
4. Gwathmey, A. T.; Benton, A. F. J. Phys. Chem. 1940, 44, 35.
5. Gwathmey, A. T.; Benton, A. F. J. Chem. Phys. 1940, 8, 431.
6. Gwathmey, A. T.; Benton, A. F. J. Phys. Chem. 1942, 46, 969.
7. Gwathmey, A. T.; Benton, A. F. J. Chem. Phys. 1940, 8, 569.
8. Leidheiser, H. Jr.; Gwathmey, A. T. J. Am. Chem. Soc. 1948, 70, 1200.
9. Gwathmey, A. T.; Cunningham, R. E. J. Chem. Phys. 1954, 51, 497; Cunningham, R. E.; Gwathmey, A. T. J. Am. Chem. Soc. 1954, 76, 391.
10. Wagner, J. B. Jr.; Gwathmey, A. T. J. Am. Chem. Soc., 1954, 76, 390.
11. Leidheiser, H. Jr.; Gwathmey, A. T. J. Am. Chem. Soc. 1948, 70, 1206.
12. Leidheiser, H. Jr.; Meelheim, R. J. Am. Chem. Soc. 1949, 71, 1122.
13. Cunningham, R. E.; Gwathmey, A. T. Adv. in Catalysis 1957, 9, 25.

RECEIVED November 17, 1982

10

Contributions of P. W. Selwood to Catalysis Research

R. P. EISCHENS
Broomall, PA 19008

This symposium on the history of catalysis is well served by inclusion of the work of P.W. Selwood. I have two reasons to be pleased to make this presentation. First, it allows attention to be focused on a momumental body of classical catalysis research. Hopefully, this attention will stimulate further efforts to exploit the methods which Selwood has pioneered. Second, as a former student, it provides an opportunity to express my indebtedness to Selwood whom I regard to be a model of what a professor should be.

Selwood began his career as an independent researcher when he joined the faculty of Northwestern in 1935. His decision to apply magnetic studies to catalysts was probably influenced by his doctoral work at Illinois where he became acquainted with the magnet and his postdoctoral work at Princeton which stimulated his interest in catalysis. In 1935, the concept of using physical methods to arrive at an understanding of catalysis was not widely accepted. Thus, it was necessary both to develop physical methods and to demonstrate how the new methods could be applied to problems of practical interest.

Selwood has written excellent thorough reviews of his work (1,2). These reviews present details of experimental methods and underlying theory. I plan to emphasize the use of magnetic studies in determining: the oxidation state of supported oxides; the degree of dispersion of supported oxides; and the structure of adsorbates chemisorbed on supported metals.

Paramagnetic susceptibility, due to unpaired electrons, is used to measure the oxidation state of transition metal cations in supported metal oxides. In many cases, the number of unpaired electrons, n,

0097–6156/83/0222–0131$06.00/0

is simply related to the magnetic moment, μ, by the equation, $\mu = \sqrt{n(n+2)}$. The magnetic moment is calculated from the experimentally determined susceptibility, χ, by the relationship, $\mu = 2.83\sqrt{\chi\ (T+\Delta)}$ where T is the temperature and Δ is the Weiss constant which is a measure of interactions between magnetic moments of paramagnetic atoms. The value of this constant is determined from the effect of temperature on susceptibility. Nickel, in the plus two oxidation state, has two unpaired electrons in its *d* orbital so the calculated moment is $\sqrt{8}$ or 2.84. Plus three nickel has three unpaired electrons so the calculated moment is 3.87.

Figure 1 shows the magnetic moment of nickel in nickel oxide as a function of the concentration of nickel in the catalysts (3). The catalysts have been prepared by supporting nickel oxide on magnesia, or on alumina, by impregnation methods under experimental conditions where the expected form of nickel oxide would be NiO. When magnesia is used as the support, the magnetic moment is consistent with divalent nickel over the entire concentration range. However, on alumina in concentrations below 5 wt%, the magnetic moment indicates that a portion of the nickel is in a higher oxidation state. Selwood has called this effect "valence inductivity." It implies that at low concentrations the supported oxide is forced from its normal oxidation state to an oxidation state which matches the structure of the support.

The data point, indicated by a square at a nickel concentration of 10.5%, is of special interest. For this sample of nickel oxide on alumina, the concentration of 10.5% was built up in a series of nine impregnations. Single impregnations were used for the other samples. The 10.5% multiple impregnation sample has a moment of 3.1. This corresponds to the moments observed for single impregnation samples having a nickel concentration of 2%.

Selwood also applied the "valence inductivity" approach in extensive studies of supported manganese oxides (4). The ignition temperature used in the final step of the sample preparation was found to be a critical factor. Ignition at 600°C is expected to produce Mn_2O_3. When manganese oxide, supported on alumina, is ignited at 600°C the manganese is found to be in the plus three oxidation state at all concentrations. Ignition at 200°C is expected to produce manganese in the plus four state. At high concentrations, the manganese oxide on alumina samples prepared by 200°C ignition has manganese in

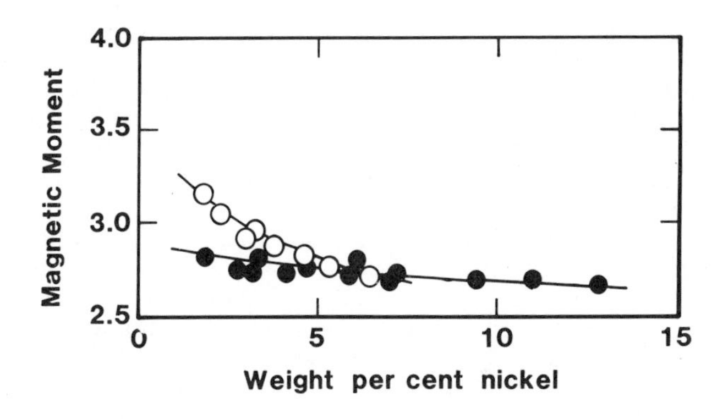

Figure 1. Magnetic moment as a function of nickel oxide supported on: O, alumina and ●, magnesia.

the plus four state. However, at concentrations below 10%, the manganese in the low ignition samples is forced into the plus three state by the epitaxial effect of alumina. When TiO_2 is used as the support, the 200^oC ignition samples are in the plus four state at all concentrations.

Valence inductivity, and the manner in which it is affected by catalyst preparation variables, is obviously a factor which must be considered in all laboratories which utilize supported oxide catalysts.

Dispersion of supported oxides can be studied by determining the effect of concentration on susceptibility at a single temperature. A susceptibility isotherm for chromium in chromium oxide on alumina at -188^oC is shown as the top portion of Figure 2 (5). This isotherm shows a sharp decrease in susceptibility as the chromium concentration is increased to 9%. At higher concentrations, a smaller decrease in susceptibility with increasing concentration is observed. The susceptibilities for supported Cr_2O_3 does not attain the value for bulk Cr_2O_3 which is about 30×10^{-6} at 188^oC.

The relationship, $\mu = 2.83 \sqrt{\chi(T+\Delta)}$ shows that changes in susceptibility could result from changes in the moment (i.e.,changes in oxidation state) and/or from changes in the Weiss interaction constant, Δ. The susceptiblity, χ, is related to the interaction constant, Δ, by the Curie-Weiss equation:

$$\chi = \frac{C}{T+\Delta} \quad (\text{or } T = \frac{C}{\chi} - \Delta)$$

where C is a constant related to the permanent atomic moment. A plot of temperature versus reciprocal susceptibility produces a straight line with an intercept equal to $-\Delta$.

The Weiss constant is a function of the number of nearest paramagnetic neighbors. For Cr_2O_3 in the corundum structure, the number of nearest paramagnetic neighbors, Z, can be related to the number of monolayers of oxide. The value of Z is three for a single monolayer. For two layers Z=6, for three layers Z=7, and for four layers Z=7.5. A plot of Z versus number of layers shows a sharp change of slope at three monolayers. Above three layers the value of Z changes slowly because the relative importance of the external layers diminishes as the number of layers increases.

The lower portion of Figure 2 shows experimentally determined values of the Weiss constant as a function of chromium concentration. The change in slope in the 9% concentration region corresponds

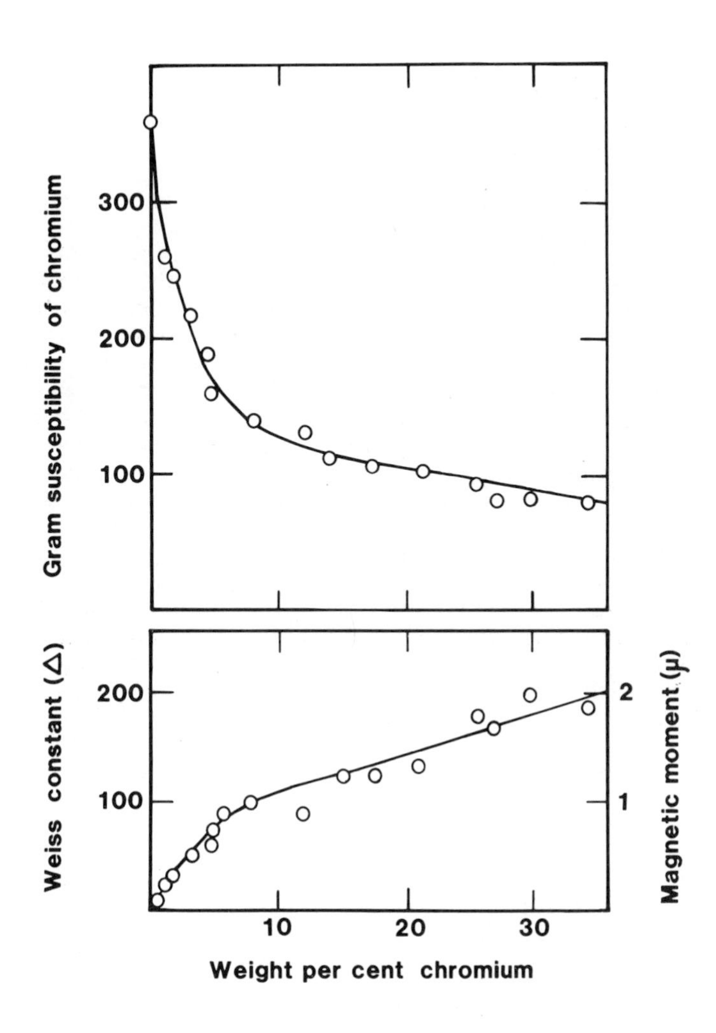

Figure 2. Susceptibility-composition isotherm for chromia impregnated on alumina (upper curve) and relation of delta and magnetic moment to chromium concentration (bottom curve).

with the change of slope in the susceptibility isotherm. As predicted by the valence inductivity concept, the magnetic moment of chromium in Cr_2O_3 is not appreciably affected by the Al_2O_3 support. Thus, the change in slope near 9% in the susceptibility isotherm can be attributed to changes in the Weiss interaction constant. On the basis of the above discussion of the value of Δ as a function of the number of oxide layers, it is reasonable to assign the 9% point on the susceptibility isotherm as the concentration where the thickness of the Cr_2O_3 particles begins to exceed three monolayers. It is interesting to note that, at a concentration of 9%, chromium oxide on the alumina could cover only about one fifth of the alumina surface even if the chromium oxide were present as a monolayer. Thus, multilayer particles of chromium oxide are formed even though large portions of the alumina are not covered.

The studies of oxidation state and dispersion, discussed above, have depended on measurements of the paramagnetic susceptibility of unpaired electrons. Determination of the number of chemisorption bonds is limited to samples which show the property of "collective paramagnetism." Collective paramagnetism is observed for ferromagnetic materials which are in small particles so that each particle is a single magnetic domain. Fortunately, particle sizes which are below 100A in diameter are routinely obtained with supported metals and these particles are sufficiently small to show the collective paramagnetism effect. An important feature of collective paramagnetic particles is that the magnetization, the magnetic moment per unit volume, is directly related to the number of atoms in the particle.

A chemical bond between the surface atom of the particle and an adsorbed species effectively removes the surface atoms from the collectively paramagnetic particle. Thus, chemisorption produces a decrease in the magnetization, M. In practice, the change in relative magnetization, $\Delta M/M$, is calibrated by comparison with changes produced by chemisorption of hydrogen on the same sample. It is well established that hydrogen is adsorbed as atoms so that each molecule of H_2 forms two chemisorption bonds. Experimentally a straight line is obtained in initial stages of plots of $\Delta M/M$ versus volume adsorbed. Thus, the number of bonds formed by an adsorbate can be obtained by comparing the slope of the magnetization-volume plot with that observed with hydrogen.

Selwood used this technique to determine the mode

of chemisorption for most adsorbates of catalytic interest (2). With hydrocarbons, a question of interest has been the degree of dissociation during chemisorption. The power of Selwood's magnetic approach to this problem is illustrated by the study of the effect of temperature on the dissociation of chemisorbed benzene. The data in Table 1 show the number of bonds formed as a function of temperature for benzene on silica-supported nickel.

TABLE I

Average Bond Number Per Adsorbed Benzene Molecule

Temperature $^{\circ}C$	25	100	150	200
Bonds Formed	5.5	6.0	10.5	18

The data in Table I show that dissociation is detected between 100° and 150°C. At 200°C, carbon-carbon bonds are dissociated as well as carbon-hydrogen.

Figure 3 shows magnetization-volume isotherms for hydrogen and H_2S on nickel-kieselguhr (6). The slope of the H_2S isotherm is twice that of hydrogen indicating that H_2S forms four chemisorption bonds. This is consistent with dissociation of the H_2S to produce,

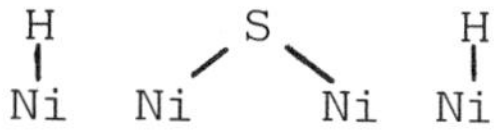

```
H         S        H
|       /   \      |
Ni    Ni     Ni    Ni
```

The magnetization-volume isotherm for CO on a supported nickel shows that on samples, in which the nickel is present in very small (less than 40A°) particles, the slope for CO isotherm is one half that for H_2 (7). This shows that CO is primarily in the linear structure, Ni-C ≡ O. Samples in which the average nickel particle size is in the range of 60A show an initial slope which is equivalent to that of H_2. At higher concentrations the slope is about one-half that of H_2. This is consistent with initial adsorption in the bridged structure:

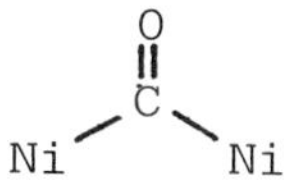

```
       O
       ||
       C
     /   \
   Ni     Ni
```

Many of the infrared studies of CO on nickel have shown the presence of both bridged and linear carbon monoxide. However, infrared spectra of CO on supported nickel have been observed which show only the

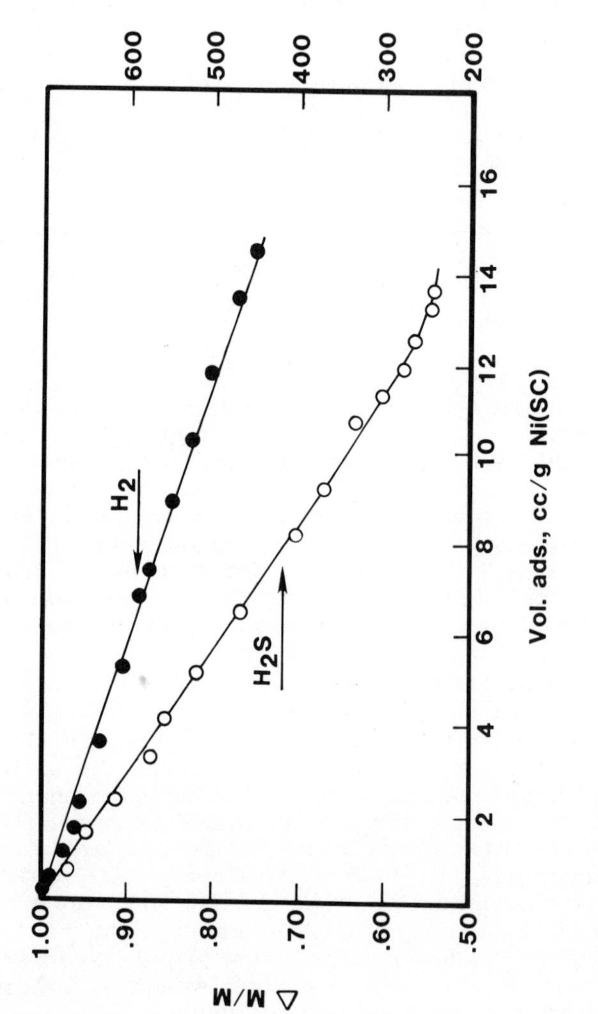

Figure 3. Magnetism-volume isotherms for H_2 and H_2S at $25^{o}C$ on nickel-kieselguhr.

linear form of adsorbed carbon monoxide (8). This is consistent with Selwood's view that the mode of CO chemisorption depends on the nature of the sample.

In addition to the three areas which I have chosen for emphasis, Selwood has contributed in other areas. For example, he was the first to utilize NMR in the study of solid catalysts. This study involved measurement of proton relaxation times for water on aluminas which supported oxides of iron, copper, and chromium (9,10,11). In these studies Selwood determined the accessibility of supported cations by comparison of relaxation times observed for cations in solution. Another important discovery involved the observation that the rate of ortho-para hydrogen conversion, over catalysts such as Pr_2O_3 and Yb_2O_3, could be changed by placing the catalyst in a weak magnetic field of only a few Oersteds (12). Changes in catalyzed reaction rates due to such weak magnetic fields raises the possibility that terrestial magnetism may be a significant factor in the navigational instincts of living organisms.

In the introduction, I indicated that I considered Selwood to be a model of what a professor should be. It is reasonable to question why I make this evaluation. I admit to the possibility that times have changed and it is now necessary for professors to be businessmen as well as scientists. With this proviso, one of Selwood's great contributions to his students was that he was available to them. His office was adjacent to the laboratory and we felt no inhibition in opening the door to request his help or to tell him about new results which appeared important to us. In addition, communication with him was facilitated because he often worked in the laboratory on mysterious projects of his own. The work which I have outlined here was clearly Selwood's. He determined the direction of the work and contributed the significant portions of the interpretations. Despite the importance of his personal contributions, Selwood was able to preserve the delicate balance by which the students were able to feel that they were partners in the effort rather than technicians working at his direction. As a final comment, Selwood's oral presentations were outstanding in their clarity and implied respect for the audience. Although I am happy to make this presentation, I regret that Selwood was not here to make it himself.

Literature Cited

1. Selwood, P.W. Advances in Catalysis 1951, 3, 28.
2. Selwood, P.W. "Adsorption and Collective Paramagnetism", Academic Press, 1962.
3. Selwood, P.W. and Hill, F.N. J. Am. Chem. Soc. 1949, 71, 2522.
4. Selwood, P.W.; Moore, T.E.; Ellis, M. and Wethington, K. J. Am. Chem. Soc. 1949, 71, 693.
5. Eischens, R.P. and Selwood, P.W. J. Am. Chem. Soc. 1947, 69, 1590.
6. Den Besten, I.E. and Selwood, P.W. J. Catalysis 1962, 1, 93.
7. Den Besten, I.E.; Fox, P.G. and Selwood, P.W. J. Phys. Chem. 1962, 66, 450.
8. Low, M.J.D.; Morterra, C.A. and Severdia, A.G. Spectroscopic Letters 1982, 15, 415.
9. Selwood, P.W. and Schroyer, F.K. Faraday Soc. 1950, 8, 337.
10. Spooner, R.B. and Selwood, P.W. J. Am. Chem. Soc. 1950, 71, 2184.
11. Hickmott, T.W. and Selwood, P.W. J. Phys. Chem. 1956, 60, 452.
12. Selwood, P.W. Nature 1970, 228, 278.

RECEIVED February 16, 1983

11

The Development of the Platforming Process—Some Personal and Catalytic Recollections

VLADIMIR HAENSEL [1]

University of Massachusetts, Amherst, MA 01103

An examination of one's career in science and technology is an interesting exercise. Despite a long passage of time, there are events that remain very clear in one's memory bank. And yet, when people ask why or how or when did you think up a particular idea, one cannot put together a sequence in either reasoning or time. Thus, it appears that ideas germinate over a fairly long period of time and then they are given a push by certain "gut feelings" which then propel the individual to design a set of experiments. Thus, it is difficult to recall specific sequences of the thinking process, but details of the experiments are very easily recalled. One can also recall very well discussions which had an input on the direction of work. Indeed, my catalytic recollections are really those of people and catalytic events.

My first introduction to catalytic reforming was in the summer of 1935. I had just graduated from Northwestern University and was to start a chemical engineering course at M.I.T. in the fall. Through the efforts of Prof. V. N. Ipatieff, who was a great friend of my father's, I got a summer job at UOP. I had worked there at no pay during previous summers to acquire experience, however, during the summer of 1935 I received $50 a month working as a technician in the catalytic reforming laboratory. In those days our usual catalyst was chromia on alumina which had just been discovered as a dehydrocyclization catalyst. The pressures were nearly atmospheric and no hydrogen was used.

When my supervisor left for vacation, Dr. Gustav Egloff came to the laboratory. To those of you who have heard of "Gasoline Gus" but have not met him, I shall give you a short description. A short, wiry,

[1] Retired from Universal Oil Products

0097-6156/83/0222-0141$06.00/0

dark complexioned individual, a champion wrestler in his days at Columbia, a prolific inventor, with some three hundred patents, and the author of a number of books and innumerable articles. He was one of the few individuals who could explain petroleum refining to a taxi driver and usually did. Egloff used to come to the UOP laboratories in Riverside at frequent intervals. On hot days, he used to put dry ice in the hollow of his hat apparently to keep a "cool head". At various technical meetings, he used to appear dressed in all synthetic clothing, and championed the conversion of petroleum to all possible products.

On that day in July 1935, "Gasoline Gus" indicated to me that, after all, the main trouble with catalytic reforming, as we were doing it then, was the formation of carbon. Young and inexperienced as I was, I realized this also, but then came the bombshell: "You have about three weeks on your own -- why don't you figure out a way of doing the reaction without all this carbon formation?" Frankly, I did not think it was impossible so I worked on it and at the end of the three weeks there was a singular lack of success. To this day, I do not know if "Gasoline Gus" was serious about this or not. This was my first brush with catalytic reforming.

Then came M.I.T. and in 1937 I was hired by UOP as a chemical engineer, working in the pilot plants. In 1939 came an opportunity to work as an assistant to Ipatieff in the newly established Ipatieff High Pressure Lab at Northwestern and obtain a Ph.D. degree at the same time while still working at UOP. In 1941 I returned full time to the UOP Riverside Laboratories to work for Ipatieff. The next five years were most instructive not only from the standpoint of technical growth, but also working for a man who was one of the greatest in the world of catalysis.

In retrospect, the greatness was that of a simple man. He knew he was good and he did not have to impress anyone. He was confident because he knew chemistry and was bound and determined to learn more. When Ipatieff first came to this country he was 62 years old. He knew very little English and the conversations with his associates were most difficult, but chemical progress, nevertheless, was very rapid, despite many comical situations that arose from the language differences. Ipatieff became a U.S. citizen in 1935 and the judge asked him if he attended church. Ipatieff said "yes". The judge persisted: "Which church?" To which Ipatieff replied: "Any church, God is everywhere." After that reply there were no more questions. Actually, Ipatieff was a very devout person. He was a member of the Russian Orthodox Church. Unfortunately, the Russian church was miles away and the services there lasted two hours and were not very prompt. The punctual and precise Ipatieff usually went to the Catholic Church a block away where the masses were frequent and quite short. In 1950 we flew to the Hague to the Petroleum Congress -- this was his first flight at age 82. As he said, there was no point doing something small, you might as well fly across the Atlantic. We boarded the plane, sat down, and at takeoff, the devout Ipatieff crossed himself and left it in God's hands. We did, however, return on the Queen Mary.

It has been said that I was Ipatieff's protegé and, frankly, knowing the man, I cannot think of a greater compliment. When he decided that I should be on my own, he let go of the reins, but was always available for discussion and consultation, and anyone would have been very foolish not to take advantage of that opportunity.

The next attempt at catalytic reforming came a number of years later. We had done some work on demethylation, then drifted into hydrocracking of kerosenes. That required desulfurization and we were delighted at how neatly kerosenes could be hydrocracked over catalysts like nickel-silica-alumina at quite low temperatures as long as the stocks were well desulfurized. One of our tests for six-membered ring naphthene content of the gasoline product was the passage over platinum on carbon at very low space velocities, wherein only these particular naphthenes could be converted into aromatics. Then we removed the aromatics by sulfuric acid treatment and established the five carbon ring naphthene content of specific fractions by a refractive index measurement.

Then came the idea to treat an entire gasoline fraction, well desulfurized, over a platinum catalyst. Various supports were made up and, as expected, we did convert a part of the naphthenes into aromatics, but the octane number increase was nothing sensational. When we moved up in temperature, the catalyst lay down and died. So we ran with hydrogen and applied a moderate pressure. The results were not particularly startling but at least the catalyst survived this ordeal. So we kept moving up and, sure enough, we did get better conversions.

All this time, we were just barely keeping ahead with our desulfurization runs to make enough charge stock for our dehydrogenation runs. One day we ran out of desulfurized stock and used a virgin straight run gasoline as our feed stock. Surprisingly enough, the catalyst did not even bat an eye, but kept right on converting. By this time we were at about 450°C and 500 psig pressure and using about 5 moles of hydrogen per mole of feed. This temperature was some 200°C higher than at that time recommended for a platinum catalyst.

In retrospect, we had wasted a lot of time cleaning up the feed, but somehow the idea of sulfur poisoning was so ingrained in our minds that the mere idea of using a poison containing stock was unthinkable. A few other things happened; we did notice that now we were converting more naphthenes and that some cracking appeared to take place. The cracking helped to concentrate the aromatics in the product and thus enhanced the octane number.

By this time we were doing about as well as had been done previously with a molybdenum oxide alumina catalyst, but with considerably less carbon formation. So now things became more serious, but not serious enough to get people very excited about it. After all, we had been using a 3% platinum on silica catalyst, and even in those days 3% platinum was pretty expensive. Platinum on silica-alumina did much better with respect to octane number but we could not control the hydrocracking very well, so we switched to alumina which had an intermediate activity. The results looked pretty good, particularly because we could run for days without much loss in activity.

At this point, I was quite fortunate in that Larry Gerhold, who was at that time manager of the laboratories, took quite an interest in the project and pointed out to me a few facts of life, one of them being that if I were really serious about this I better start cutting down on the platinum inventory on the catalyst. This we did. It is surprising what you can do when you have to. We devised all sorts of schemes for incorporating the platinum into the alumina, but one observation was quite critical, and that was that when we made the alumina from aluminum nitrate it was not so good a catalyst as one made from aluminum chloride. This was a real puzzle, until we observed that less washing of the cake made from aluminum chloride made a still better catalyst. Then we found a slight acidity in the exit gas from the unit and the picture began to gel. If chlorine is lost, but is active, then fluorine could be more active and more stable. The first fluorided catalyst gave us one of the highest octane products we had ever obtained.

Thus, things really began to look up. You can tell that you have something good and this, at the time, looked extremely good. This catalyst not only gave a high octane number product, it also had durability, at least in the relatively short tests that we were able to carry out. Within a year we had about one hundred people working on the project -- a large pilot plant crew, an engineering staff and a substantial research effort, primarily devoted to catalyst manufacture.

It would be only fair to name all the people who contributed to the project, but two persons gave it the greatest push -- they were Larry Gerhold and Bob Sutherland. It has been often said that the most difficult sale is within your own company. Platforming suffered the same fate and it was largely as a result of the efforts of these individuals that the project was not abandoned.

One of our troubles was deactivation, in fact, the pilot plant operation did show that we did accumulate carbon on the catalyst. If the initial rate of carbon formation were maintained, the run would not last more than a month or so. This would mean regeneration, and instead of a relatively simple operation, we would have to go to a more complex plant involving a cyclic regeneration system and all its difficulties that, at that time, were most formidable. Fortunately, as we learned, the carbon formation did not continue but leveled out at an acceptable value.

Nevertheless, our troubles mounted. The catalyst requirement was a very pure alumina substrate and that meant that we could not use a commercially supplied material which contained iron and other impurities which seriously impaired the activity. The initial larger scale catalyst manufacture was done by purifying anhydrous aluminum chloride by redistillation and then hydrolyzing it, followed by filtration, washing, drying, pilling, impregnation, etc. Furthermore, misgivings began to arise. We were considering a catalyst which would cost about $10.00 per pound, compared to the usual catalyst cost of about 15-60 cents a pound. The sales department was not too fond of the new process -- it was too radical a departure from the processes they were used to selling. The service department, which oversees the commercial operations, was also rather lukewarm about the whole business, and advocated more pilot plant testing and some modifications. At this darkest hour, the project was rescued by Mr. David Harris, president of UOP. He decided to go ahead. It is amazing how many people fall in line when the boss makes a decision!

It must have been a difficult decision. Mr. Harris took over the management of UOP in 1945 after it was given to the ACS by the former owners, the major oil companies. Very little cash was available and, at the time of the gift, the cupboard was pretty bare. Even more disturbing was that the basic patents in thermal cracking were expiring, cutting off additional revenue, and, furthermore, the company was also facing some 60 million dollars worth of legal suits against it.

To Mr. Harris, a very successful businessman, this was quite a challenge. His number one job was to organize the company operation on a more business-like basis and then settle the legal suits. As he said: "only the lawyers get rich", and carried this message in his visits to the opposing parties in the suits, and did his very best in staving off possible bankruptcy. (Actually, Mr. Harris, in his remark about the lawyers, did not foresee the dilemma of a lawyer who had to call a plumber to unplug a sink. After the work was done in about 3/4 of an hour, the lawyer was presented with a bill for 75 dollars. The lawyer exclaimed: "You know, this amounts to 100 dollars an hour, and even I, as a lawyer, do not make that much per hour", to which the plumber replied: "I did not either, when I was a lawyer.")

When it came to research and development, Mr. Harris knew it was the one substantial asset he must keep intact and make profitable. To him, Platforming represented a substantial investment but the potential was great and he trusted his advisors. His gamble did pay off and within ten years UOP, already free of legal suits, was sold to the public for about 75 million dollars, thus providing a real bonanza for the Petroleum Research Fund administered by the ACS.

Although Platforming was a very substantial contributor to the welfare of the company, the overall efforts in other areas, such as catalytic cracking, alkylation and other refining processes, also picked up and resulted in considerable income. During the first ten years after the process was first announced, Platforming units were licensed at the average rate of one every two weeks, so that by 1960 some 230 units were in operation, representing a capacity of some 1.5 million barrels per day. At the present time, the total licensed Platforming capacity is in excess of 4 million barrels per day with nearly 500 units in operation. We have catalyst manufacturing facilities in the United States, Europe and Japan.

Needless to say, anytime something like this comes along you get a fair number of competitors, so that the total volume of gasoline produced in platinum containing catalytic reforming units is very large. In fact, at the present time one would be hard pressed to buy gasoline which did not contain a portion processed over a platinum catalyst.

You might wonder how the name "Platforming" was coined. In retrospect, platinum reforming shortened to Platforming was a clever condensation and was thought up by Horner Eby, one of the three chemists that comprised our group at that time. It was clever because it was descriptive and also implied elevation, and we did mean elevation in octane number and yield. Our public relations department, whose domain was to think up names, took a dim view of the name we used and thought up a number of alternatives, but, like most forced-to-think-up-ideas, they fared poorly and the name had stuck for lack of better ones.

The name "Platformate" implied the product from the Platforming process and we used it extensively. The Shell Company asked us for the rights to use it in their advertising campaign. The greater mileage due to the use of Platformate was quite real and arose from the fact that when you have more aromatics present the gasoline is more dense and, on a per gallon basis, more total BTU's are produced on combustion.

None of this did we foresee in the spring of 1949. The first public announcement of the new process was very inauspicious. The "soft sell" was decided upon, probably from necessity, because we could hardly afford an all-out sales campaign. Ed Nelson, a vice-president of UOP and a great supporter of the process, had been asked to deliver a general paper on recent trends in petroleum technology at the Western Petroleum Refiners' Association Meeting in April of 1949 in San Antonio. In this paper, as one of five items, he included several paragraphs generally describing Platforming.

But this proved to be enough. The superintendent of the Old Dutch Refinery at Muskegon, Michigan, Elmer Sondregger, heard the paper and stopped at UOP's offices on the way home from the meeting to plead that we install the first unit of the new process at his refinery. He had a thermal reforming unit which he would make available for revamping to Platforming. The offer was accepted, and the next six months were busy ones around UOP. All records were broken in carrying out the design, purchasing and construction in this short period. Many laboratory and pilot plant runs had to be made to provide specific answers to problems which arose. Probably the most critical item was the manufacture of commercial quantities of a catalyst which had previously been made only in small laboratory batches.

By November, the construction was completed, catalyst supplied and the unit was ready to start. I was delighted to be asked to attend the start-up that fateful Saturday afternoon. I had been with the company some 12 years and during that time, except for a few scientific accomplishments and a number of patents, I had not really contributed to the company's welfare. So this was the first opportunity to justify some of the investment the company had in me. The unit started up beautifully, but disaster -- at least it seemed like a disaster -- struck in about three hours on stream, and the unit had to be shut down.

What do you do when a disaster, such as this obviously was, really strikes? You retire to a bar. There was only one topic of conversation. In the middle of it, Howard Nebeck, a most capable chemical engineer, left without finishing his drink and retired to his room to do some sketches on the redesign of the reactor.

The problem was excessive overheating of the outer shell. I still remember the frequency with which George Thompson, the UOP operator in charge of the unit, was applying the temperature chalks to the outer shell. When the highest temperature chalk promptly melted on the shell, he pulled the switch and the unit went down. I should explain that the catalyst was contained in an inside insulated reactor, and the high temperature insulation was supposed to protect the carbon steel shell from the combined effects of high temperature and high hydrogen partial pressure. Apparently the insulation was not insulating properly. Howard Nebeck's redesign that night involved a new liner configuration and, seven days later, the unit was modified and brought back on stream, this time to run for about nine months on the same batch of catalyst.

The process proved to be exactly the right solution to Old Dutch's problems. Where previously the gasoline from the refinery had been piling up in its tanks with no customers, everyone seemed anxious to try new "platinum treated" gasoline. This market required additional feed stocks and in order to meet this sudden need, the area around Muskegon was scoured for all available naphthas. Among the stocks fed to the unit was a very high sulfur bottoms product from light gasoline sweetening rerunning, and a used chlorinated cleaning solvent. Some of these stocks didn't run very well, but the unit did manage to survive and some of the limitations of the process were quickly defined.

From the start, there was quite a stream of visitors to see the unit in operation. We even had a sample of "catalyst" in a gallon container as an exhibit in the office of the plant. The volume of the sample shrank in proportion to the number of visitors, who undoubtedly just wanted a souvenir. I would have hated to have them be misled by an analysis of that particular souvenir.

Despite extensive publicity and a good performance, we did not start up the second unit for about two years. In a way, this could have been a good thing. It gave us a chance to consolidate our position by learning a great deal more about the process and, at the same time, make some substantial improvements. In addition, the competition was lulled into a false feeling of security and we gained the most important element -- time. The chief competitor was fluid hydroforming and considerable improvements were being made in the process. I remember well a verbal battle at the World Petroleum Congress at The Hague between the proponents of fluid hydroforming and ourselves. We had a good thing and we knew it, but they felt the same way and by sheer mass they could and did swing a lot of weight. The next year settled the issue, and sales of Platforming units increased very rapidly. Here again, we were fortunate in getting some of the majors interested enough to sign up, and once you do that it is just like getting support from Mayor Daley. You graduate into the big time.

Here again, I must emphasize that throughout the whole time the rat race continued, primarily due to the pressures exerted from within the company, from people like Gerhold, who weren't satisfied with the status quo. Our catalyst manufacturing method was rather complex and costly. The new method, developed from the very ingenious work by Jim Hoekstra, made all the difference in the world by providing for a much better catalyst support, made by a continuous process of manufacture. But even this proved to be a difficult thing to put on the marketplace, since refiners who had the old catalyst in their units, did not really want to change, so we had to maintain duplicate facilities for a period of time. The performance in the field was good enough to attract imitating competitors.

One disturbing occurrence really shook us. A Platforming unit at Bell Oil and Gas Company was started and ran very well for a few weeks -- then reports started coming in that all was not well. This is typical, you are slow to hear of the good performance, but the poor ones reach you quickly. The ailment was quite real. The temperature drop across the first reactor declined from the normal $100^{o}F$ to about $30^{o}F$ and the second reactor took over the reaction. However, this also started to lose activity and the heaters could not keep up with the demand for additional heat to maintain the overall activity. Finally the unit was shut down and the catalyst from the first of the three reactors was returned to us for examination. Here again, we were fortunate in having astute people on our staff, like Jack Murray and Ed Bicek of our Physical and Analytical Research. This group developed new analytical procedures which identified arsenic as the poisoning factor, and traced it to the feed stock. The arsenic was present to the extent of only about 30 parts per billion. The platinum in the catalyst just loved the arsenic, even in these low concentrations, and we were showing good patterns of platinum arsenide.

If the unit in the field showed deactivation, we should be able to do so in the laboratory, so we added a few parts per million of an oil soluble arsenic compound to a clean feed and, despite a fairly long run, could not show any poisoning effect. Finally, we realized that in a laboratory unit the surface to volume ratio of the hardware was many orders of magnitude greater than in a commercial unit, and we were undoubtedly picking up the impurity selectively on the walls of the hardware prior to entry into the catalyst bed. As a last resort, we suspended a minute particle of arsenic in the gas phase of the preheater and, sure enough, we deactivated the catalyst within hours.

The story of the arsenic poisoning troubles did reach the industry before we had demonstrated a solution to the problem, and some of our newly established competitors were publicizing their platinum catalyst as poison resistant. This created a minor uproar within UOP, and some of the people who got on the Platforming bandwagon by edict were again voicing their misgivings. Before too long the whole thing died down, including the claims of the competitors. I suspect that they had fallen into the same trap that we had by relying upon a laboratory test.

Despite all this the poisoning was real -- and the only way out was a pretreatment of the feed and this was readily accomplished by a treating step prior to feeding the stock to the Platformer. Thus, we became quite aware of the effect of possible poisons, and as the octane number requirements became more severe we learned to install more sophisticated pretreating facilities.

Platinum is a scarce material, and the success of the process depended on development of an economical process for recovering it from spent catalyst. This was very well handled by Herb Appel. On the other hand, catalyst modifications have proven to be a continuing problem.

In retrospect, it is amazing how one does <u>not</u> learn from history. When Universal was deeply involved in synthetic cracking catalysts, new catalyst modifications were always tested at the same conditions that were developed for the initial catalysts, and it is no wonder the new catalysts did not stand much of a chance. We fell in the same trap with the modified catalysts in the Platforming series and a number of good compositions were overlooked for a time because they were tested at the conditions found best for the reference catalysts. Nevertheless, advances were made and, from the initial R-4 catalyst,, as it was called, we have gone through a substantial number of commercial catalysts so that now we are in the R-30 series with the most sophisticated bimetallic catalysts, the new developments coming largely through the efforts of Ernie Pollitzer and John Hayes and their group.

Throughout all this work, one fact has emerged most clearly. You can think up all sorts of catalyst compositions, but they are of no avail unless they are properly tested. The proper testing implies meticulous attention to the design and details of the test and proper interpretation of the results. It is people like Rod Donaldson and a number of other most capable chemical engineers who have made this work meaningful and useful for further catalyst development.

At the same time, the search continues for the understanding of how these catalysts can be characterized by means other than the catalytic act. We are now finding that such characterization appears to be a reality, in other words, through the work done by the group headed by Hertha Skala, we have developed new insight of the relationship between the metallic components and the substrate.

At this point, I would like to digress to mention an important segment of the research effort. In the development of the R series of catalysts, we have been very fortunate in having on our staff a number of technicians who, despite the lack of formal technical degree, were able to contribute substantially to the development of new catalysts. In fact, a number of our commercial catalysts do bear the distinct mark of contribution by these people. We like to think that our method of operation, which includes their participation in our discussion sessions and providing for them an opportunity to try out their own ideas, has a direct bearing on their growth. We, in turn, benefit and we recognize their contributions by taking them off hourly pay and putting them on a salary basis, commensurate with professional status.

Thus, it is quite apparent that any technical development is only as good as the interest you can arouse in people. There are those you work for, those with whom you work, and those who work for you. The more unusual the idea the more difficult is the sale, but the people who get sold on the unusual idea are the real top-notchers, and once these critical people are sold, the input on their part is tremendously important for the final technological success.

In retrospect, I can only say that it was, and still is, a wonderful experience. To me one of the greatest thrills is to see a large scale unit in operation which is based in part on your own efforts. This thrill is greater than all the advancements, financial gains and honors that may come your way.

RECEIVED December 1, 1982

12

Otto Beeck and His Colleagues in Catalysis

J. NORTON WILSON[1]
Berkeley, CA 94709

O. A. Beeck and a talented team of collaborators (Smith, Wheeler, Ritchie) were pioneers in the application of evaporated, porous, metal films as "clean" surfaces for the study of chemisorption, heats of adsorption and the catalytic hydrogenation of ethylene. They discovered also the development of preferred orientation in such films. With another team of gifted investigators (Otvos, Stevenson, Wagner), Beeck also made early important contributions to the study of acid catalysis in hydrocarbon reactions by means of both stable and radioactive isotopes together with mass spectromentry. Beeck was not only a creative scientist but also a warm and generous person and an able organizer and administrator. The highlights of these investigations and some reflections about the people involved are presented.

Though I never had the privilege of working on catalysis with the late Dr. Otto Beeck, I knew most of his colleagues in that field as friends and talked on many occasions with them, and occasionally with him, about their work. For what follows, I have drawn upon my memory, recent conversations with J. W. Otvos, A. W. Ritchie, Fred Rust and A. E. Smith, the published literature and a brief biography that I wrote in 1962 (1).

I first met Beeck in 1943, when he hired me to work at Shell Development Company, in Emeryville, California, as a physical chemist in his Physics Department. Over the ensuing years, I came to know him fairly well as both colleague and friend and developed a strong admiration for him, not only as a man of vision and a creative scientist, but also as

[1] Retired from Shell Development Company

0097–6156/83/0222–0153$06.00/0

Figure 1. Photograph of Otto Beek taken during the 1940s.

a warm, outgoing and empathetic person and a strong and inspiring leader who used a loose rein very effectively. In his later years he developed a remarkable knack for putting together creative and effective research teams of people whose talents, knowledge and expertise combined well together for the project at hand. Many of the research results for which he is well known were accomplished through such teams, many of whose members deserve a large share of the credit. Nevertheless, the importance of his role as instigator, organizer and leader is not to be minimized. He played another role that is always important in a large industrial organization: that was as a promoter and defender of the work to higher management. To characterize him briefly, he was a very active and selective human catalyst.

During the years I knew him, he was a very busy man. His administrative responsibilities were substantial, not only as Head of the Physics Department but also as an Associate Director of Research with broader concerns. He served the National Research Council as an active member of its Committee on Catalysis and the National Advisory Committee for Aeronautics as a prominent member of its subcommittee on Lubrication, Friction and Wear. Those were heavy responsibilities during the years of World War II. He was a member of the Advisory Council of the American Institute of Physics and was an Associate Editor of the Journal of Applied Physics. He was active also on the Advisory Committee of the American Petroleum Institute's Research Project 44, which was a very active one during the war. Needless to say, he travelled a lot. In his spare time he led an active personal life. He lived on a small ranch in the countryside to the east of San Francisco Bay; there he bred and rode horses; he was also a skier and had a lively interest in photography and other visual arts. His wife was an accomplished sculptress and they had many friends in the world of the arts.

His career came to an untimely end on July 7, 1950, when he died from a massive coronary infarction shortly after a completely unexpected heart attack. He was then not quite 46 years old. In his brief career, however, he accomplished a great deal.

Achievements

He was author or co-author of 46 publications, including twenty-four on heterogeneous catalysis and related topics and five on the surface chemistry and physics involved in lubrication, friction and wear under conditions of high loading or of start-up. Many of his publications presented not only significant experimental facts but also some instructive insights; some of them were still being cited in 1980 and 1981. Two papers, which bear his name as co-author, were written by his colleagues over a year after his death.

He lectured from time to time at universities served as chairman for many specialized technical meetings, including several of the Gordon Research Conferences. He played a significant role in the organization and early development of those justly renowned Conferences; indeed, they grew out of the series of annual conferences held at Gibson Island on Heterogeneous Catalysis in which Beeck was a vigorous participant and developed friendships with many leaders of catalysis research in both industry and academia.

Among Beeck's important contributions to the progress of chemical research was his early and sustained encouragement of the application of spectrometric techniques to problems of chemical analysis. The existence of the modern analytical instrumentation industry is an expression of the fact that rapid and accurate chemical analyses are vital to modern technology. Beeck was directly responsible for the recruitment of R. Robert Brattain to lead a program in infrared spectrometry for analytical research; these efforts, together with those of other collaborators, led to the development of rapid methods for multicomponent analysis of hydrocarbon mixtures by infrared and ultraviolet spectrometry. These methods provided important savings in manpower and time during the development of processes for the manufacture of butadiene for synthetic rubber and of toluene and other high-octane components of fuels during the war. Similar applications of mass spectrometry to hydrocarbon analysis were developed early by David P. Stevenson with Beeck's support.

Important contributions to the development of such methods came from many individuals in other industrial organizations during that period. One of the important pioneering groups in the field, however, was the one at Emeryville associated with Otto Beeck.

Early History and the Road to Catalysis

Born in Stettin, Germany, Beeck received his

higher education in the Free City of Danzig, now the Polish city of Gdansk. There he earned a degree in engineering in 1928 and a doctorate in physics only two years later. His thesis involved the measurement of cross-sections and ionization potentials of noble gas atoms as determined from a study of their collisions with molecular beams of alkali ions and atoms. He continued that work as a research fellow and instructor in physics at Caltech from 1930 to 1933. He published a total of twelve papers in that difficult field, a testament to his experimental skill and productivity.

During that period, his name came to the attention of Dr. E. C. Williams, who was then Director of the young and growing laboratory at Emeryville. He talked with Beeck about the intellectual challenges posed by the phenomena of heterogeneous catalysis and invited him to come to Emeryville to see whether the disciplines of physics would be useful in a laboratory whose primary mission was to explore and exploit the chemistry of hydrocarbons and their derivatives.

Beeck accepted the challenge and came to Emeryville in 1933, a depression year when good jobs were scarce. His first studies of catalysis made use of the molecular beam techniques with which he was familiar; he impinged beams of hydrocarbon molecules on thin strips of transition metals heated in high vacua. In 1934 and 1935 he published a couple of interesting papers (2,3) on the catalytic decomposition of paraffins under these conditions. An interesting by-product of that work was a thorough investigation of the Knudsen accomodation coefficients and their variation with temperature for a number of simple paraffins and olefins. From these data he calculated the first reliable values for the specific heats of many of these compounds in the range C_2 to C_7, and he correlated these numbers with the corresponding infrared spectra (4,5).

By now he had realized that, with the vacuum techniques available at the time, a small surface could not be kept free of contamination long enough for catalytic studies. He therefore decided to use evaporated metal films as model catalysts. This was by no means a new idea; papers describing results with such systems had been published, especially in the European literature, since the late 1920's. The results, however, were not mutually consistent; there was clearly room for improvement. Around this time he acquired an assistant, a bright young chemist named Fred Rust, who later went on to a distinguished career in free-radical chemistry. They began to study the hydrogenation of ethylene over evaporated films of nickel. They learned that such films were so active that the reaction was diffusion-limited; the ethane produced interfered with the transport of reactants to the active surface in a static system.

An important development occurred in 1936; Beeck was authorized to organize and lead a Physics Department and he started to recruit people. One of his first recruits was Albert E. Smith.

Al Smith had an interesting background. In the early 1930's, he was a graduate student in physics at the University of California in Berkeley and working in E. O. Lawrence's new Radiation Laboratory. He had hopes of going on to a career in nuclear physics. He had taken the graduate courses in physics, all the advanced mathematics he could lay his hands on, many courses in chemistry and had a master's degree to attest to his scholastic achievements. He was well versed in the laboratory arts and metal-working and was an accomplished glass-blower; indeed, he had earned his way through school by part-time work in that capacity. In late 1935, however, the economic exigencies of the time required him to interrupt his doctoral work and earn some money to assist his family. He applied for a job at Shell Development Company and was accepted. After a couple of short-term assignments he was asked by Mirko Tamele, of cracking catalyst fame, to develop a spectroscopic method to measure the concentrations of porphyrins and other condensed polyaromatics in heavy petroleum fractions. This was relevant to the then-current debates about the origins of petroleum. In those days one could not buy a spectrometer off the shelf, so Smith designed and built one, with photographic recording of absorption spectra in the visible and near ultraviolet. While he was busy applying this tool to the porphyrin problem, Beeck beckoned with an opportunity to build an electron diffraction machine for the study of surfaces. Though Tamele was reluctant to let him go, Smith finally joined the new Physics Department. Guided by Germer's publications, he proceeded to design an electron diffraction apparatus, including a precisely stabilized high voltage supply and the vacuum system. The system was built in the laboratory shops, with some specialized work by Smith himself and it worked. At Beeck's instigation, he set out to use it to study the structure of the films formed on metal surfaces under "boundary lubrication" conditions in the presence of lubricants containing certain polar additives.

In 1937, while this work was in progress, Ahlborn Wheeler joined the department. "Bud" Wheeler was a new Ph.D. from Princeton, where he had worked with Henry Eyring. Though he had a keen appreciation for good experimental design and technique, his primary leanings and aptitudes were theoretical. During his years at Emeryville, he devoted a lot of effort to an analysis of the effects of catalyst porosity on the apparent kinetics and activation energy of catalytic reactions, with attention to the effects of both pore shapes and pore-size distributions. Many of his findings were communicated informally from time to time at Gordon Research Conferences; they were later summarized in an important monograph (6). His first assignment, however, was to take over Rust's responsibilities in the metal film work. Rust was glad of the chance to return to his primary interest: exploratory research on homogeneous reactions.

In the metal film work, the immediate need was to overcome the problem of diffusion limitation. This was accomplished by the invention of an ingenious glass turbine which could be mounted inside the vacuum system and used to circulate the reactant gas mixtures at a high rate over the catalytic film surface. Normal to its axis, it carried an encapsulated rod of soft iron, so that it could be driven from outside the vacuum system as a kind of synchronous induction motor by a pulsed 60-cycle magnetic field. This device must have been very difficult to build and balance, but it was made to work. While it was being developed, they carried out a study of the chemisorption of nitrogen on iron films at 23 and 100 C and found it to be an activated process. By treating the saturated films with acid and finding ammonium ion in the resulting solution, they determined also that it was a dissociative adsorption (7). Then they returned to the study of ethylene hydrogenation on nickel.

Comprehensive Studies of Ethylene Hydrogenation

For this work, Beeck and Wheeler used a more elaborate vacuum system than had been used in the earlier exploratory work. They used large McLeod gauges for pressure measurements, with which noncondensible gas pressures of 10^{-7} Torr. could be detected and a "sticking vacuum" implied 10^{-8}Torr. or less. Higher pressures were measured with a mercury manometer and an accurate cathetometer. The reactor system was isolated from the pumps and gauges by means of cold traps. To obtain reproducible results, they found it necessary to outgas the metal filaments before evaporating the films; this was done by heating them electrically under vacuum to a temperature just below that at which evaporation rates became appreciable. It was necessary also to bake out the cylindrical glass reaction tube at 500° C for some time before depositing the film on its inner surface and to control the temperature of the substrate while depositing the film. With these precautions and use of the circulation turbine they could obtain reproducible reaction rates that were not diffusion-limited. They found that the reaction rate was accurately first order in the partial pressure of hydrogen but independent of that of ethylene; this implied that ethylene was an inhibitor as well as a reactant and they could now define a rate constant. Chemisorption of hydrogen or of carbon monoxide was too fast to follow with a McLeod gauge, even at low temperatures. These processes therefore had very low activation energies, contrary to earlier literature results for conventional supported nickel catalysts.

Now they made a surprising discovery. Guided by a recent European publication, they evaporated a nickel film in a low pressure of purified inert gas instead of in vacuum. The resulting film was about ten times as active as the same mass of vacuum-deposited film, even though its chemisorptive surface area was only about twice as great. These effects could be repeated; there must be a structural difference. Beeck immediately invited Al Smith to join the investigation.

Smith undertook an exhaustive study of the nature of these films. To do so, he had to build his own specialized vacuum system because he had to deposit his films on flat plates in order to examine them by electron diffraction. He also had to be able to transfer them into the electron diffraction camera without exposure to atmospheric contamination. For pressure measurements he, too, used a McLeod gauge because the ionization gauges one could build at that time were prone to outgassing, which would contaminate the films. He made films of a great many transition metals, in addition to nickel.

He studied films grown in vacuo and in inert gas ranging in pressure from 10^{-5} to about 20 Torr. All were found to have the usual metal crystal structure. As film thickness increased, the films, even those grown in vacuo, developed an increasing tendency to grow crystallites with a preferred orientation. For face-centered cubic metals, such as nickel, the orientation was such that the 110 planes were roughly parallel to the substrate surface. For body-centered metals, such as iron, the 111 planes became so oriented. This finding, of course, gave essentially no information about which crystal planes were exposed to the gas phase.

The major observable distinction of the gas-grown films was that preferred orientation developed much more rapidly as thickness increased. The optimum pressure for this effect was one to two Torr; at higher pressures, the preferred orientation was increasingly suppressed.

Smith studied also the sintering of the films, either by annealing at temperatures above that of deposition or by growing them on substrates held at higher temperatures. One by-product of that work was one of the first and perhaps the first example of what we now call epitaxial growth of single-crystal films. By evaporating palladium on to the cleavage plane of a sodium chloride crystal held at 350°C, he made a single crystal film of the metal with its 100 plane parallel to the backing.

These observations, of course, guided the ongoing work by Beeck and Wheeler on chemisorption and catalysis. In 1940, Beeck, Smith and Wheeler published a landmark paper describing their findings in the Proceedings of the Royal Society (8). It stimulated similar work elsewhere, because it showed that evaporated metal films, properly made, were reproducible, porous and pretty uniform throughout their thickness. On unsintered films, given that a molecule of hydrogen dissociatively occupies two surface chemisorption sites, a molecule of carbon monoxide occupied only one, whereas a molecule of ethylene occupied at least four. Sintering of the films before use introduced complications into both chemisorption and catalysis.

The work continued thereafter but with little further participation by Smith, who went on to other problems. By self-training, he later became a distinguished X-ray crystallographer. The main burden of the later laboratory work was carried by Wheeler, with assistance at various times from various people, including W. A. Cole, W. H. Thurston, J. W. Givens and A. W. Ritchie. Walt Ritchie's story is an interesting one. For economic reasons, he had to interrupt his college career when he had taken only a couple of courses in chemistry. He joined the project as a laboratory assistant. He was a bright young man, however, and expressed an interest in what was going on. Wheeler lent him text books, encouraged him to work the problems and even offered to correct them. When Beeck heard about this, he had several talks with Ritchie and encouraged him in his self-education. In time, with continued learning and broader laboratory experience, Ritchie was promoted to professional status. He was a co-author of some of Beeck's later publications.

The first important development after the 1940 paper was a simple but elegant calorimeter for measuring the heat of adsorption of fractions of a monolayer on evaporated metal films (9). For both hydrogen and carbon monoxide on nickel, the magnitude of the initial heats showed that strong chemical bonds to the surface were being formed. The heats decreased monotonically as coverage increased, and more rapidly as monolayer coverage was approached. This result implied some kind of interaction among chemisorbed species; it also implied that chemisorbed hydrogen atoms were mobile over the metal surface. Curiously, the curves for heat of adsorption versus coverage for hydrogen on oriented and unoriented nickel films were essentially superposable, despite the large difference in catalytic activity per unit surface.

A significant observation was made during measurements of the heats of adsorption of ethylene: after partial coverage of the surface, a residual pressure of ethane was detected in the gas phase. Ethylene was hydrogenating itself! This fact, together with the number of sites covered by a chemisorbed ethylene molecule, led to the concept of the so-called "acetylenic complexes". The chemisorption of ethylene was dissociative to form chemisorbed hydrogen and a strongly adsorbed, less hydrogen-rich residue. During the catalytic hydrogenation of ethylene under steady-state conditions, the surface was presumed to be covered largely by such hydrogen-poor residues, whose removal by hydrogenation then became the rate-determining step. This would account for the observed kinetics.

A second important development was the extension of the catalytic and thermochemical measurements to most of the transition metals. Most of this had been done by 1945, when Beeck published a review article entitled "Catalysis - a Challenge to the Physicist" (10). A second review was issued in 1947 (11). In it, Beeck mentioned that Wheeler had been able, by means of an approximate quantum-chemical calculation, to account for the shapes of the curves relating heat of adsorption to fractional coverage. Unfortunately, no details of that calculation were ever published.

Almost every year from 1940 to 1949, Beeck attended the Gordon Research Conference on Catalysis, sometimes accompanied by Wheeler. There he usually presented a report of their recent progress, so the catalysis community was apprised of new developments. A paper on the adsorption of hydrogen on sintered nickel films was published in 1948 (12). It revealed that the usual very rapid adsorption was accompanied by a slower, activated process. This slower step was attributed to the penetration of hydrogen into regions that had been closed off by thin barriers formed by sintering. The finding resolved, at least in part, the difference between the results obtained initially with evaporated films and those reported for supported nickel catalysts formed by conventional chemical means.

In 1948, Beeck published his third review article on surface catalysis; it summarized the findings and practical applications of all the preceding studies by him and his colleagues since 1934 (1). For example, in the early work with platinum foils (3), it was found that carefully dried hydrocarbons did not decompose rapidly on such surfaces until a temperature of 1600°C was reached, at which point the resulting fragmentation of the reactant was severe. Addition of traces of water vapor led to decomposition to simpler products at much lower temperatures. He attributed the promotional effect to dissociation of the water on the surface to produce chemisorbed H. and OH. radicals which reacted with the impinging hydrocarbon. This finding led to a patent (14).

He mentions also that in the early work with evaporated nickel films, it was found that acetylene in a mixture with ethylene was hydrogenated selectively and almost completely before the ethylene was attacked. This finding led to a useful method for removing traces of vinylacetylene from butadiene for synthetic rubber manufacture during World War II. Also, during that war, he assigned A. E. Smith to use X-ray diffraction to study the reason for the rapid deactivation of hydroforming catalysts. The problem was traced to a rapid transformation, under reaction conditions, of the high-area alumina support to alpha alumina with a much lower area. A method was found to stabilize high-area alumina supports by the incorporation of small amounts of alkali ions, especially lithium (15,16).

Before the war, using one of the earliest samples of tritium from the Radiation Laboratory at Berkeley, he found the rate of isomerization of n-butane to isobutane, over aluminum chloride promoted with water, was proportional to the rate of exchange of hydrogens between the hydrocarbon and a catalyst promoted with tritiated water. This observation may have been part of the stimulus for the more detailed studies with isotopic tracers of acid catalysis after the war.

Final Publications on Metal Films

1950, the year of Beeck's death, was also an important publication year. A monograph describing the chemisorption results appeared in Advances in Catalysis (17). Beeck also attended the Faraday Society Discussion on Heterogeneous Catalysis, an international meeting, where he presented three papers: a summary review by himself (18), a paper with Ritchie on the reaction rates with a variety of metal catalysts (19) and one with Cole and Wheeler (9) describing the adsorption calorimeter and the results obtained therewith. A number of stimulating experimental facts, ideas and correlations were presented, including the famous "volcano-shaped curve" in which the logarithm of the isothermal rate constant for hydrogenation, covering four orders of magnitude, was plotted against the lattice spacing for eleven transition metals. The curve peaked at rhodium, the most active catalyst, and fell off sharply on either side. That correlation was marred, however, by the presence of several significantly out-lying data points. A better correlation was a plot of log k versus the initial heat of adsorption of either hydrogen or ethylene. Rhodium exhibits the lowest heat while tungsten, the least active catalyst, has the highest. Strangely, the activation energy for the overall hydrogenation rate constant was stated to be 10.7 kcal/mol for all the metals tested, including both oriented and unoriented films of nickel. Unfortunately, few experimental data were presented to support that statement, nor was there a discussion of how the effects of sintering were circumvented in the measurement of activation energies.

Some previously unpublished results were presented about the rate of removal by hydrogenation of pre-adsorbed ethylene from a covered film surface.

The rate of removal from a rhodium surface was much higher than from the surfaces of less active metals, in conformity with the kinetic model. Morever, the product obtained from the rhodium surface was largely ethane, whereas from the less active surfaces of other metals, saturated oligomers of ethylene were obtained also. This showed that the "acetylenic complexes" could polymerize on the surface, thus providing a route toward the known phenomenon of carbonization of catalyst surfaces during hydrogenation.

The format of that meeting was such that each author was required to submit a complete manuscript of his paper well in advance of the meeting. Preprints of the papers were sent to all expected participants before the meeting so that they could prepare their discussions. At the meeting itself, each author was given only a short time to introduce and summarize his paper, so that most of the time could be devoted to discussion. Alex Oblad told me an interesting story about that. As the meeting started, the discussion following the first few papers was rather desultory. During a break, Beeck, who was scheduled to come on next, suggested to Oblad, who was to follow him, that they liven things up. Beeck would devote most of his allotted time to a spirited attack on Oblad's work and Alex, in turn, would reciprocate. Alex agreed; the resulting exchange led to a lively discussion which continued through the meeting, as the published proceedings attest.

Also in 1950 a paper was published describing adsorption isobars for hydrogen on nickel films from 20°K to room temperature (20). These results were achieved with the aid of the newly developed Collins helium cryostat.

It is extremely fortunate, I believe, that this final burst of papers appeared before Beeck's untimely death. At the same time, it is regrettable that a great deal of the experimental detail underlying this large and stimulating body of work was never published and, presumably, never will be.

Acid Catalysis

The history of Beeck's involvement in the study of acid catalysis was very different from the foregoing. Here his role was primarily to assemble a talented research team to attack a broad objective. That objective was to use stable isotopes, which became available after the war, to study mechanisms of catalysis; the idea may have been stimulated by conversations with D. P. Stevenson. The team comprised Stevenson, an accomplished mass spectrometrist and a broadly experienced physical and structural chemist; John W. Otvos, a physical chemist and spectroscopist with background in reaction kinetics and radiotracers, and Charles D. ("Chuck") Wagner, an organic chemist with strong physical leanings; his primary role was to synthesize hydrocarbons with carbon-13 or deuterium in known molecular positions.

Acid catalysis was a natural field for such a group to study since it involved both skeletal rearrangements and probable hydrogen transfers. So far as I know, there was little formal research direction of this work by Beeck; experiments were planned informally among the three investigators and products were usually analyzed mass spectrometrically by Stevenson. Beeck dropped in at the laboratory from time to time to learn what was going on and to offer encouragement and informal suggestions.

This work resulted, over the period 1948-1952, in eight publications dealing with the isomerization of paraffins and their hydrogen-exchange reactions in acidic systems. The interfaces involved were either solid, with catalysts based on aluminum chloride, or liquid, with concentrated sulfuric acid. The results provided strong support for the utility of the carbonium-ion model and new information about the properties of the postulated transient intermediates.

First came a thorough investigation by Stevenson of the nature of aluminum halide catalysts and their activity for the isomerization between methylcyclopentane and cyclohexane (21,22). In agreement with industrial lore and with slightly earlier publications by others on related reactions, it was found that small amounts of water were essential to catalytic activity; pure hydrogen halides, on the other hand, were not effective promoters. The active species were proposed to be $Al_2X_{6-x}(OH)_x$; this proposal was suported by X-ray powder diffraction data obtained in collaboration with A. E. Smith. The equilibrium constant and its temperature coefficient were re-determined and the thermodynamic parameters for the reaction were computed. Many of the byproducts were also determined by mass spectrometry.

Concurrently, a series of papers began to appear on the isomerization and hydrogen-exchange reactions of hydrocarbons with acid catalysts. Over slightly hydrated aluminum bromide at room temperature, propane containing carbon-13 at one end (C_3H_8-1-C^{13}) was found to isomerize toward a statistical mixture with

C_3H_8-2-C^{13} at a rate comparable to that of the isomerization of n-butane to i-butane under the same conditions (23). Under similar conditions, n-butane-1-C^{13} was found to isomerize to a statistically weighted mixture of the isomers with the carbon-13 in all possible positions (24). The data showed that the activation process was purely intramolecular and suggested that the activated complex was capable of rearranging among all possible isomers before reverting to the unactivated hydrocarbon.

A study was then made of the hydrogen-exchange reactions of all seven of the monodeuterated C_2,C_3 and C_4 hydrocarbon isomers at room temperature over a water-promoted aluminum chloride-alumina catalyst. Contrary to expectation, no exchange of deuterium between the catalyst and any of the deuterated hydrocarbon isomers was detected, even though most of the hydrogen in the reaction systems was incorporated in the catalyst. On the other hand, a variety of intermolecular exchanges between hydrocarbon molecules was observed, though no evidence was found for disproportionation reactions (25). Some of these reactions were identified by the use of doubly-labeled hydrocarbons, with carbon-13 in one position and deuterium in another. Under the conditions studied, some of the exchange reactions were very fast, with activation energies of approximately zero and half-times of the order of two minutes or less. Others, such as the intermolecular exchange between the primary and secondary hydrogens of propane, were much slower. A substantial number of general observations were made, but detailed mechanistic conclusions were deferred. An important difference between isomerization and hydrogen exchange was signalled by the fact that the activity of the catalyst for isomerization declined by a factor of 10 during the course of the series of experiments, whereas the activity for the exchange reactions did not diminish noticeably.

This important short communication was accompanied by another on hydrogen exchange reactions of the butanes in concentrated sulfuric acid (26). In this case, i-butane was found to exchange its primary hydrogens with the hydrogens of sulfuric acid at a significant rate, the half-time at 25°C decreasing from 1700 minutes in 91.5 percent sulfuric acid to 48 minutes in 98.3 percent acid. The reaction displayed a short induction period which could be eliminated by adding 0.1% of isobutene to the starting hydrocarbon. The induction period was thus attributed to the necessity for trace oxidation of the starting paraffin to form olefin, leading to a carbonium ion, before reaction could proceed.

The overall kinetics were found to be pseudo-first order, as is required for most exchange reactions, regardless of mechanism (27,28). The logarithm of the apparent rate constant was found to vary linearly with the Hammett acidity function.

A reaction with the same rate constants was found to be the exchange of *tertiary* hydrogens between isobutane molecules, one of which was labelled with deuterium in the tertiary position and the other with carbon-13 in a primary position. Nevertheless, exchange of the tertiary hydrogen of isobutane with the hydrogens of the sulfuric acid was not observed at all, except with 98.3 percent acid, in which case significant oxidation of the hydrocarbon was shown by discoloration of the acid and liberation of sulfur dioxide. With the *n*-butanes, neither the primary nor secondary hydrogens showed any significant reaction in 96 percent acid.

These findings had important implications for the properties of the hypothetical carbonium ions. Though several qualitative generalizations were drawn, detailed discussion was deferred for longer communications that followed.

Preparation and characterization of the isotopically labelled hydrocarbons were described in two papers (29,30); a third described a method for determining the isotopic purity (32) of perdeuterosulfuric acid. With this foundation, the behavior of isobutane was discussed (31) in terms of a carbonium-ion chain reaction in which the chain-carrying step is the transfer of the tertiary hydrogen atom, carrying its electron pair, from an isobutane molecule to the tertiary carbon of a tertiary carbonium ion to terminate the life of one ion and generate another. The lifetime of a given carbonium ion is long enough so that all, or nearly all of its primary hydrogen atoms are exchanged with hydrogens of the acid. Then followed a substantial paper (32) in which a detailed description and discussion was given of the hydrogen exchange and isomerization reactions of a number of C_5 to C_7 alkanes and several cycloalkanes in the presence of sulfuric acid. Again the kinetics were found to be pseudo-first order and the products were accounted for in terms of the carbonium-ion mechanism. A number of generalizations were adduced concerning the many details of the hydrogen-exchange reactions; these could all be accounted for by the mechanism.

After this work was completed, the team disbanded and its members pursued other avenues of research. Stevenson returned to mass spectrometric research and other forms of spectrometry. He later assumed administrative responsibilities, first as Manager of the new Chemical Physics Department and later as Director of the General Science Division. Otvos turned to research applications of radiotracers and later worked with Wagner on radiation chemistry. He subsequently succeeded Stevenson as Manager of the Chemical Physics Department and later became Manager of Analytical Research. Wagner went on to pioneering work in radiation chemistry and also assumed supervisory responsibilities. He later achieved widespread recognition for this work in X-ray photoelectron spectroscopy.

Concluding Remarks

Though Otto Beeck's research career was a relatively short one, his achievements were diverse and substantial. In his interpretation of the high catalytic activity of oriented metal films, however, he espoused and actively promoted one thesis that was not shared by Smith or Wheeler and has not stood the test of time: that was to suggest that the planes of preferred orientation were also those exposed to the gas phase. A more physically acceptable interpretation of the observed effects can now be made, at least in principle. Nevertheless, the overall significance of the work that he inspired and led in his chosen fields of investigation was such that his successors in those fields are in his debt, whether they realize it or not.

Literature Cited

1. Wilson, J. N. "Dr. Otto A. Beeck", 1962. A biographical sketch, available in the files of the Center for History of Physics, American Institute of Physics, 335 E. 45th St., New York, N.Y. 10017. Contains a complete list of publications.
2. Beeck, O. A. Phys. Rev. 1934, 45, 331.
3. Beeck, O. A. Nature 1935, 136, 1028-9.
4. Beeck, O. A. J. Chem. Phys. 1936, 4, 680-89.
5. Beeck, O. A. J. Chem. Phys. 1937, 5, 268-73.
6. Wheeler, Ahlborn. "Advances in Catalysis"; Frankenburg, W. G.; Komarewsky, V. I.; Rideal, E.K.Eds. Academic: New York, 1951; 3, p.250-327.

7. Beeck, O. A.; Wheeler, A. J. Chem. Phys. 1939, 7, 631-2.
8. Beeck, O. A.; Smith, A. E.; Wheeler, A. Proc. Roy. Soc. (London) 1940, A177, 62-89.
9. Beeck, O. A.; Cole, W. A.; Wheeler, A. "Heterogeneous Catalysis"; Discussions of the Faraday Society; Aberdeen University Press: Aberdeen, 1950, 8, 314-21.
10. Beeck, O. A. Revs. Mod. Phys. 1945, 17, 61-71.
11. Beeck, O. A. Record Chem. Progress 1947,8,105-.
12. Beeck, O. A.; Ritchie, A. W.; Wheeler, A. J. Coll. Sci. 1948, 3, 505-10.
13. Beeck, O. A. Revs. Modern Phys. 1948, 20,127-30.
14. Beeck, O. A.; Burgin, J.; Groll, H. P. A. "Activating and Maintaining the Activity of Dehydrogenation Catalysts", U.S. Patent 2,131,089 (to Shell Development Company).
15. Smith, A. E.; Beeck, O. A. "Thermally Stable Aluminas for Catalyst Supports", U.S. Patent 2,454,227, 1948, (to Shell Development Company).
16. Smith, A. E.; Beeck, O. A. "Lithium-Alumina Catalysts", U.S. Patent 2,474,440, 1949 (to Shell Development Company).
17. Beeck. O. A. "Catalysis and Adsorption of Hydrogen on Metal Catalysts", Advances in Catalysis, Frankenburg, W. G.; Komarewsky, V. I.; Rideal, E. K., Eds., Academic: New York, 1950, 2, p. 151-195.
18. Beeck, O. A.; "Heterogeneous Catalysis", Discussions Faraday Soc., Aberdeen University Press: Aberdeen, 1950, 8, p. 118-28.
19. Beeck, O. A.; Ritchie, A. W.; ibid., p. 159-66.
20. Beeck, O. A.; Given, J. W.; Ritchie, A. W. J. Coll. Sci. 1950, 5, 141-7.
21. Stevenson, D. P.; Beeck, O. A.; J. Am. Chem. Soc. 1948, 70, 2890-94.
22. Stevenson, D. P.; Morgan, Jane H. J. Am. Chem. Soc. 1948, 70, 2773-7.
23. Beeck, O. A.; Otvos, J. W.; Stevenson, D. P.; Wagner, C. D. J. Chem. Phys. 1948, 16, 255-6.
24. Otvos, J. W.; Stevenson, D. P.; Wagner, C. D.; Beeck, O. A. J. Chem. Phys. 1948, 16, 745.
25. Wagner, C. D.; Beeck, O. A.; Otvos, J. W.; Stevenson, D. P. J. Chem. Phys. 1949, 17,419-20.
26. Beeck, O. A.; Otvos, J. W.; Stevenson, D. P.; Wagner, C. D. J. Chem. Phys. 1949, 17, 418-19.
27. Wilson, J. N.; Dickinson, R. G. J. Am. Chem. Soc. 1937, 59, 1358-61.
28. McKay, H. A. C. Nature 1938, 142, 997-8.

29. Wagner, C. D.; Stevenson, D. P. J. Am. Chem. Soc. 1950, 72, 5785.
30. Wagner, C. D.; Stevenson, D. P.; Otvos, J. W. J. Am. Chem. Soc. 1950, 72, 5786.
31. Otvos, J. W.; Stevenson, D. P.; Wagner, C. D.; Beeck, O. A. J. Am. Chem. Soc. 1951, 73, 5741-6.
32. Stevenson, D. P.; Wagner, C. D.; Beeck, O. A.; Otvos, J. W. J. Am. Chem. Soc. 1952, 74,3269-84.

RECEIVED October 29, 1982

13

Ernest W. Thiele: A Pioneer in Defining the Role of Diffusion in Heterogeneous Catalysis

RAVI RANDHAVA
Xytel Corporation, Mt. Prospect, IL 60056

Randhava: Dr. Thiele, scientists have gradually developed an appreciation of the important role that diffusion plays in heterogeneous catalysis. Your paper in 1939 provided the theoretical foundation for later developments in this research area; the widely used "Thiele Modulus" attests to the importance of your contribution. Can we talk about this for a few minutes.

Thiele: I'd have to say the whole thing started out at MIT. I was working on my doctor's thesis at that time studying the steam-carbon reaction. Some very peculiar results in kinetics made me think that maybe porous carbon had something to do with the problem. All this was in the early twenties—you have to remember that at that time no one in the oil industry had ever heard of catalysis. For that fact, it was only after Ipatiev wrote his big paper that interest in catalysis caught on in the US.

After leaving MIT, I joined Standard Oil of Indiana in 1925. The question of porosity stayed in my mind, since many catalysts are porous. I kept thinking about it for several years and gradually evolved my theory. What we are talking about here is quite straightforward. Small catalyst particles behave very differently from large particles. In small grains, the catalytic action is proportional to the mass of the catalyst; in large grains, to the external surface.

All my work on kinetics and the modulus was essentially done in my spare time. My paper was published in 1939. Soon after that, I found out that two other researchers, Damkohler in Germany, and Zeldovich in Russia, had also been working on the same problem and had recently published their results. It just goes to show you that no one is indispensible; if you don't do something, someone else will. The approaches that the three of us followed differed markedly, but we all arrived at essentially the same result: there is a modulus, an expression depending on the diffusivity of the reactants, the rate of reaction, and the size of the grains, which determines the effectiveness of the catalyst mass; if this modulus is low, all the mass is effective; if it is high, only an outer layer is effective.

0097–6156/83/0222–0173$06.00/0

Randhava: What else was going on at Standard Oil in those days?

Thiele: Keep in mind that I spent almost 35 years at Whiting. At one time or another, I was engaged in just about every aspect of petroleum refining, including a great deal of practical work on catalysts.

The Whiting refinery used to be the biggest facility of its type in the world, when I joined in 1925. Executives came up through the ranks of chemical engineers. The general atmosphere towards people in research was not particularly good, but it improved as you went higher.

I always said that in those days the lab used to be like a worm. There was little or no organization. Now things are carefully departmentalized. A good deal of my time at Whiting was spent on cat cracking. Everyone was interested in that field. Houdry got Sun Oil involved and they developed a fixed bed process. Standard Oil of New Jersey went ahead and set up a powdered oil process.

Most of the early work that I did at Whiting was elementary in nature—things like studying the physical properties of oil. The first noticeable project that I got involved with had to do with delayed coking where you take heavy oil and heat it to a high temperature so that the remains are solid. When I arrived the refinery was using the Burtron process for thermocracking. We set up our first semi-works plant in 1930. We didn't think that we could get away with anything as small as a pilot plant for delayed coking then. Of course, nowadays people have more experience in designing and building smaller units and things have changed.

In addition to cat cracking and coking, we did a lot of work on ultraforming. This was a hot area and we got caught up in a wide variety of pilot plants and catalysts. We also got involved in alkylation using sulfuric acid, a pretty messy process. I'd have to say that over the years at Standard Oil there was literally no end to the number of projects that we handled.

I joined Whiting as a chemical engineer, later became a group leader, assistant director of research, and finally their associate director of research. At one time I was in charge of the high pressure lab where I had 70 people, including 10-12 PhD's working for me. Later on they divided up my job and gave it to about three people.

Randhava: Let's go back a few years. Where did you go to school?

Thiele: We can start even before that. My parents came to the U.S. from Germany. I was born in 1885 in Chicago, went to Catholic schools in the city, and received an AB degree from Loyola University after studying subjects such as English, philosophy, Latin, and Greek. From there it was on to the University of Illinois at Urbana-Champaign, where I graduated in 1919 with a BS in chemical engineering.

Randhava: What made you decide to go to MIT?

Thiele: After leaving the University of Illinois, I spent about 6 months as an analyst for Swift & Co. This was followed by a couple of years with People's Gas Light and Coke in Chicago. I was stationed in the lab doing many types of work, and did not receive any formal training, but did gain some useful experience in manufacturing gas. However, even though I had graduated with honors from Illinois, to give you a feeling for how little I knew then, I had no idea that you have to pour reflux back into a column. That's when I realized something was missing and made the decision to go to MIT for my doctor's degree.

Randhava: What happened at MIT?

Thiele: Things were quite exciting there. I started out in the fall of 1922, and got both my master's and doctor's degrees at MIT. My thesis work was on the kinetics of the steam-carbon reaction. I did that under Prof. R. T. Haslam. I specially remember the oral exam for my doctor's degree. Dr. Lewis, the head of the department, was there, along with an instructor, Prof. C. S. Robinson. They were not interested in any meaningless ceremonies and didn't ask me any questions. Lewis remarked that McCabe had an idea for computing columns. I proposed a nomographic technique, showed it to McCabe, and he showed me his method. I knew we had a good thing, drew up most of the figures, and finally left. McCabe had his troubles with department reviewers, but the paper was eventually published.

Randhava: Dr. Thiele, what are your feelings about scientists like you publishing papers?

Thiele: I've written and published my share of papers. However, I want to point out that my papers have always been a side issue for me. Most of my life was in business. You musn't expect an enormous number of papers from me because I was not primarily an academic man. What I went after and got a lot of were patents, over 30 in my lifetime.

One of my significant patents was on delayed coking. One of my later patents relates to lowering the pour point of steam cylinder oils. We used to take the dirtiest stuff such as low grade asphalt to lower the pour point. As soon as I became associate director of research at Standard Oil, my flow of patents came to a halt.

Randhava: To what extent do you think we should rely on industry for funding of research?

Thiele: It seems right to rely on industry, but the truth is, there's many projects in which, if you expect industry to do it, it will never get done. The government must do many things. Industry doesn't look as far ahead into the future. And there's very large projects that won't work out with industry funding. You can see what's happening with regard to synthetic fuels. Neither government nor industry can make it work. In this case it might reasonably be left to the private sector. I think there's been a tradition of looking a long way ahead that's decreasing.

Randhava: Weren't you involved with nuclear work at one time?

Thiele: Yes, I was. When the project was very young, I was the only engineering technician who worked on the Standard Oil Development Company committee which was investigating the atom bomb. We worked with Arthur Compton on that committee and eventually turned over all our findings to General Grove who ran the Manhattan Project.

At about that point from 1942 to 1943, I was given a special project to design and start-up a heavy water separation plant in Trail, British Columbia. I didn't have much faith that they would get this job done in time for the war. The work was extra confidential. One of the techniques we used involved a layer of catalyst between every two bubble trays. You had to get heavy water in the right phase—it resembled distillation.

After I returned to Whiting, I got a call from the U.S. Army to go to DuPont to see why their heavy water plant was not working properly. I eventually gave them my opinion about what they could do better. Needless to say, I enjoyed telling DuPont what to do. Sometime during the forties, I also got involved in a project for evaluating nuclear propulsion of aircraft. In 1949 I was engaged as a consultant by the Congressional Joint Committee on Atomic Energy to conduct a major investigation into the possible loss of uranium at Argonne National Laboratory. Some of the senators, especially the one from Iowa, were making a big fuss about this problem. I was paid directly from Senate funds to do this work. A comment and editorial on my report hit the front page of the New York Times.

Randhava: What was going on in the catalysis scene during the war?

Thiele: A lot of major projects were launched, many of them related directly to the war effort. I attended many meetings and seminars. The interesting thing was that the government relaxed many of the anti-trust laws. This meant that all of us from different companies could freely get together, exchange ideas, and pool our experience. All the big names would be there: Hensel and the Russians from UOP, people from Texaco, Union Oil, Standard Oil of New Jersey, and many others. Some of the meetings were held in various locations in North Carolina, California, and Mississippi.

Randhava: What happened after you left Standard Oil of Indiana?

Thiele: I thought it was a little early to retire when I left Standard Oil in 1960. I joined the staff of Notre Dame as a visitng professor and spent about 10 years there. Teaching is demanding work, so at age 74 I decided to retire.

I wrote a few more papers and reviewed many others. Hopefully I have been able to keep a few bad ones out of the literature. Catalysis continues to be an exciting field with its share of excellent people: Brunauer, Emmett, Wei, Varma, and Carberry to name a few.

Randhava: What about the work you did with Geddes?

Thiele: When you go to a refinery, it's 85% distillation. You are dealing with complex mixtures and we both thought that some research on multicomponent distillation was in order. I had been thinking about this area for a long time and had worked out a theory in my head. Like with the modulus, two other people had rival ideas. When our paper was first published, it seemed too long and there were very few applications. I think it was ahead of its time. Later, when computers became available, our work proved very useful.

Randhava: What has happened with the modulus since you presented it in 1939?

Thiele: The concept gradually became very popular and is now considered a citation classic. According to the Institute for Scientific Information my work on the modulus was referenced a total of 227 times during the period 1961-1977.

Twenty-five years after the paper was published, a graduate student at MIT wrote to say that there was a mistake in one of the equations. I checked it out, and sure enough, he was right. Nobody else had caught this error. Later on, Wheeler did some additional work on my modulus and Hogan and Watson introduced the effectiveness factor.

Randhava: Dr. Thiele, on behalf of ACS and the entire scientific community, thank you for sharing your thoughts with us. We specially admire your description of your major academic achievements, such as your breakthrough paper on the Thiele modulus, as "side issues", all done only in your spare time.

RECEIVED February 16, 1983

14

Ahlborn Wheeler–Catalytic Scientist

G. ALEX MILLS

University of Delaware, Center for Catalytic Science and Technology, Newark, DE 19711

Ahlborn Wheeler is widely recognized in the catalytic community for his pioneering contributions to catalysis and particularly his development of an understanding of the quantitative relationship between the physical properties of catalysts and their activity and selectivity. The fundamental relationships which he developed have stood the test of time. They are recognized as his personal intellectual contributions. At the same time, he worked closely with others and published the results of outstanding research with some fifteen different collaborators while at Princeton University, Shell Oil Co., the DuPont Co., Houdry Process Co., Aerojet-General Corp., and Lockheed Missile and Space Co.

Wheeler began his catalytic career at an unusual time and place - in the early 1930's at Princeton (B.S., M.S. and Ph.D.). Princeton was at that time famous as <u>the</u> exciting U.S. center for catalysis. The concept of active sites had been proposed by Taylor and were discussed vigorously. Deuterium had just been discovered by Urey at Columbia. Wheeler published a remarkable series of papers with Eyring, Pease, and others on the relative rates of combination of H_2 and D_2 with ethylene and acetylene as well as the absolute rates of combination of H_2 with halogens.

Significantly, he turned his attention to questions on physical adsorption of gases on surfaces. He published the results of his work on relative adsorption of H_2 and D_2 and, with Otto Beeck and E. A. Smith, on the activated adsorption of nitrogen on iron and the significance of this to NH_3 synthesis.

Research on catalyst surfaces at Shell was at a period of time of great excitement, fostered greatly by the enthusiasm of Beeck. Studies on the detailed

0097–6156/83/0222–0179$06.00/0

structure of metal surfaces utilized a new technique of examining the relationship of the orientation of atoms in metal films and their catalytic behavior. This new concept was to open the way for a series of research investigations on the nature of sorption of H_2 on nickel and on the significance of heats of adsorption.

Following a decade at Shell, Wheeler in 1946 joined the DuPont Co. where for the next eight years he plunged into practical research on a vast array of catalytic problems concerned with the application of catalysis to organic reactions. However, it was at this time in the early 1950's that he made what many regard as his greatest individual contribution - his original and comprehensive treatise on "Reaction Rates and Selectivity in Catalyst Pores". Fortunately, the results of his intellectual and original tour-de-force was published in two classical book series, Advances in Catalysis, (1951) edited by Frankenburg, Komarewsky and Rideal, and in Catalysis, (1955) edited by Emmett.

At this time it had become possible to determine experimentally total surface area and the distribution of sizes and total volume of pores. Wheeler set forth to provide the theoretical development of calculating the role of this pore structure in determining catalyst performance. In a very slow reaction, reactants can diffuse to the center of the catalyst pellet before they react. On the other hand, in the case of a very active catalyst containing small pores, a reactant molecule will react (due to collision with pore walls) before it can diffuse very deeply into the pore structure. Such a fast reaction for which diffusion is slower than reaction will use only the outer pore mouths of a catalyst pellet. An important result of the theory is that when diffusion is slower than reaction, all the important kinetic quantities such as activity, selectivity, temperature coefficient and kinetic reaction order become dependent on the pore size and pellet size with which a pellet is prepared. This is because pore size and pellet size determine the degree to which diffusion affects reaction rates. Wheeler saw that unlike many aspects of heterogeneous catalysis, the effects of pore structure on catalyst behavior can be put on quite a rigorous basis, making predictions from theory relatively accurate and reliable.

Wheeler examined the mechanism of diffusion and flow in catalyst pores and their consequence on reaction rates in pores. He always was concerned and anxious to translate the results of his sophisticated mathematical calculations into significant practical terms. The attached figure illustrates his classical estimate for the % of internal catalyst surface available for a number of important industrial reactions. The influence of pore structure on other significant events was calculated, such as pressure and temperature gradients to be expected in catalyst pellets. Special attention was drawn to the effect on activity and selectivity of catalyst poisons which adsorb selectively on pore mouth locations.

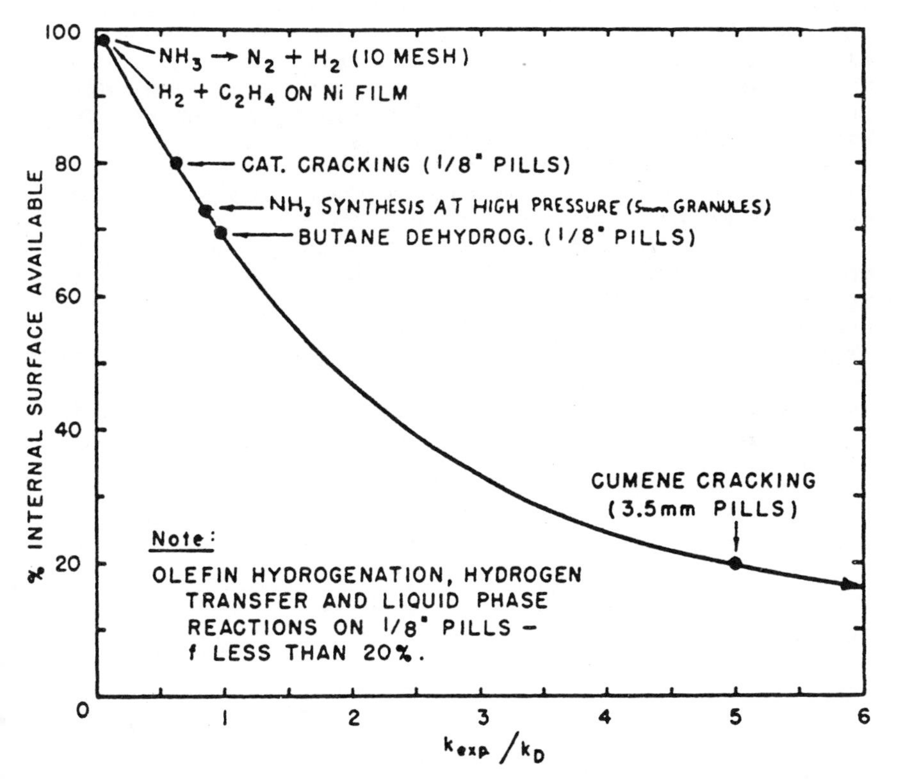

Dependence of fraction of surface available on the ratio $k_{exp.}/k_D$

Figure 1. Wheeler's classical estimate for the percent of internal catalyst surface available for a number of important industrial reactions.

Wheeler's research at Houdry involved him in petroleum refining and petrochemicals. Again, he became involved early in a new field, namely the homogeneous activation of molecular hydrogen (by now hydrogen was an old friend). The unusual capability of aqueous solutions of cobalt cyanide to act as a catalyst served to broaden the field of catalytic action of metal ions and served to construct a bridge to metal surface catalysis. In more recent years Wheeler continued his concern to establish the fundamental relationship between the physical system and catalytic behavior. Research published while at Aerojet-General and Lockheed deal with intermolecular forces in gases, rates of sublimation, trace contaminants adsorption and sorbent regeneration as well as the performance of fixed bed of catalyst reaction with poisons in the feed. Each research endeavor is characterized by a deep concern for the fundamental phenomena occurring at surface and what can be derived based on chemical principles.

As mentioned, Ahlborn made individual contributions by looking deeply into mathematical relationships between the theory of physical events and their consequence to catalysis. At the same time he enjoyed discussions with everyone concerned with applied catalysis, and he was happy to cooperate. A congenial co-worker, "Bud" Wheeler conducted his arguments with gentlemanly persistence. As to personal style, it was well known that it was almost impossible to get him into the laboratory on time in the morning and likewise nearly impossible to get him to leave in the evening. When he was interested in a problem, it became all pervasive to him. Perhaps the one exception noted at Gordon Research Conferences, were periodic absences during trout fishing season.

RECEIVED November 17, 1982

15

Early Infrared Studies of Adsorbed Molecules at Beacon

R. P. EISCHENS
Broomall, PA 19008

When Professor Davis asked me to describe the early phases of infrared efforts at Beacon, he emphasized that he did not want a review of published work. Rather, he wanted to learn the origin of ideas and the nature of problems encountered before it was evident that infrared would provide a feasible approach to surface studies. He was also interested in the personalities involved in establishing a research environment which made possible a twenty-five year period of fundamental work in an industrial laboratory. I shall try to fulfill this request because, in later years, as I have learned more about research policies, I more fully appreciate the rare and valuable opportunities enjoyed by my colleagues and I at Texaco's Beacon Laboratory.

When the infrared work was started, Dr. W.E. Kuhn was overseeing Texaco research and Dr. W.J. Coppoc was the Director responsible for the research area which included our Physical Research Section. Wayne Kuhn and Joe Coppoc have been active in the ACS and are well known and respected. They had assigned the late Dr. Louis C. Roess to establish and supervise a fundamental research group. His personal research interests were mainly in the area of electronics. He had received his B.S. degree in Electrical Engineering and his Ph.D. in Physics.

When I joined Texaco in 1948, one of the company's major research efforts was in the production of synthetic fuels by an iron catalyzed Fischer-Tropsch reaction. At that time there was fear of a worldwide oil shortage. The oil fields in Arabia had been discovered but their potential was not yet realized. Because of this company interest, my first projects were studies of CO chemisorption on iron. The most significant of these studies were carried out in collaboration with Dr. A.N. Webb.

0097–6156/83/0222–0183$06.00/0

Our infrared work stemmed from the 1949 Gordon Research Conference on Catalysis. At this conference, I had listened to a spirited discussion of the question of whether cracking catalyst acidity was due to Lewis or Bronsted acid sites. Prior to this discussion, I had been only vaguely aware of the importance of catalyst acidity even though chemisorption of basic nitrogen compounds was being used as an index of cracking activity. After the conference, it occurred to me that ammonia would retain the NH_3 structure on a Lewis site and would be converted to NH_4^+ on a Bronsted site and that infrared might be able to identify these structures. I went to J.E. Mapes who was working with one of the infrared spectrophotometers available to our group. Mapes agreed to cooperate in an attempt to determine whether infrared could differentiate between the two possible forms of chemisorbed ammonia. Success in these experiments encouraged us to apply infrared to other chemisorption systems. Professor Davis may be disappointed to learn that the infrared work evolved from an effort to answer a limited, specific question rather than from a keen insight of the potential application of infrared to a broad range of catalysis studies.

Mapes and I did not have an in situ sample cell. Our sample was placed in a tube to which salt windows were attached with sealing wax after the sample had been dried and ammonia chemisorbed. Despite this primitive technique, with its danger of converting Lewis sites to Bronsted sites by inadvertent exposure to water vapor, the infrared spectrum indicated that most of the ammonia was in the Lewis configuration (1).

During the early stages of the ammonia work we received important assistance which was made possible by Texaco's sponsorship of a fellowship at Columbia University. Part of the Columbia work was devoted to infrared spectra of minerals. To minimize scattering of the infrared beam, the Columbia workers ground the minerals in a mortar and pestle and used the smaller particles obtained by sedimentation. Mapes and I obtained details of this procedure and we used it with our cracking catalyst sample. It is likely that our ammonia experiment would have been unsatisfactory, due to excessive scattering, if we had not had access to this method of preparing the sample.

We used a Perkin-Elmer Model 12 instrument. This instrument had an interesting background which also contributed to the success of the ammonia work. It was one of the first commercially available infrared units. When this instrument was purchased, spectra were recorded on a photographic film. Dr. Roess had used the instrument to develop the, now widely used, electronic system involving a chopped beam and alternating current amplification. This made it possible to record spectra on a paper chart. Roess had also designed an attachment which made it possible to subtract the background without affecting absorbance of the bands. The operator would first obtain the background spectrum of the catalyst prior to chemisorption. After the adsorbate had been added, the operator would follow the background with a stylus as the second spectrum was being scanned. This gave the advantage of convenient and accurate background subtraction.

After the ammonia results encouraged further effort, our previous interests caused us to select the CO-iron system for the next infrared work. This was an unfortunate choice which almost led to early termination of our infrared studies of metal adsorbents. Our CO-iron chemisorption studies had been limited to iron samples produced by reducing unsupported iron oxide. We had experienced no difficulty in obtaining satisfactory chemisorption of CO on this unsupported iron. Since it was apparent that smaller metal particles would be essential in the infrared work, we attempted to produce small iron particles by reduction of supported iron oxide. We did not anticipate the difficulty of reducing supported iron. Several months of effort produced spectra which, with stretches of imagination, might have been attributable to chemisorbed CO, but which were not reproducible. We tried the CO-platinum system after we became discouraged with iron. This led to infrared studies of CO on a variety of metals. At a later time, we were able to obtain reproducible spectra of CO on iron. However, except for one study of vapor phase corrosion inhibitors (2), sample preparation problems discouraged us from expending effort on iron.

About a year prior to the start of our infrared work, I attended a lecture by Professor W.A. Patrick, of Johns Hopkins University, at a meeting of The Catalysis Club of Philadelphia. Professor Patrick described the properties of a silica which was produced by burning silicon chloride. A unique property of this silica was that its high surface area was attributable to small, non-porous spheres. With the hope that this type of silica would eliminate the need for obtaining small particles by the sedimentation procedure, a search was made throughout the Beacon laboratory. This led to Cabosil which was being tested as a grease thickener. Prior to the pressed disc technique, the infrared transmitting properties of Cabosil proved to be an important advantage. More significantly, Cabosil had the unanticipated advantage of being the support which is least likely to affect the properties of supported metals.

When Mapes transferred to another position, he was replaced by W.A. Pliskin whose Ph.D. training had been devoted to theoretical infrared work. This began a productive collaboration which extended over a period of ten years. As the infrared work progressed, we encountered numerous unexplained effects such as minor differences in band positions and band shapes. We did not attempt to explain such effects unless we had a well defined experimental approach which gave promise of clarifying them. There were also some major "loose ends" which should have been pursued further.

Anamolous bands were observed when impure zinc oxide was exposed to oxygen. Our work with zinc oxide had involved a study of the heterolytic cleavage of adsorbed hydrogen, $ZnO + H_2 \rightarrow ZnH + ZnOH$ (3). This had been observed with a commerically available zinc oxide, Kadox-25, which was produced by burning metallic zinc. We wanted to work with a sample having a higher surface area so we prepared zinc oxide by decomposing zinc oxalate. The oxalate had been made from zinc nitrate and ammonium oxalate. Unlike Kadox-25, our laboratory prepared zinc oxide produced bands in the 1700-2200 cm^{-1} region when exposed to oxygen (4). These bands are shown in Spectrum B of Figure 1. Spectrum A is the background. The bands in Spectrum B were reversible. Their intensity was a function of the oxygen pressure. Analysis of our zinc oxide showed that it was contaminated with 0.2 wt% nitrogen and 0.4-1.0% carbon.

It first appeared as though the bands were produced by the reaction of adsorbed oxygen with the nitrogen and carbon impurities. This interpretation was weakened when it was found that treatment with O_2^{18} failed to produce shifts in the frequencies of the bands. It was abandoned when the bands appeared after exposing the impure zinc oxide to chlorine because it was not reasonable to attribute the bands

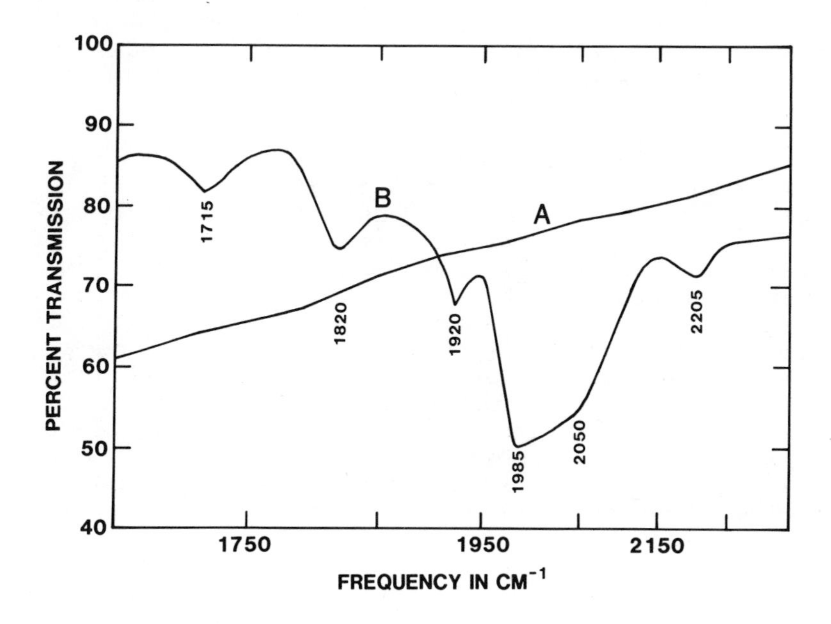

Figure 1. Effect of oxygen on ZnO (N,C): (A) at 30 °C under vacuum; (B) in 20 cm O_2 (4).

to chlorine-carbon and chlorine-nitrogen species. The bands attributable to carbon-oxygen species were shifted, as expected for a C^{13} carbonyl, when C^{13} labeled oxalate was used to prepare zinc oxide.

The published explanation of the above experimental results was that adsorption of electronegative adsorbates on impure zinc oxide produces new functional groups by reaction of the impurity with lattice oxygen within the bulk of the zinc oxide. In later unpublished work at the University of Denver, Professor Smith prepared zinc oxide which contained only nitrogen as an impurity. Exposure of this zinc oxide to oxygen produced the bands attributable to nitrogen-oxygen vibrations in Spectrum B. However, our qualitative explanation which implies that adsorption causes reaction within the bulk zinc oxide has been neither supported nor challenged by more sophisticated interpretations.

Another loose end involves efforts to use dipole-dipole interactions between adsorbed CO's to determine which crystal faces are exposed on particles of supported platinum (5). The band in the 2060 cm^{-1} region, attributable to CO chemisorbed on platinum, shifts toward higher frequency as surface coverage is increased. A shift in this direction would be expected to be produced by dipole-dipole interactions. However, by itself, the shift is not adequate for study of dipole-dipole effects. It could also be due by surface heterogeneity, which causes CO to be adsorbed preferentially on sites where bonding is strongest, or by induced heterogeneity whereby the initially adsorbed CO modifies the platinum so subsequently adsorbed CO is more weakly held. Because of these difficulties, the CO interactions were studied by using intensity data derived from co-adsorption of normal $C^{12}O$ and $C^{13}O$.

In the most simple case where only a pair of adsorbed CO's are considered, dipole-dipole interactions would lead to coupling and produce an in-phase mode $\overset{O}{\underset{C}{\uparrow}}\ \overset{O}{\underset{C}{\uparrow}}$, and an out-of-phase mode, $\overset{O}{\underset{C}{\uparrow}}\ \overset{O}{\underset{C}{\downarrow}}$. The intensity of the in-phase band would be equivalent to that of two isolated CO's. However, the intensity of the out-of-phase band would be close to zero because the two dipole changes would cancel. For reasons that are not obvious, the frequency of the in-phase mode for a coupled $C^{12}O$-$C^{13}O$ pair falls near the frequency of isolated $C^{12}O$'s and the frequency for the out-of-phase mode falls in the region expected for isolated $C^{13}O$'s. When a mixture of $C^{12}O$ and $C^{13}O$ is adsorbed, bands are observed which are only slightly shifted from the expected frequencies. However, dipole coupling causes the higher frequency band in the $C^{12}O$ region to have a high relative intensity because the coupled pairs contribute to this intensity and do not significantly contribute intensity to the low frequency C^{13} region.

Since the dipole-dipole interaction has a repulsive effect which will cause the CO's to stay as far apart as possible, it is possible to calculate a measure of the intensity ratio as a function of surface coverage. This is accomplished by considering all neighbors, rather than pairs, and assuming that the distance between adsorbed CO's is established by the distance between platinum atoms on the simple crystal faces. Calculated curves of the intensity ratio of the high and low frequency bands for a 67% $C^{12}O$ and 37% $C^{13}O$ mixture are shown in Figure 2. The experimental points, indicated by circles, start at the non-coupling value of 1.7 and follow the [110] calculated curve to a coverage of 0.5. Unfortunately, at higher coverages the experimental values do not coincide with any of the calculated curves.

A determination of faces exposed on supported palladium has been made by comparing frequencies of adsorbed CO's with those observed on single crystals (6). There is merit to this approach to study of face exposure for supported metals. However, the Kugler-Boudart method is not suitable for supported platinum where only small frequency differences are expected. Dipole-dipole interactions deserve further study and refinement.

Carbon is deposited when alumina is exposed to hydrocarbons at elevated temperature. An in situ study of carbon deposition showed that bands are produced in the 1580 cm^{-1} and 1430 cm^{-1} regions (7). These bands appear to be due to the asymmetric, C (↙↖ O O), and symmetric, C (↙↘ O O) vibrations of carboxylate groups. Figure 3 shows these bands as observed after exposure of alumina to acetylene at 250°C. Carboxylate bands are observed even though the system is under reducing conditions. The production of carboxylates is therefore attributed to hydrolysis rather than oxidation. Hydrolysis would produce an oxidized moity, carboxylate, and a reduced moity, hydrogen. Hydrogen is not detected by infrared because it is not retained on

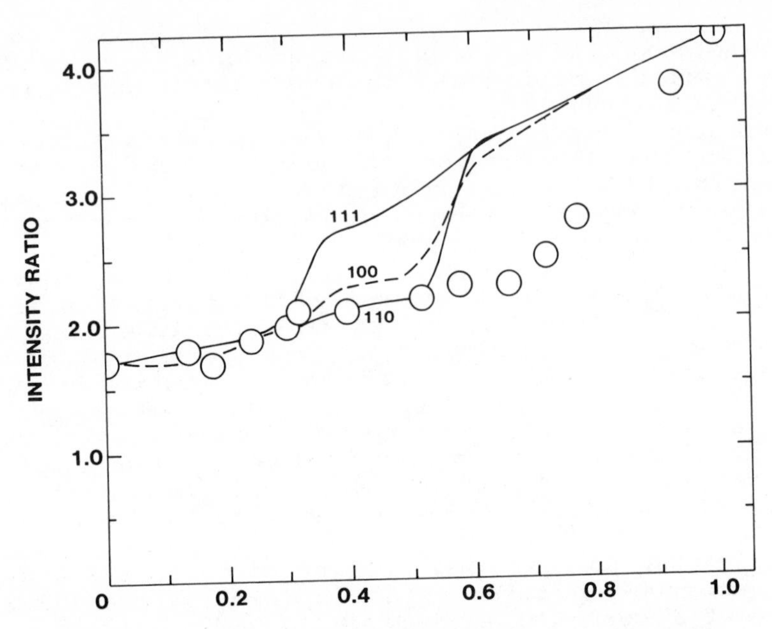

Figure 2. Comparison of experimental intensity ratios (circles) and calculated intensity ratios as a function of surface coverage for 63% ^{12}CO samples (5).

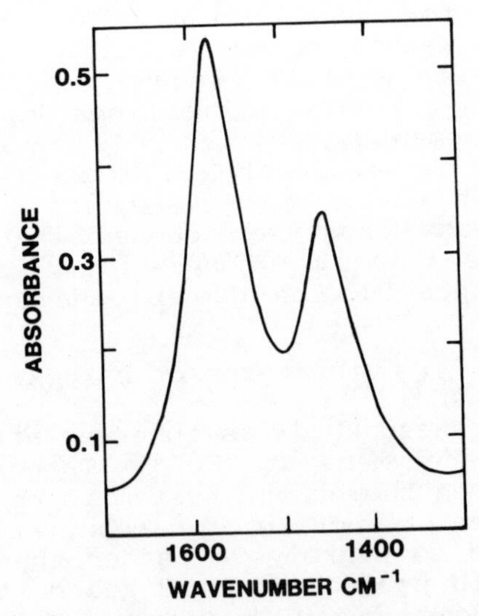

Figure 3. Carboxylate bands formed by treating alumina with acetylene at 250 °C (7).

the surface. An overly simplified interpretation of the infrared results might lead to the conclusion that the alumina surface has oxidizing properties. Carbon deposition is a limiting factor in many of the most important catalytic processes. Prior to the in situ infrared experiments, there was a consensus that carbon on catalysts is a mixture of poorly defined hydrocarbon species in which the hydrogen/carbon ration is a function of the severity of experimental conditions. The conclusion, based on Figure 3, suggests that carboxylate formation may be an additional important factor in catalyst deactivation which merits further study.

Other loose ends which I might have discussed are the missing asymmetric stretching band for carboxylates on reduced metals (8), the reactive surface carbon which is produced by self hydrogenation of acetylene on nickel (9), and simultaneous infrared and magnetic studies of chemisorption (10). In some cases termination of projects was due to reaching the limit of what we could effectively accomplish. In other cases it was due to a shortage of time because of the numerous worthwhile projects which were open to us in 1952.

The latter point is worth further examination. There had been moderate activity in applying infrared to study of minerals at the time our infrared work was started. A paper, which had major implications to surface studies was published in 1937 by Buswell, Krebs, and Rodebush (11). These workers reported an infrared study of the drying of montmorillinite. They demonstrated that infrared could detect and distinguish adsorbed water and surface hydroxyls. Despite the fact that this work was published in a readily available journal, it did not receive the attention it merited from catalysis researchers. At Texaco, we did not learn of the Buswell paper until we were preparing to write a review paper in 1957 (9).

If we return to the questions expressed by Professor Davis, it can be concluded that our early infrared work was not influenced by published literature. Ideas which stemmed from the Gordon Research Conference and the lecture by Professor Patrick at the Catalysis Club of Philadelphia were of critical importance. The information gained through informal channels made possible by Texaco sponsorship of the Columbia fellowship was vital at a critical stage of our work. However, the most significant factor was the research environment which Roess established for his group.

He instilled the concept that the objective of our fundamental work was to advance knowledge in fields of special interest and was not to be justified by the promise of immediate competitive advantage. Young researchers were allowed maximum freedom but were subjected to continuous knowledgeable and skeptical questioning by Roess and by their colleagues in the group. Roess had a sophisticated appreciation of the fact that the pace of research is often slow and he displayed extreme patience during difficult non-productive periods such as our early work with iron. The Physical Research Section included about ten Ph.D.'s whose research interests covered a broad range of petroleum related projects. During the period when Pliskin and I were applying infrared to systems of catalytic interest, Francis and Ellison were establishing the factors important to reflection infrared in studies related to lubrication (12) and Lewis was originating the application of x-ray absorbtion edge studies to determination of the oxidation state of supported metals (13). Although Roess did not personally participate in the research being carried out in the Physical Research Section, his evaluation and understanding of the significance of the projects was critical to their success. Support of the Physical Research Section at the higher levels of administration by Dr. Kuhn and Dr. Coppoc was a necessary factor in making it possible to have an extended fundamental effort. Their confidence in Roess was undoubtedly important in their decision to provide this support.

Literature Cited

1. Mapes, J.E. and Eischens, R.P. J. Phys. Chem 1954, 68, 1959.
2. Poling, G.W. and Eischens, R.P. J. Electrochem. Soc. 1966, 113, 218.
3. Eischens, R.P.; Pliskin, W.A. and Low, M.J.D. J. Catalysis 1962, 1, 180.
4. Smith, D.M. and Eischens, R.P. J. Phys. Chem. Solids 1967, 28, 2135.
5. Hammaker, R.M.; Francis, S.A. and Eischens, R.P. Spectrochimica Acta 1965, 21, 1295.
6. Kugler, E.L. and Boudart, M. J. Catalysis 1979, 59, 201.
7. Ludlum, K.H. and Eischens, R.P. Division of Petroleum Chemistry, ACS meeting, New York, New York, 1976.

8. Eischens, R.P. and Pliskin, W.A. Proceedings Second International Congress on Catalysis, Paris, 1960.
9. Eischens, R.P. and Pliskin, W.A. Advances in Catalysis and Related Subjects 1958, 10, 1.
10. Mertens, F.P. and Eischens, R.P. "The Structure and Chemisorption of Solid Surfaces", G.A. Somorjai, Ed.,Wiley, New York, N.Y., 1969.
11. Buswell, A.M.; Krebs, K. and Rodebush, W.H. J. Am. Chem. Soc. 1937, 59, 2603.
12. Francis, S.A. and Ellison, A.H. J. Opt. Soc. Am. 1959, 49, 131.
13. Lewis, P.H. J. Phys. Chem. 1963, 67, 2151.

RECEIVED February 9, 1983

16

The Fixed Nitrogen Research Laboratory

PAUL H. EMMETT
Portland State University, Department of Chemistry, Portland, OR 97207

The Fixed Nitrogen Research Laboratory was created in the summer of 1919 (1) for the expressed purpose of helping the country to establish a synthetic ammonia industry that could serve to assure us a supply of nitrate fertilizers and explosives that are so essential to our existence as a nation. World War I had emphasized the possibility of our being cut off by a submarine blockade from our only source of nitrates, Chile. The laboratory was imminently successful in reaching its goal by laying a foundation for a proposed and successful synthetic ammonia industry. At the same time it has been described as a textbook illustrating the development of a catalytic process and the carrying out of the necessary steps to ascertain the nature and characteristics of the catalyst surface and to get some idea as to the mechanism of the catalysis. It is, therefore, quite in order to review briefly work on ammonia synthesis and related reactions involved in the preparations of the 3:1 $H_2:N_2$ gas mixture and in converting it into ammonia.

The writer had no personal contact with the laboratory until the summer of 1926. At that time the name was changed to "Fertilizer Investigations Unit, Bureau of Chemistry and Soils." The laboratory continued toward its overall goal and so will be treated as the Fixed Nitrogen Laboratory to about 1940, the last year for which the writer has a complete list of its publications. The review will cover all catalytic work done at the laboratory during this time plus a few papers of work done at the laboratory but published in 1941 and 1942.

Though he has not located a detailed special report that gives the complete list of directors and assistant directors of the laboratory, the writer

0097-6156/83/0222-0195$06.25/0

has put together a little information that gives one a feeling for the quality of leadership with which the laboratory was blessed. Dr. Arthur B. Lamb was the first director and served for a year with Drs. R.C. Tolman and W.C. Bray as assistant directors. In 1920, Professor Tolman took over as director and served for two years. At the end of his term the laboratory was officially transferred from the War Department to the Department of Agriculture. Dr. F.G. Cottrell became the director. He served a number of years (probably until about 1934) when the director became a well-known physicist, C.H. Kunsman. In the 1925-26 period the highly respected Dr. S.C. Lind was assistant director.

As a first step in catalyst development, Larson and his group designed a testing unit that would enable them to test the activity of catalysts at pressures up to 100 atms., temperatures up to 550°C and space velocities as high as 45,000 volumes of gas per volume of catalyst per hour. Of course, the supplementary purifiers of the 3:1 gas were also developed. The synthesis gas could be used dry or could be bypassed through a high pressure saturator to add any desired partial pressure of water vapor (6,7,8).

In summary then, the group at the Fixed Nitrogen Research Laboratory by 1925 had shown that doubly promoted iron catalysts containing about 3% aluminum oxide and one percent potassium oxide were entirely satisfactory for commercial use and would, if operated on pure gas, have a very long life. Actually, many similar commercial catalysts are said to retain their activity for more than 5 years.

The American Process for Ammonia Synthesis

As pointed out above, within the first six years, not only was an active catalyst developed and evaluated but an "American Process for Ammonia Synthesis" was designed and described (9).

It was tested in several small 3 to 10 ton per day plants but was never operated intact on a large scale commercial basis. Instead the large commercial units became adaptations of the Haber process, the Cassale Process, or the Claude Process for ammonia synthesis. The American Process was originally described as using electrolytic hydrogen. This was much too expensive for commercial use in America at that time. Instead hydrogen was obtained from the reaction of steam with coal or later from the steam re-forming of natural gas. The American Process also incorporated a rough clean-up catalyst designed to remove final traces of impurities from the synthetic gas before the latter came in contact with the main body of catalysts. This feature too became unnecessary in commercial units as more effective purification of the synthesis gas was introduced.

One of the more valuable contributions of the laboratory to the synthetic ammonia industry was the supplementary information furnished relative to details essential to proper plant designs. For example, the equilibrium concentration of ammonia for synthesis reaction

$$N_2 + 3H_2 = 2NH_3 \quad (1)$$

were determined at temperatures of 200 to 700°C and pressures up to 1,000 atms. (10)

They also measured the compressibility of nitrogen, hydrogen and 3:1 hydrogen-nitrogen mixture up to 1,000 atmospheres and 400°C (11-14). Likewise, the solubility of these gas mixtures at high pressure in water and liquid ammonia were measured. (15-19)

One of the problems that had to be overcome in building commercial units was the deleterious effect of high pressure hydrogen-nitrogen-ammonia mixtures on the steel walls of the reactors. Corrosion tests were run on various steels to permit a selection that would be satisfactory. (20) Also in rector design it was always kept in mind that the walls should be cooled by the incoming gas to keep the wall temperature as low as possible.

Dr. Lamb was a surface and catalytic chemist who had spent a postdoctorate with Haber - one of the leaders in the German synthetic ammonia industry. Dr. Tolman was among the leading theoretical physical chemists of his day. Dr. Cottrell was noted for his invention of the Cottrell precipitator for removing particular matter from gas streams. Dr. Lind, of course, was best known for his work on radium and for kinetics of reactions involving radiation. These directors and their very able personnel got the laboratory off to a flying start. Within six years active iron catalysts had been developed and tested, an American Process for synthesizing ammonia had been designed and a start was made on basic research.

Enough has been said about the chronology of the work. From here on it will be discussed in terms of the two main subdivisions:

1. Catalyst Development.
2. Basic work on the Iron Ammonia Catalysts and on the Equilibrium Constant for the Water Gas Shift Reaction.

Catalyst Development

Priorities on the idea of using a combination of an irreducible oxide such as alumina and an alkaline oxide such as potassium oxide to prepare a superior catalyst are a little uncertain. Mittasch and his co-workers in Germany are generally given credit (2,3) for developing the doubly promoted catalyst containing a few percent alumina and a smaller percent of potassium oxide. On the other hand, Larson in the United States, in describing the reason for some of their work at FNRL, stated in an article in 1924 (4) "although there was nothing in the literature to indicate that combinations or mixtures of two or more substances would produce a better promoter for the iron catalyst, it appeared to be a logical course of experimentation. The results of these experiments disclosed the fact that no single substance effected an improvement in the catalytic properties of iron even remotely compared to that produced by a combination of two or more properly selected substances." He then went on to relate that aluminum oxide alone as a promoter for iron yielded a catalyst capable of producing 8% ammonia under standard test conditions (450°C, 5,000 space velocity and 100 atmospheres pressure); one containing only K_2O yielded 5% ammonia and an iron catalyst promoted with both alumina and potassium oxide formed 14% ammonia. It therefore seems that Larson and his co-workers at the Fixed Nitrogen Laboratory were unaware of the development of a doubly promoted (alumina plus potassium oxide) catalyst at the time they were carrying on their work. Several patents covering doubly promoted and related catalysts were issued to Larson in the period 1925-1934. Table I shows data published by Larson and Brooks in 1926 (5) illustrating the superiority of doubly promoted catalysts and calling attention to the fact

Table I

The Results Obtained for Mixed Promoters, Consisting of Potassium Oxide and an Oxide of Aluminum, of Silicon, and of Zirconium

Promoters	AMMONIA 30 Atmos. Percent	AMMONIA 100 Atmos. Percent
1.01% Al_2O_3	5.02	9.00
0.35% K_2O + 0.84% Al_2O_3	5.82	13.60
0.61% ZrO_2	4.88	7.72
0.96% K_2O + 2.76% ZrO_2	5.43	12.73
0.51% SiO_2	4.67	7.49
9.57% K_2O + 0.75% SiO_2	5.33	10.90

that silica and zirconium oxide resemble alumina as promoters, the high activity toward ammonia synthesis being obtained only when potassium oxide is also added.

In connection with the design of reactors and plants, it should be pointed out that one of the contributions of the laboratory was the training of a group of chemists and engineers who helped later to build and operate successful commercial synthetic ammonia plants.

Basic Study of Ammonia Catalysts and Water Gas Shift Catalysts

Professor H.S. Taylor, who was considered during his lifetime to be the dean of American Catalytic Chemists, is quoted as remarking that the work at the Fixed Nitrogen Research Laboratory could be considered as the basis of a textbook on heterogeneous catalysis. It included catalyst development, reactor design and basic catalyst studies. These latter included the usual basic studies of the physical and chemical adsorption of reactants, the kinetics of the reaction, studies of the catalyst surface and finally, mechanism studies and conjectures. In line with this suggestion we shall now give a compact summary of the basic work.

The Physical Adsorption of Nitrogen on Iron Catalysts Usually physical adsorption measurements on a catalyst are made in conjunction with chemisorption measurements. In the present instance they had a slightly different objective. We wished to find a method for measuring the area of the internal surface of the catalyst particles so that we would always be able to judge whether an improvement in activity was due to a quantitative extending of the surface or a qualitative change. Benton (21) had shown that a nitrogen adsorption isotherm made near the boiling point of nitrogen gave an S-shaped curve with two "kinks" in it which he suggested might correspond to a monolayer of adsorped nitrogen for lower pressure "kink" and two layers for the higher pressure "kink". We decided to follow this up and made a large number of isotherms for nitrogen, CO, CO_2, methane, oxygen, and argon near their respective boiling points (22). A few such curves are shown in Figure 1. It turned out that the "kinks" that Benton had observed disappeared when one made corrections for the imperfection of nitrogen at its boiling

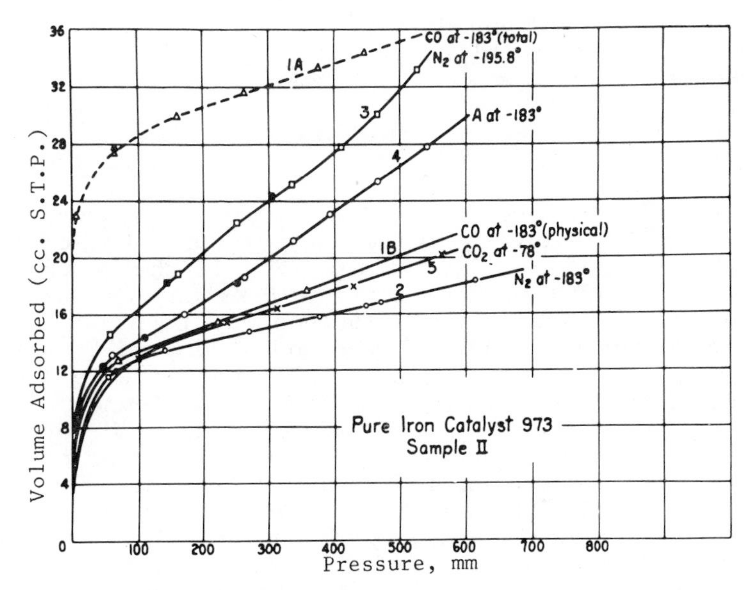

Figure 1. Adsorption isotherms for a pure iron synthetic ammonia catalyst for various gases near their boiling points. Curve 1A is for physical plus chemical adsorption of CO. Curve 1B is for physical adsorption occurring at -183°C after the evacuation of the samples at -78°C for an hour. The solid symbols are for desorption. (Reproduced from Ref. 22. Copyright 1937, American Chemical Society.)

point. However, a point corresponding to the beginning of the long straight portion of the isotherm gave adsorption values for the various gases that agreed with each other nicely and were considered to be monolayers. The multiplication of the number of molecules on a catalyst at this point (which we designated as point B) by the area of each molecule (calculated from molecular size deduced from the density of the liquid adsorbate) gave an area for the catalyst. A little later Brunauer and Teller added a theoretical interpretation to the S-shaped curves of Figure 1 to yield an equation for the adsorption isotherms (23) where x is the relative pressure

$$\frac{x}{V(1-x)} = \frac{1}{V_m C} + \frac{(C-1)x}{V_m C} \qquad (1)$$

defined as the pressure of a particular adsorption measurement divided by the vapor pressure, Po, of the adsorbate at its boiling point; V_m, is the volume of gas in a monolayer; and C is a constant related to the heat of adsorption. Fortunately point B, that we had selected as a monolayer and V_m proved to agree with each other. We applied the equation to all our miscellaneous adsorption isotherms and found it to work satisfactorily. The method of measuring surface areas of catalysts or other finely divided or porous solids has become known as the BET method and the equation as the BET equation. The method is now 44 years old and still seems to be universally accepted for measuring catalyst surface areas. It has proved to be a very useful and valuable byproduct of the ammonia research at the Fixed Nitrogen Laboratory.

Chemisorption of Nitrogen and Hydrogen It is generally agreed that at least one of the reactants of every catalytic reaction must be chemisorbed on to the catalyst surface. Such data were, therefore, taken on typical iron ammonia catalysts. Results show that three different types of chemisorption of hydrogen exist on a singly promoted iron catalyst: type C (24) in the temperature range -195°C to -130°C, type A between -78°C and 0°C, and type B from 100°C to 500°C (25). Apparently type B adsorption is involved in ammonia synthesis. It takes place even below the temperatures ordinarily employed in ammonia synthesis (450° to 500°C) and is, therefore, not rate determining in the synthesis.

Curves for the rates of adsorption of nitrogen are shown in Figure 2 (26). It was pointed out that

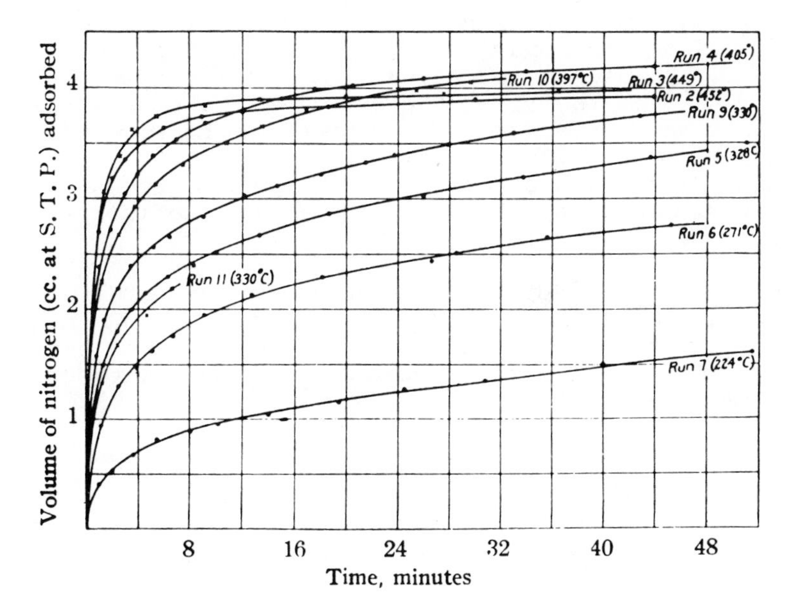

Figure 2. Rates of chemisorption of nitrogen at one atmosphere pressure on 16.46 grams of doubly promoted catalyst 931 as a function of temperature. (Reproduced from Ref. 26. Copyright 1934, American Chemical Society.)

the observed rate of nitrogen adsorption at 450°C was approximately equal to the rate of ammonia synthesis. It was, therefore, suggested that the rate controlling step in the synthesis was the chemisorption of nitrogen. This has been confirmed by later work and is generally accepted today (3).

Other information in regard to the properties of the adsorbed nitrogen was obtained. Rate determinations at several temperatures showed an energy of activation of 16 kilocalories per mole. The adsorption at a given temperature varied as the 1/6th power of the nitrogen pressure. Very strong evidence was presented to show that nitrogen was chemisorbed (23) as atoms and not as molecules (27). The evidence has to do with the fact that the amount of CO chemisorbed on the iron surface at -78°C was not influenced by the prior adsorption of nitrogen at 450°C. Thus, for example, the chemisorption of CO proved to be the same on a catalyst before and after the chemisorption of nitrogen (Table II).

Kinetics Rate measurements were made both on the decomposition of ammonia (28) and on the high pressure synthesis (29). The decomposition of ammonia on a doubly promoted catalyst was found to obey the equation

$$\frac{-d(NH_3)}{dt} = \frac{k_d(NH_3)^{0.6}}{(H_2)^{0.9}} \qquad (2)$$

The energy of activation of the decomposition reaction is about 45 kcals per mole of ammonia. For reasons that are still not too well understood, the kinetics over a singly promoted catalyst over the temperature range 400 to 450°C are the direct inverse of the above equation the rate of decomposition of the ammonia being proportional to $(H_2)^{0.9}$ and inversely proportional to ammonia to the 0.6 power. On a singly promoted catalyst the energy of activation appears to be zero over the 400 to 450°C temperature range.

Rate measurements on the synthesis were made at pressures of 30 to 100 atmospheres, temperatures of 370 to 450° C and space velocities of 22,000 to 145,000. An especially designed reactor was used in an endeavor to cut down on thermal effects.

Before these kinetic data were published, several papers relating to the kinetic calculations had appeared. Temkin and Pyzhev (30) proposed three

Table II

Effect of N_2 Chemisorption on CO Chemisorption

Catalyst	N_2 Chemisorption cc	CO Chemisorption, cc		Decrease in CO Chemisorption cc
		True	On N_2 Chemisorption	
973	3.6	25.2	24.4	0.8
954	13.3	14.2	14.2	0.0
931	8.3	45.8	45.8	0.0

(a) N_2 chemisorption 450°C.
(b) CO chemisorption at -183°C.

equations to explain the rate of nitrogen adsorption as well as the rate of desorption. They suggested that

$$\theta = 1/f \quad na_o P' \qquad (3)$$

$$V = k_a P e^{-g\theta} \qquad (4)$$

$$W = k_d e^{h\theta} \qquad (5)$$

where θ is the fraction of the surface covered with nitrogen and P′ is the 'virtual pressure' of nitrogen as given by the equilibrium relation

$$P' = (NH_3)^2/(H_2)^3 K_{equil.}$$

V is the rate of nitrogen adsorption and W is the rate of desorption. By assuming that the slow step in the decomposition of ammonia is the rate of nitrogen desorption and that the slow step in the synthesis is the rate of adsorption of nitrogen, one then has a basis for predicting the kinetic expressions for synthesis and decomposition of ammonia over the iron catalysts.

For example, the application of these equations to the decomposition of ammonia would be expressed as

$$W = k_d e^{h/f \ln/a_o \frac{(NH_3)^2}{(H_2)^3}} = k_d \left[\frac{(NH_3)^2 a_o}{(H_2)^3 K}\right]^{h/f} = k \left[\frac{(NH_3)^2}{(H_2)^3}\right]^{\beta} \qquad (6)$$

If β has a value of 0.3 the equation is in perfect agreement with the kinetic expression representing the data of Love and Emmett given above (23).

Nothing more will be said about the synthesis data except to point out that one numerical change has to be made in the kinetic equation according to a much later paper by Anderson and Tour (32). This has yet to be applied to the FNRL data. The Temkin-Pyzhev equations without this added refinement represent the data for 3:1, 1:1, and 1:3 $H_2:N_2$ fairly well.

Shortly after the paper by Temkin and Pyzhev appeared, Brunauer, Love and Keenan (31) derived a general equation for the adsorption and desorption rates for nitrogen. They also showed that their equation lead to equations (3), (4), and (5) of Temkin and Pyzhev over a certain intermediate pressure range. The overall rate of adsorption (difference between the rate of adsorption and the rate of desorption) was given by the equation

$$dv/dt = k_a P_{N_2} V_m e^{-\frac{Jv}{V_m RT}} - k_d V_m e^{-\frac{Bv}{V_m RT}} \quad (7)$$

which is identical to the one derived from equations (3), (4), (5) of Temkin and Pyzhev is v/V_m is taken as θ. The constants $k_a V_m$, $k_d V_m$, J/V_m and B/v_m were evaluated from their rate data and found to have values of 0.02, 0.000957, 2100 calories and 800 calories respectively. The equilibrium value for adsorption can be calculated from the equation

$$\ln P_{N_2} = \ln k_a/k_d + \frac{(B + J)v}{V_m RT} \quad (8)$$

B/B + J is equal to β in equation (6). The numerical equation for the kinetics of ammonia decomposition over the same catalyst for which Brunauer, Love and Keenan determined the rate constants for adsorption and desorption would be

$$\frac{-d(NH_3)}{dt} = \frac{k_d (NH_3)^{0.55}}{(H_2)^{0.85}} \quad (9)$$

This is in approximate agreement with the experimental results obtained by Emmett and Love (equation (2).

Brunauer, Love and Keenan showed an extraordinary result from equation (7). By inserting the value for the constants derived from the experimental rate data they were able to calculate the equilibrium volume of nitrogen adsorbed as a function of pressure. The results are shown in Table III. There is remarkable agreement between the observed and calculated equilibrium adsorption values, an agreement which is a tribute not only to the theory of the equations but to the reproducibility of the rate and equilibrium data.

Studies of Solid Phase In any catalytic study, one of the first questions asked is whether or not the catalysis occurs by way of an identifiable intermediate solid phase. Compounds of the approximate composition Fe_4N, Fe_3N and Fe_2N were known. Can any

Table III

Adsorption Isotherm of Nitrogen on Catalyst 931 at 396°

p,mm	V(obs.),cc at STP	V(calc'd),cc at STP
25	2.83	2.88
53	3.22	3.22
150	3.69	3.70
397	4.14	4.15
768	4.55	4.45

of these nitrides serve as intermediates in the catalytic synthesis of ammonia?

To answer this question the free energy of formation of Fe_4N was determined (33). By passing ammonia-hydrogen mixtures over the catalyst at various temperatures and $NH_3:H_2$ ratios and examining the solid phases by x-ray analysis, it was shown that the phase relations presented in Figure 3 were obtained. At 450° C, for example, as the ammonia-hydrogen ratio is increased at a total pressure of one atmosphere, only adsorbed and dissolved nitrogen is picked up by the catalyst until the ratio reaches 0.3:0.7. Then at this composition of gas the iron is converted into Fe_4N.

If one combines the equation for Fe_4N formation with that for ammonia synthesis, one obtains the value for the dissociation pressure of Fe_4N to be about 4500 atmospheres at 450° C. Therefore, the Fe_4N cannot be an intermediate in the ammonia synthesis.

Wyckoff and Crittenden (34) used x-ray line broadening to measure the crystal size of a reduced iron catalyst before and after heating it to 600°C. They showed that the promoted catalysts were stable to this temperature whereas the pure iron catalysts or those promoted only with K_2O sintered badly and grew large crystals. This work was done before the development of a method for measuring surface area. It predicted properly, however, the stabilizing effect of both aluminum oxide and two promoters, aluminum oxide and potassium oxide.

Wyckoff and Crittenden (35) also showed that it was possible to form the compound FeO by adding metallic iron to molten Fe_3O_4. They determined the crystal structure of FeO for the first time.

<u>Surface Studies on Iron Catalysts</u> In order to ascertain the surface coverage of the catalyst with promoter, chemisorption measurements were made for CO at -195° C and CO_2 at -78°C (36). The results are shown in Figures 4 and 5. The sum of the CO chemisorption volume and the volume of CO_2 chemisorbed turned out to be equal approximately to the volume of nitrogen required to form a monolayer over the entire catalyst. It was suggested, therefore, that CO could be used to measure the metallic content of a catalyst surface and CO_2 chemisorption to measure the K_2O or K_2O = Al_2O_3 coverage. Love and Brunauer (37) later showed that when the K_2O is less than the stoichiomatric amount required to form $K_2Al_2O_3$ the carbon

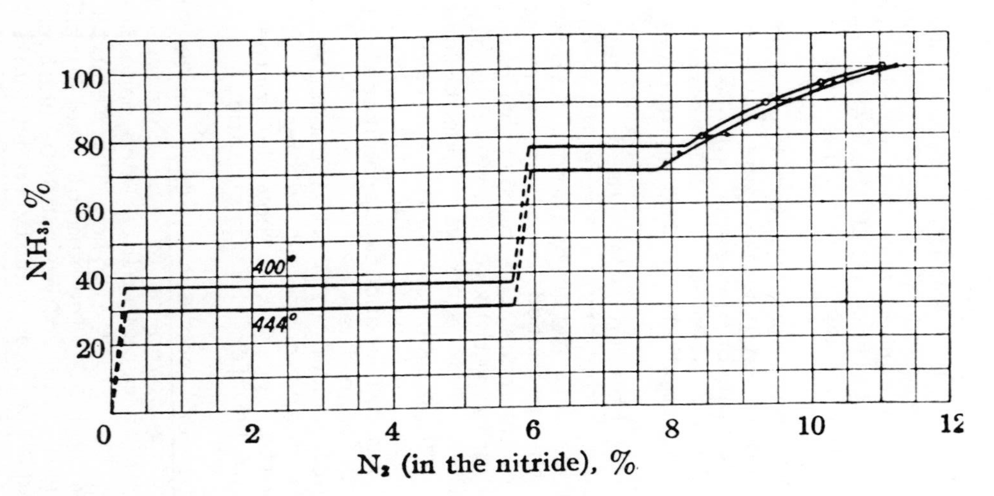

Figure 3. Equilibrium diagram showing the variation of the percentage nitrogen in the iron nitrides with the percentage ammonia in the ammonia-hydrogen mixtures. (Reproduced from Ref. 33. Copyright 1930, American Chemical Society.)

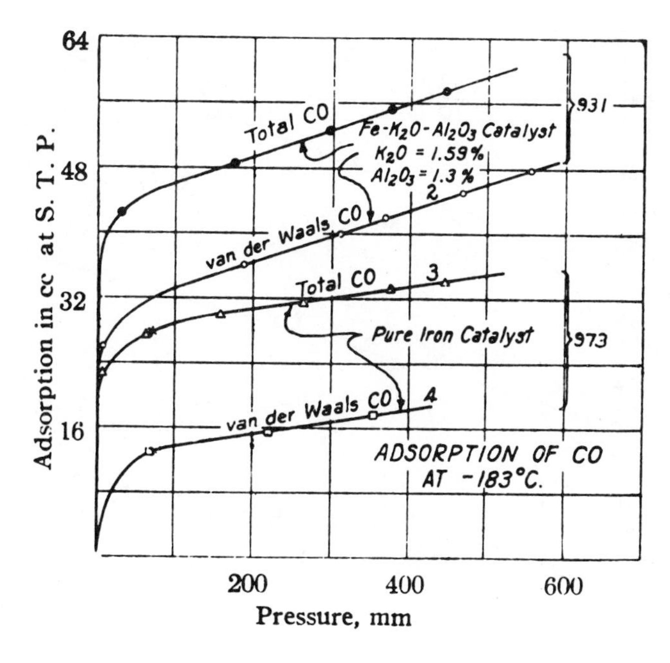

Figure 4. Comparison of the total and the van der Waals adsorption of carbon monoxide for doubly promoted catalyst 931 and pure iron catalyst 973. (Reproduced from Ref. 36. Copyright 1937, American Chemical Society.)

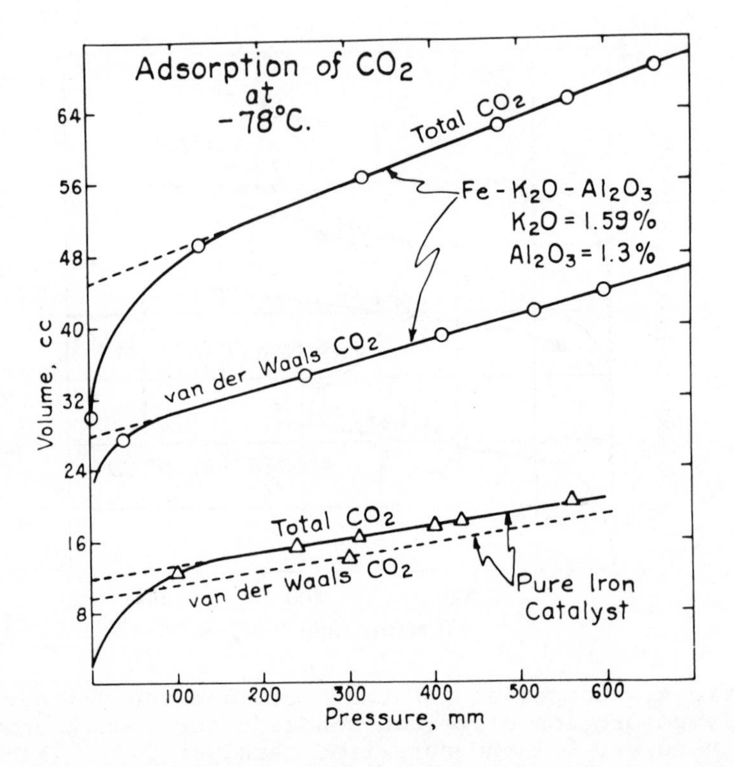

Figure 5. Comparison of the total and van der Waals adsorption of carbon dioxide on a doubly promoted and on a pure iron catalyst (42).

dioxide chemisorption measures only the portion covered by alkali. Thus the method used on a singly promoted catalyst (Al_2O_3 as the sole promoter) can measure the fraction covered with metallic iron but tells nothing directly about the portion covered by alumina.

Almquist and Black (38) studies the poisoning of iron catalysts by water vapor. In experiments conducted at 1 atmosphere pressure, they showed that quantities of water vapor 0.3% as large as those required to convert the iron to Fe_3O_4 would nonetheless form oxygen-Fe bonds on the surface. The amount of surface oxygen increased approximately with the square root of the pressure of water vapor. If the poison was added as O_2 it all was converted to H_2O and emerged from the catalyst in that form and not as O_2. The catalyst recovered its initial activity if the poison was shut off and the catalyst was treated for an hour or two with pure dry $3H_2:N_2$ gas. Oxygen is, therefore, a reversible poison.

Almquist proposed a very interesting interpretation of these poisoning results. He suggested (39) that the active iron atoms on the surface of the catalyst had higher free energy values than the normal iron atoms and therefore would be converted to oxide at much lower values of $H_2O:H_2$ - such as are used in the poisoning experiments. Thus for the normal iron atoms and for the active iron

$$3/4Fe + H_2O = 1/4Fe_3O_4 + H_2 \quad K_1 = H_2/H_2O = 5 \text{ at } 444^\circ C \tag{3}$$

$$3/4Fe_{act} + H_2O = 1/4Fe_3O_4 + H_2 \quad K_2 = H_2/H_2O = 76/0.04 = 1875 \tag{4}$$

$$\text{therefore } 3/4Fe_{act} = 3/4Fe \quad K_3 = 375 \tag{5}$$

Hence the free energy to be associated with 3/4 gram atom of iron is 8600 calories or 11,500 calories per gram-mole of iron. This calculation assumes of course, as pointed out by Almquist, that one assigns the same free energy to the Fe_3O_4 in bulk and to that formed on the surface. He would thus picture the iron surface of a promoted catalyst as having a spectrum of activities ranging from high activities for a few sites down to values of 11,500 calories per atom of iron for the lower limit of sites capable of being poisoned by 0.04% water vapor in hydrogen. By measuring the amount of oxygen taken up at each partial pressure of water vapor in the gas feed, one could obtain a plot of the excess free energy of the active surface atoms against the number of such atoms.

Frankenburg (40) has pointed out that the same type of calculations can be extended to the iron-iron nitride system. Thus the same iron atoms capable of picking up oxygen to form surface iron oxide with 0.04% water vapor would be capable of picking up nitrogen at a few atmospheres pressure to form surface iron nitride even though the dissociation pressure of the bulk Fe_4N is about 4500 atmospheres.

Explanation of the Cause of Two Values for the Water Gas Shift Reaction

In the early thirties, two values for the equilibrium constant for the water gas shift reaction

$$H_2O + CO = CO_2 + H_2 \qquad (6)$$

existed. One was obtained by analyzing the effluent gas from a catalyst when a stream of water vapor and CO was passed slowly over it. The other value was obtained by combining the equilibrium constants for the two equations

$$Fe + H_2O = FeO + H_2 \qquad K_1 = H_2/H_2O \qquad (7)$$

and $$FeO + CO = Fe + CO_2 \qquad K_2 = CO_2/CO \qquad (8)$$

give $$H_2O + CO = H_2 + CO_2 \qquad K_3 = \frac{(CO_2)(H_2)}{(CO)(H_2O)} \qquad (9)$$

A literature search revealed that values for the constant K_1 obtained by about half the people who had measured the equilibria seem to be about 40% too small. The others, including flow experiments of our own, were close to what we believe to be the correct value. We were able to show (41) that the phenomenon of thermal diffusion accounted for the disagreement among various experimental results. Briefly whenever the boat containing the iron-iron oxide was placed in a quartz tube about 2 cm in diameter, thermal diffusion would cause the water vapor in the cold part of the tube, at the furnace exit, to be too high so that the $H_2:H_2O$ measured would be about 40% too low. In the apparatus that we used the quartz tube was 2 cm in diameter at one end of the furnace and 2 mm at the other. By reversing the direction of flow we would obtain the correct result (circulation through the exit containing the 2 mm tube) or the erroneous result (circulation through the 2 cm exit). The values for the equilibrium constants for reaction (7) and hence of (9) were therefore firmly established.

In summary one may conclude that the Fixed Nitrogen Research Laboratory and its continuation as a section of the Bureau of Soils over the period of 1919 to 1940 did an outstanding job in laying a foundation for a prosperous and well-built synthetic ammonia industry and for doing its share to develop and study the detailed characteristics and mode of operation of the various catalyst systems involved.

Literature Cited

1. Clarke, Sister Margaret Jackson. "The Federal Government and the Fixed Nitrogen Industry 1919-1926"; Thesis, Oregon State University:Corvalis, OR, June 1977; p. 126.
2. Mittasch, A. "Advances in Catalysis", 1950, Vol. III, p. 81-101.
3. Boudart, M. Catalysis Reviews 1981, 23, 1-12.
4. Larson, Alfred T. Ind. and Eng. Chem. 1924, 16, 1002.
5. Larson, Alfred T.; Brooks, A.P. Ind. Eng. Chem. 1926, 18, 1305.
6. Larson, Alfred T.; Brooks, A.P. Chem. Met. Eng. 1922, 26, 555.
7. Tour, R.S. Chem. Met. Eng. 1922, 26, 588.
8. Larson, Alfred T.; Karrer, S. Ind. Eng. Chem. 1922, 14, 1012.
9. Larson, A. T. Chem. Met. Eng. 1924, 30, 948.
10. Larson, A. T. J. Am. Chem. Soc. 1924, 46, 367.
11. Bartlett, C. F. J. Am. Chem. Soc. 1927, 49, 65.
12. Cupples, H. L. J. Am. Chem. Soc. 1929, 51, 1026.
13. Bartlett, E. P.; Hetherington, H. C.; Kvalnes, H. M.; Tremearne, A. H. J. Am. Chem. Soc. 1930, 52, 1363.
14. Bartlett, E. P.; Hetherington, H. C.; Kvalnes, H. M.; Tremearne, A. H. J. Am. Chem. Soc. 1930, 52, 1374.
15. Wiebe, R.; Gaddy, W. L.; Heins, C. I. J. Am. Chem. Soc. 1933, 55, 947.
16. Weibe, R.; Tremearne, T. H. J. Am. Chem. Soc. 1933, 55, 975.
17. Weibe, R.; Gaddy, V. L. J. Am. Chem. Soc. 1935, 57, 1487.

18. Weibe, R.; Tremearne, T. H. J. Am. Chem. Soc. 1935, 57, 2601.
19. Weibe, R.; Gaddy, V. L. J. Am. Chem. Soc. 1937, 59, 1984.
20. Vanick, J. S. Trans. Amer. Soc. for Steel Treating 1922, Sept.
21. Benton, Arthur F.; White, T. A. J. Am. Chem. Soc. 1932, 54, 1819.
22. Brunauer, S.; Emmett, P. H. J. Am. Chem. Soc. 1937, 59, 1553.
23. Brunauer, S.; Emmett, P. H.; Teller, E. J. Am. Chem. Soc. 1938, 60, 309.
24. Kummer, J. T.; Emmett, P. H. J. Phys. Chem. 1952, 56, 258.
25. Emmett, P. H.; Harkness J. Am. Chem. Soc. 1935, 57, 1631.
26. Emmett, P. H.; Brunauer, S. J. Am. Chem. Soc. 1934, 56, 35.
27. Brunauer, S.; Emmett, P. H. J. Am. Chem. Soc. 1940, 62, 1732.
28. Love, K. S.; Emmett, P. H. J. Am. Chem. Soc. 1941, 63, 3297.
29. Emmett, P. H.; Kummer, J. T. Ind. Eng. Chem. 1943, 35, 677.
30. Temkin, M.; Pyzhev, V. Acta. Physicochem. 1940, 12, 3279.
31. Brunauer, S.; Love, K. S.; Keenan, R. C. J. Am. Chem. Soc. 1942, 64, 751.
32. Anderson, R. B.; Tour, H. L. J. of Phys. Chem. 1959, 63, 1982.
33. Emmett, P. H.; Hendricks, S. B.; Brunauer, S. J. Am. Chem. Soc. 1930, 52, 456.
34. Wyckoff, R. W. G.; Crittenden, E. D. J. Am. Chem. Soc. 1925, 47, 2866.
35. Wyckoff, R. W. G.; Crittenden, E. D. J. Am. Chem. Soc. 1925, 47, 2876.
36. Emmett, P. H.; Brunauer, S. J. Am. Chem. Soc. 1937, 59, 310.
37. Love, K. S.; Brunauer, S. J. Am. Chem. Soc. 1942, 64, 745.
38. Almquist, J. A.; Black, C. A. J. Am. Chem. Soc. 1926, 48, 2814.
39. Almquist, J. A. J. Am. Chem. Soc. 1926, 48,2829.
40. Frankenberg, W. "Catalysis"; Rheinhold Publishing Co.: New York, NY, 1955; Vol.3, p. 203.
41. Emmett, P. H.; Shultz, F. J. Am. Chem. Soc. 1933, 55, 1376.
42. Emmett, P.H., Advances in Colloid Science, 1942, 1, 1-36.

RECEIVED February 18, 1983

17

Einstein in the U. S. Navy

STEPHEN BRUNAUER
Clarkson College, Potsdam, NY 13676

The figure of Albert Einstein has fascinated his contemporaries; he was not only the greatest scientist of our age, but also the best-loved and most admired man among all scientists. Because of his stupendous contributions to science, one of which resulted in the atomic age, his name became known, through newspapers, magazines, radio and television, not only to the educated laymen, but also to those who had less education than the present audience; and even those who know the name of no other scientist know the name of Einstein.

Several biographies of Einstein were published, which dealt with almost every aspect of his life, but one aspect of his life - in my opinion an important aspect - was not discussed in any of them but one, and in that only very briefly and to some extent misleadingly. This aspect is the story of how Einstein, a lifelong pacifist, helped during World War II to fight the Nazis through his work for the U.S. Navy.

The best and most complete biography of Einstein was written by Ronald W. Clark, with the title Einstein, the Life and Times, and it was published by the World Publishing Company in 1971. Clark devoted less than two pages in his 631-page book to Einstein's work for the Navy Bureau of Ordnance, and even that is partly erroneous, based on George Gamow's book My World Line. This is not Clark's fault; he wrote what scant information he received, and apparently no one referred him to the person who could have given him both more and more accurate information, namely, to me. At the time Mr. Clark wrote his book, I was a professor in the Department of Chemistry at Clarkson College of Technology.

First I have to tell you the background of how Einstein and I became acquainted. At the time of

0097–6156/83/0222–0217$06.00/0

Pearl Harbor, I was a research chemist in the U.S. Department of Agriculture. Soon after that, I applied for a commission in the Navy. After a long drawn-out fight with the Navy, which included one rejection, I won the fight, and received my commission as a full lieutenant (equivalent to a captain in the Army) on September 2, 1942. After that it took more than a month until I located a billet in the Bureau of Ordnance and was called in for active duty.

Mr. Clark, following Gamow's book, wrote about the "Division of High Explosives" in the Bureau of Ordnance, but there was no such thing. The Bureau had a "Research and Development Division (Re)," the division had a section called "Ammunition and Explosives (Re2)", and the section had a subsection called "High Explosives and Propellants (Re2c)." I was assigned to Re2c. It had two other reserve officers in it when I joined, and we divided the work among ourselves. One became head of propellant research, I became head of high explosives research, and the third, who was a lieutenant j.g., became my assistant and deputy. I was, on the basis of my broad experience in the field, excellently qualified for my assignment. I knew the names of two high explosives: TNT and dynamite. With that knowledge, I became head of high explosives research and development for the world's largest Navy!

But I was young and learned fast; furthermore, the staff kept on growing as the war progressed. I acquired two groups of civilian scientists; one headed by one of the speakers at this meeting, Raymond J. Seeger; another of tonight's speakers, Harry Polachek, was in this group; the other group was headed by Gregory Hartmann, who eventually became Technical Director of the post-war Naval Ordnance Laboratory at White Oak. I also had a few officers. Besides the people directly under me, I had very many other scientists working for us indirectly. The great majority of the civilian scientists doing war research was organized into the National Defense Research Committee (NDRC), which had two divisions doing research on high explosives: Division 2, headed by Professor E. Bright Wilson of Harvard University, which worked on underwater explosives in Woods Hole, Massachusetts, and Division 8, headed by Professor George Kistiakovsky, also of Harvard, which worked on explosives in air in Bruceton, Pennsylvania. This is a long introduction to my meeting Einstein, but I believe it was necessary to see the set-up to understand better what I will say from this point on.

The top people of the Army, Navy and the two civilian divisions had occasional joint conferences to discuss their research on high explosives. At one such conference, the name of Einstein was mentioned by somebody. That gave me an idea. I asked the Army people whether Einstein was working for them. The answer was no. Then I asked the civilians whether Einstein was working for them, and the answer again was no. Why? "Oh, he is a pacifist," was the answer, "furthermore, he is not interested in anything practical. He is only interested in working on his unified field theory." I was not satisfied with these answers. Like those who gave the answers, I was ignorant of the fact that Einstein had changed his pacifist views publicly since Hitler's ascension to power; nor did I know that Einstein was interested in practical things. The first biography of Einstein, that of Philipp Frank, was to appear only four years later. Nevertheless, I felt that Einstein could not be a pacifist in a war with Hitler, nor did I believe that he would be unwilling to contribute his efforts to this war. So I decided that I would try to get Einstein for the Navy.

In the second week of May, 1943, I wrote a letter to Einstein, asking his permission to visit him in Princeton. The gracious consent came by return mail. The visit took place on May 16. After the pleasant preliminaries, I asked Einstein whether he would be willing to become a consultant for the Navy in general, and for me, in the field of high explosives research, in particular. Einstein was tremendously pleased about the offer, and very happily gave his consent. He felt very bad about being neglected. He had not been approached by anyone to do any war work since the United States entered the war. He said to me, "People think that I am interested only in theory, and not in anything practical. This is not true. I was working in the Patent Office in Zurich, and I participated in the development of many inventions. The gyroscope too." I said, "That's fine. You are hired." We both laughed, and agreed that Einstein would talk the details over with Dr. Frank Aydelotte, the Director of the Institute for Advanced Study, where Einstein was employed.

Already on the next day, both Einstein and Aydelotte wrote separate letters to me, and it is worth quoting both letters in full. Both letters came on the stationery of the Institute. The following was Einstein's letter:

May 17, 1943

Dear Lieutenant Brunauer:

I have your kind letter of May 13 and have discussed with Dr. Aydelotte, Director of the Institute for Advanced Study, the matter of my cooperation with the Research and Development Division of the Navy. Dr. Aydelotte approved heartily of my participating in your research operations. He and I both feel that the individual contract would be most suitable, and I agree fully with the arrangements outlined in the enclosed letter from Dr. Aydelotte.

I very much enjoyed your visit and look forward with great satisfaction to this association with you in research on Navy problems. I shall expect to receive from you in due course the contract and information about the work which you wish me to undertake, and I hope that I shall be able to make some useful contribution.

In this connection, I should like to raise one question: Would it in any way interfere with my usefulness to the Navy if I should spend a part of the summer in a cottage at Lake Saranac? I do not know whether it will be possible for me to take a holiday away from Princeton in any case, and certainly if my usefulness to the Navy would be increased by remaining in Princeton I should be most happy to do so. If, however, it would be equally convenient for you, I think I could probably work to better advantage in the more agreeable climate of Lake Saranac during the hot months of summer.

Yours very sincerely,
(signed) A. Einstein

How clear from this letter is Einstein's joy over the fact that he was finally drawn into the war research! "I very much enjoyed your visit and look forward with great satisfaction to this association with you in research on Navy problems." I think it is obvious that I enjoyed the visit at least as much as he. This was my first opportunity to meet the man whom I considered one of the two greatest scientists of all times (the other was Newton). And how clearly the letter shows Einstein's humility, asking the permission of a simple Navy lieutenant to spend the summer at Lake Saranac. I am sure that Einstein couldn't know at that time that I was a scientist, my field being very far from Einstein's interests.

He could have written simply that during the summer he may be reached at Lake Saranac, but no - he asked the permission of the Navy lieutenant. Naturally he received it from me, but he didn't use it. He stayed at Princeton. The letter of Dr. Aydelotte is also interesting and I quote that also in full.

May 17, 1943

Dear Lieutenant Brunauer:

Professor Einstein has told me of his conversation with you and showed me your gracious letter of May 13th suggesting arrangements under which he may be of assistance to the Navy for theoretical research on explosives and explosions.

In talking over the matter with Professor Einstein he and I have both come to the conclusion that probably the best arrangement would be for the Navy to make an individual contract with him on the basis of $25 per day, Professor Einstein to let you know at intervals the amount of time he has actually spent on Navy problems. I think it is important to leave in the arrangements for an assistant in case a great deal of routine work should be necessary, although Professor Einstein cannot tell at this time whether or not he will need the services of an assistant.

I take the liberty of writing to you simply to say that the Institute for Advanced Study cordially approves of this arrangement with Professor Einstein and looks forward with pride to having him undertake this service for the Navy.

Believe me,
(signed) Frank Aydelotte,
Director

This gracious letter shows that Dr. Aydelotte was doubly happy about my offer: for the sake of Einstein and for the sake of himself and the Institute. The most amusing part of the letter for us today is the consultant fee of $25 per day for the world's greatest scientist. It was a ridiculously small fee even at that time. As to the assistant mentioned in the letter, it was never needed because no routine work was ever assigned to Professor Einstein.

The originals of the two letters are in the Navy or in the Archives, but I had photocopies made of them, and had them framed. This is the only thing I had on a wall of my office, wherever I worked, and one of the greatest joys in my life has been that I was able to make Einstein happy. And so I became, using some exaggeration suggested by a friend, Einstein's "boss" for three years.

The news of my successful visit spread like wildfire in the Bureau of Ordnance. Officers, from ensigns to admirals, came to me with the question, "is it true that Professor Einstein is working for us?" When they found out that it was true, it settled the matter of the outcome of the war in their minds. The U. S. Navy and Einstein were an unbeatable combination.

Einstein's security clearance was obtained very quickly, and the contract was signed on May 31. Soon after that, I made my second trip to Einstein, taking to him for consideration one of the toughest problems that puzzled us at that time. The problem was whether the detonation of a torpedo should be initiated in the front or in the rear. The three most important characteristics of the shock wave produced in a detonation are the peak pressure, the impulse or momentum of the shock wave, which includes the duration of the shock, and the energy released in the explosion. If in a torpedo the detonation of the high explosive is initiated at the forward end, one obtains the highest peak pressure. If the detonation is initiated in the rear end, one obtains the highest momentum. The energy developed is the same, regardless where the explosion is initiated. So the question was which of the three main characteristics causes the most damage. If it is the peak pressure, the explosion should be initiated at the front end of the torpedo; if it is momentum, it should be initiated at the rear end, and if it is the energy, the location of the initiation does not make any difference.

Einstein was thinking about the problem for about ten minutes, and finally chose momentum and gave the reasons. But a few days later I received a letter from him telling me that he gave much furthur thought to the matter, and changed his conclusion. He decided that the energy developed in the explosion was the most important factor, and gave his reasons. Very expensive experiments performed much later showed that he was right. Of course this subject was highly confidential during the war, but I hope that now - thirty-five years later - it is declassified.

This is a good example of the problems we took to Einstein, and this one example should suffice. He always gave very careful thought to the problems we took to him and always came up with a reasonable answer. I alluded to some misleading statements in Clark's excellent book. These he took from George Gamow. Gamow, a brilliant theoretical physicist, was also one of my consultants during the Second World War. According to Gamow's story, he was the Navy's liaison man with Einstein; he took the research we did to Einstein, who listened with interest and praised the work. The implication is that Einstein only "listened," but made no contribution, and this is false. Less important is the implication that he was the only contact with Einstein. He claimed that he visited Einstein every two weeks, which is not true; I visited Einstein about once in two months and that was more frequent that Gamow's visits. Raymond Seeger and many others also utilized Einstein's services.

I mention here two men. A young man, who worked on torpedoes in the old Naval Ordnance Laboratory, asked my permission to consult Einstein about his research. Naturally, I gave my permission. My co-workers and I considered this young man very brilliant, but we did not suspect that he would be the first man, and to date the only one, to receive two Nobel Prizes in the same subject, physics. His name is John Bardeen. Another man was Henry Eyring, who is one of the greatest physical chemists of the country and the world. Eyring was then a professor at Princeton University, but he had never met Einstein. He and his brilliant group of young co-workers worked on a high explosives project for us. I introduced Henry to Einstein, and our walk in Einstein's garden became one of the great experiences of Eyring's life.

If I were asked to state what specific contributions were made by Einstein to our high explosives research, I would have to say this. New and more effective high explosives were developed during the war, and they were used by the Navy and the Army (which then included the Air Force) against Germany, Japan and their allies. (I found out later that at least the underwater explosives, possibly others, were also used in the Korean and the Vietnam War.) But these developments were the results of the efforts of large groups of people, including Einstein. It is impossible to assess the contributions of the individuals within the groups. The new developments resulted from team work, and

Einstein was a member of the team - three of tonight's speakers were members of the team. But it is easy to assess the value of a different type of contribution of Einstein - his contribution to morale. It was uplifting to know that "Einstein was one of us."

I learned from Clark's book that in July 1943, i.e., soon after Einstein joined the Navy, he wrote to his friend Bucky, "So long as the war lasts and I work for the Navy, I do not wish to begin anything else." But we were unable to supply him with enough work to occupy him full time. Whenever I visisted him, I found that the tall and long blackboard in his study was filled from the left to right end, and from top to bottom, with long, complicated equations, written with neat, small-sized symbols - obviously work on the unified field theory. Nor were my bimonthly or more or less frequent visits spent on business alone. After the business came the conversation. We talked about progress of the war, about interesting items in the news, about history, philosophy, about personal experiences, and about a great variety of other things. Einstein had a wonderful sense of humor; he loved to make witty remarks and tell humorous anecdotes. He laughed heartily at his own jokes and also at mine. His well-known wisecrack, "I am in the Navy, but I was not required to get a Navy haircut," was born in one of our conversations. I am very sorry now that I did not make detailed notes after each trip to Einstein. But that was the busiest time of my life; I worked seven days a week and twelve hours or more every day, as did many others. During three and a half years, we had three days off, the three Christmas days.

The great mathematician G. H. Hardy called Einstein "good, gentle, and wise," and it would be difficult to find better adjectives for him than these. But I would add one more, "humble." You could see that in his letter to me, which I read to you, but that is only one example. In all my visits, I received the impression of a genuinely humble person. On one occasion, he gave me one of his books as a present. It was <u>The Meaning of Relativity</u>. On the empty page under the cover he wrote in his beautiful, small, clear letters only this much: A. Einstein, and under it the year, 1945. I was disappointed that he did not write more, but I attribute it to his humility, to his great modesty.

C. P. Snow visited Einstein only once in his life, in the summer of 1937. He wrote a long essay about his visit. He found Einstein a sad man and a pessimist. During the eight hours they spent together, he heard Einstein's famous laughter only once. Einstein had good reasons to be pessimistic then; that was the time of the rapid rise of Hitler, and the western powers did nothing about it. But since Snow's essay appeared twelve years after Einstein's death, it created the impression that Einstein was always like that. This is not so. During my visits, while the war lasted, Einstein was gay and ebullient; in those visits he laughed heartily and often. He had good reasons for that too; we were on the way to eradicating Hitler and his Nazi system, and Einstein was - by his work in the Navy - one of the eradicators.

Although Einstein's third and last contract as "Consultant for Research on Explosives" ran from July 1, 1945 (before the end of the war) to June 30, 1946 (nine months after the end of the war), there was no need to consult him after the end of the war. Hiroshima shook up many people, and Einstein more than anyone else. I visited him twice after the war; last time in April 1946. Einstein's mood changed - he was worried about the fate of mankind. I expressed the deep gratitude of the Navy, the Bureau of Ordnance, and especially of my own for the privilege of working with him, and he in turn thanked me for getting him into the war research, which gave him great satisfaction. When we said goodbye to each other, I was deeply moved, and perhaps he was moved too.

That was the last time I saw him. Our paths diverged after that. Our aim was the same: the prevention of a third world war, but our paths were different. I stayed in the Navy for four and a half more years to build up a new organization for high explosives research and development. Einstein's path was complete disarmament and the establishment of a world government, and he exerted all his effort, and all his influence to achieve those ideals. As we know, he failed. I believe that Einstein knew that his efforts were doomed to failure; he was a prophet way ahead of his time. But the "conscience of the world," as Einstein was called, could not but fight to the end for what he believed, however hopeless the fight was.

This is the story of Einstein in the Navy in a nutshell. It is incomplete for two reasons: I myself could have said more if the time allotted to my talk had been longer. What is more important, doubtless others could add their experiences to mine. Some day a more complete story will emerge about this important part of Einstein's life. But even this short history is far more complete than anything you can find in print to date. Thank you for your attention.

Editor's Note: In 1979, on the occasion of the one hundredth anniversary of Einstein's birth, the Washington Academy of Sciences held two commemorative meetings. Stephen Brunauer was one of the invited speakers. The talks were subsequently published in the Journal of the Washington Academy of Sciences, 69 #3, 108-113, 1979 (Lancaster Press, Lancaster, PA, journal publisher).

RECEIVED April 5, 1983

The History of the BET Paper

EDWARD TELLER
Hoover Institution, Stanford, CA 94305

In January, 1935, while working at University College in London, I got a letter from my friend, George Gamov, inviting me to join him as Professor of Physics at The George Washington University in Washington, D.C. In September of that year, my wife and I arrived in this country where we have lived ever since.

The first six years in this country were quiet and productive years for me. I taught at the University, worked with George Gamov on some relatively early question of nuclear physics and worked with a great number of other people on the physics of molecules, including some properties of solids.

Washington, D. C., then as now, had many laboratories. Quantum mechanics at that time was a relatively novel theory which in principle could solve all the questions of atomic and molecular interactions. However, many people who had obtained their degrees even a few years earlier were unfamiliar with it. My first class on the subject at The George Washington University consisted of a dozen or so "older" students and was given in the late afternoon to accommodate their schedules.

This schedule also happened to agree with my way of life at that time. I was hardly ever in bed before midnight, but much more pertinent, I was hardly ever out of bed before ten in the morning. Incidentally, during those years I never managed to visit the Capitol or any of the House or Senate buildings. Now that I reside on Stanford campus in California rather than just off Connecticut Avenue, I wish I could say the same thing.

One of my students in these first classes was a fellow Hungarian, Stephen Brunauer. He worked in a place that was unofficially called "the fixed-nitrogen lab," more formally named the Bureau of Chemistry and Soils.

More than once, I have been asked the flattering question: why have Hungarians played a disproportionately important role in the physical sciences? My standard answer is that this disproportion is the result of an optical illusion--Hungarians seem to be more visible. I have also offered the theory that

0097–6156/83/0222–0227$06.00/0

they carried away from Hungary the characteristics of people who have escaped from a shipwreck--who have become more dedicated to life having narrowly been saved.

The most colorful explanation was offered by the most important Hungarian immigrant scientist: the aerodynamicist, Theodore von Kármánn, who played the leading role in the development of the United States air force in World War II. He gave away the "secret:" we are not Hungarians at all but Martians who are the vanguard of the planned conquest of earth and have settled first in a little known region (Hungary) to assume a superficially human appearance.

I particularly like this explanation because Theodore von Kármánn included in this unique fifth column not only scientists but also Zsa Zsa Gabor. I am also glad that through the recent movie about an Extra Terrestrial (whose initials I also carry) von Kármánn's theory has been given a less imperialistic turn.

One result of the Hungarian invasion was that two years after my arrival, Steve, of whom I saw quite a lot, told me a peculiar fact. Multilayer adsorption of gases had been studied, and the dependence of the amount of the material adsorbed on the pressure of the gas adsorbed looked similar in practically all examples.

Adsorption to a monolayer had been explained most successfully by Irving Langmuir. This very success may have deflected attention from the study of multilayer adsorption. In monolayer adsorption, saturation is approached when the monolayer has been filled. But soon after this has occurred, additional adsorption sets in, thus producing a curve that has been called S-shaped although it is necessary to lay the S on its side in order to reproduce the graph.

Steve suggested that the adsorption of additional layers is due to a force from the solid acting over longer distances rather than the well-known contact forces which Langmuir used to explain monolayer adsorption.

Actually, in 1937 we knew quite a lot about the interactions over longer distances. Forces sufficiently strong to explain the multilayer adsorption appeared to be completely excluded. But Steve insisted that the observations were there, and that his explanation would account for them. If I did not believe what he proposed, then I should suggest an alternative.

At the time, my ignorance of adsorption was superlative. I did not even know what the word <u>chemisorption</u> meant. But Steve's challenge was exciting and seemed justified.

The obvious initial step in multilayer adsorption would be that, while the Langmuir adsorption depends on the interaction between the adsorbing surface and the gas molecules, the adsorption of a second layer would depend primarily on the attraction between molecules already adsorbed in the first layer and additional molecules of the second layer. It was, of course, equally clear that the adsorption of the third layer would depend on the molecules in the second layer, and that the same relationship could be continued indefinitely.

Indeed, if the pressure of the gas reached the vapor pressure of liquified gas at the temperature in question, then the layers would build up successively with essentially equal occupation. The adsorption would continue into actual condensation on the surface, thus establishing infinitely many adsorbed layers.

If we were dealing with a pressure below vapor pressure, than the amounts adsorbed in the second, third and higher layers would form a geometric series which could be easily summed. The first layer would not form a term in the geometric series and would be more strongly populated.

Therefore, it seemed that an explanation of multilayer adsorption was quite possible. Furthermore, no new constants were needed in order to explain the experimental data. The two relevant factors are: the attraction that the molecules experience in the first layer (which had already appeared in Langmuir′s equation); and the interaction between the various layers of adsorbed substance. This latter constant could be derived from the vapor pressure at each temperature.

These ideas were simple and perhaps should have been discovered at an earlier time. But at this point, Steve got deeply interested in them. He suggested that the whole question should be discussed with Paul Emmett, the most experienced man in surface chemistry who worked at the fixed nitrogen laboratory.

All the rich material of observation of surface adsorption was available for the discussion among the three of us. We soon noticed that the multilayer adsorption theory put the interpretation of adsorption curves on a more solid footing, thereby making it possible to determine the area of the surfaces of substrates, such as charcoal, more accurately.

Indeed, these surface areas could in principle be determined from the Langmuir equation, provided the interaction between the substrate and the adsorbed molecules was considerably stronger than the interaction between the adsorbed molecules themselves. If one kept the pressure well below the vapor pressure, then Langmuir′s curve should apply and the adsorbed quantity would be directly proportional to the surface.

The trouble was, however, that the surface of the substrate may not be quite uniform. This would give rise to variation in the strength of adsorption. At low temperatures, the Langmuir saturation could occur long before the buildup of further layers. But at the same low temperatures, even small differences in the interaction between the substrate and the various molecules at different locations could seriously distort the Langmuir curve.

If, on the other hand, the experiments were carried out at higher temperatures, then the Langmuir saturation is less pronounced because even before the first layer is complete, the second layer begins to get adsorbed. Differences in the adsorption at various locations would be less pronounced in the second than in the first layer, because the first layer already would have had a smoothing effect on the surface configuration.

The real advantage of the multilayer theory is, however, that it can utilize both low and high temperature data and in the end give much more reliable values for surface areas.

Of course, like the Langmuir theory, the multilayer adsorption theory is not universally valid. For instance, if the surface has holes or clefts, then only a limited number of layers may be adsorbed. We were fully aware of this exception and tried to sift the experimental material accordingly.

Another difficulty was left unanswered. The hypothesis that the second layer is exposed to the same forces as an additional layer in the liquified form of the adsorbed gas is not valid unless the first layer is already rather completely developed. Similarly, one may have doubts about the third layer being adsorbed on the second. We were aware of this difficulty, but while we mentioned it, we did not resolve it.

Actually, our proposed adsorption equation turned out to be rather more practical than I had expected. A measure of its success is that the names of the authors have been practically forgotten, and the equation bears the honorific title of BET. The whole adventure was, for me, one of undiluted pleasure, not only on account of the success but also on account of the harmonious cooperation which led to the original publication.

This is not quite the end of the story, however. About thirteen years later, a young colleague, Bill McMillan, and I finally returned to the question that the BET theory failed to answer: To what extent can the assumption that the second layer is built up as though the first layer were complete (and the third as though the second were complete, and similarly all the other layers) be justified?

The answer was, of course, that the assumption was not justified. Because of the incompleteness of the layers, changes in the BET equation were necessary. Actually, it turned out that the required changes are small and can be expressed in a surprisingly simple form.

The changes turned out to be the same as the changes that would have arisen if Steve Brunauer's original hypothesis had been correct. They are similar in nature to those that would be produced if the substrate interacted with the distant gas molecules according to a law where this interaction decreased with the inverse third power of the distance of the molecule from the surface. While these changes were in the direction of the original Brunauer suggestion, these corresponding imaginary forces would be so small that in themselves, they would not begin to explain the presence of the multilayers.

The truly amusing (and at the same time somewhat

disappointing) point comes with a comparison of the results of the two studies. The theory of the van der Waals forces predicts that the substrate interacts with distant molecules, producing a potential energy that decreases with the inverse third power with the distance from the surface. Therefore, the changes that Bill McMillan and I had to introduce did nothing more than point up an additional interaction that behaves in the same way as the van der Waals interaction.

Still this interaction is distinct from the van der Waals interaction. Different substrates have different interactions in the van der Waals interaction, whereas the term that Bill and I found is necessarily the same for all multilayers of a given gas no matter on what substrate they were adsorbed.

This later work was incomparably more difficult than the original development of the BET theory. It was also incomparably less practical. Indeed, its effect on the explanation of surface adsorption phenomena was practically nil.

But in this case as in many others, scientific work turned out to be a pure pleasure. Whether this pleasure is derived from the ease of the work, or its abundant practical results, or from the satisfaction of answering a difficult complex question in relatively few words does not greatly matter. This most positive aspect points up the similarity of science and art. In neither field can work and pleasure be distinguished.

On almost the last day of 1979, Steve Brunauer, Paul Emmett and I (together with our wives) got together to celebrate our own paper with a most quiet but even more enjoyable dinner. This was done at the suggestion and expense of a friend of the BET theory. I had just returned a few hours earlier from Taiwan. Instead of going to bed to suffer with jet lag, I met my two old friends whom I had not seen for many years. I am happy to report that having a happy time with friends who share pleasant memories of the past seems to be an excellent cure for jet lag.

RECEIVED December 22, 1982

CATALYTIC CRACKING

The Carbonium Ion Mechanism of Catalytic Cracking

HERVEY H. VOGE[1]
Sebastopol, CA 95472

Development of the carbonium ion mechanism of catalytic cracking in 1940-1950 is reviewed. Studies of the thermal and catalytic cracking of pure hydrocarbons played a key role, and were complemented by earlier hypotheses of Whitmore regarding the probable carbonium ion intermediates in low-temperature acid-catalyzed hydrocarbon reactions. Workers in several laboratories arrived at similar conclusions at about the same time.

Catalytic cracking of high-boiling petroleum fractions is the largest industrial catalytic process in terms of material processed and catalyst required. It has become the central process in petroleum refining. The history of catalytic cracking shows that it developed largely empirically, and benefited from basic understanding only in later stages. Modern catalytic cracking was preceded by an early cracking process employing $AlCl_3$ as a catalyst. This process was tried in 1913-1915 by McAfee (1) but had no extensive use. The chemistry of cracking with $AlCl_3$ was not examined closely.

The present catalytic cracking process arose from clay treatments of petroleum fractions. A strong impetus was given by the French chemist Eugene Houdry about 1924-28 (2). In a series of exploratory experiments Houdry observed that the gasoline produced when higher-boiling petroleum fractions were heated in the presence of certain acid-treated clays gave improved performance in racing cars. This led eventually to the Houdry fixed-bed cracking process, first put into use in a plant in 1936. Then, as now, the advantage of catalytic cracking did not derive from the increased rate of cracking, but rather from the greater value of the products as compared to those from the older thermal cracking processes. The Houdry process was

[1] Retired from Shell Development Company

0097–6156/83/0222–0235$06.00/0

promoted in the USA by the Houdry Process Corp. However the competitive creativity of the Oil Industry rapidly modified the process, and the fixed-bed process was fairly soon replaced by moving-bed and fluidized-bed processes.

Research at Shell

In the early development of catalytic cracking several loosely cooperating industry groups were formed. At Shell we collaborated with Standard Oil Development Co., The Texas Co., Standard Oil Co. of Indiana, Universal Oil Products and the M. W. Kellogg Co. This group was first organized as the "Catalytic Refining Agreement", and later operated cooperatively under "Recommendation 41 of the Petroleum Administrator for War". The collaboration ended shortly after the war, but by that time the fluidized-bed cracking process had been well established (3). A leading role in the creation of this process was played by Standard Oil Development Co. (a part of what is now Exxon). Because of the basic catalysis research that had been done earlier at Shell, especially by O. Beeck, A. Wheeler, M. W. Tamele, and others, it was suggested about 1939 that Shell should look into the more fundamental aspects of catalytic cracking. B. S. Greensfelder and I concentrated on the reactions of catalytic cracking; G. M. Good joined us later.

Petroleum above the gasoline boiling range is composed of very many hydrocarbon compounds, including some that also contain sulfur and nitrogen atoms. To ascertain the nature of the cracking reactions, Greensfelder wisely chose to study the behavior of representative pure compounds. In our joint work more than 50 pure hydrocarbons were tested over a typical solid cracking catalyst, a silica-zirconia-alumina material that had been made experimentally by UOP. Later various silica-aluminas were used with similar results. Comparisons were also made with thermal cracking, and with cracking over catalysts of decidedly different character, such as activated carbon. From the modern viewpoint these researches were done using primitive analytical methods, and much care and expense was devoted to identifying reaction products.

The pure hydrocarbon studies were extended by using mixtures of several compounds and by studying the effects of oxygen, nitrogen, and sulfur compounds. These studies were valuable for commercial applications of catalytic cracking, aiding the selection of feedstocks, choice of operating conditions, use of recycle cracking or staged cracking, etc. But we were especially anxious to know what was taking place on the catalyst surface.

Mechanism

In the early years of catalytic cracking there was

considerable speculation about the mechanism. The catalysts then used were non-crystalline solids of high surface area (200 - 500 m^2/g). Usually they were composed of silicon dioxide plus 13 - 25% aluminum oxide. They were not obviously acidic, and it was only much later that direct evidence of the acidic nature of the surface after proper dehydration was obtained. One early proposal ascribed cracking to strong adsorption of adjacent carbon atoms on the solid surface. Another ascribed activity to active carbon formed on the catalyst surface. There were other theories, usually rather vague and without much evidnece to back them up.

As we learned more about the hydrocarbon reactions, some similarities to reactions catalyzed by strong acids at much lower temperatures became evident. An additional and different impetus to understanding came from an interlude of thermal cracking studies. At the time we were interested in the thermal cracking of normal paraffins (waxes) for production of alpha olefins. Thermal cracking of n-hexadecane gave products in close agreement with the Rice-Kosiakoff theory of free-radical chain reactons (4). According to that theroy a reactive (and isomerizing) free-radical intermediate propagates the chain, and through progressive breakdown stages about 4 moles of product hydrocarbons form from each mole of hexadecane that cracks. In catalytic cracking we had observed 3.5 moles of hydrocarbon products per mole of hexadecane cracked, and the number was almost independent of conversion level from 11 to 70% conversion. This suggested progressive breakdown in a chain reaction involving a reactive intermediate.

The papers of Whitmore (5) proposed a carbonium ion as the intermediate in low-temperature acid-catalyzed reactions such as alkylation and polymerization; the same intermediate could perhaps be involved in catalytic cracking.

Some of the similarities between catalytic cracking and low-temperature acid-catalyzed reactions include: high reactivity of olefins compared to paraffins, removal of complete side chains from alkylaromatics, high reactivity of compounds containing tertiary carbon atoms, selective formation of isoparaffins, olefin isomerization reactions, hydrogen transfer reactions, and hydrogen-deuterium exchange reactions. There are of course important differences, most of which can be explained by the different thermodynamic equilibria that prevail at 500°C and 20°C.

As soon as we applied the concept of the carbonium ion intermediate, many previous observations could be correlated. A few simple rules about formation, isomerization, and cracking of the hypothesized ions explained most of the experimental data on hydrocarbon cracking. Quantitative prediction of the products from n-hexadecane cracking was possible with the aid of only one additional assumption, as noted in the paper of Greensfelder, Voga and Good (6). For details of the carbonium ion mechanism as applied to cracking the reader is referred to a review chapter by the present author (7). Greensfelder, in another chapter (8), extended the basis for the theory by calculating relative energies of formation and reaction of various simple gaseous carbonium ions, using thermodynamic data for hydrocarbon dissociation energies, and ionization potentials for gaseous free radicals. It has to be recognized that the carbonium ions of cracking exist adsorbed on the surface of a solid, but there is reason to expect some relationship between relative energies there and in the gaseous state.

Acidic Nature of Catalysts

Belatedly good evidence for the strong acidity of suitably dehydrated catalyst was obtained. Tamele (9), reporting work done in collaboration with O. Johnson, L. B. Ryland, and E. E. Roper, noted that evidence for acidity was obtained for silica-alumina from ammonia adsorption, and that the catalyst acidity could be titrated with butylamine in benzene. Still later, Benesi (10), using the Hammett indicators, showed that silica-alumina catalyst had an acid strength greater than that of 90% sulfuric acid.

Another sign of the importance of catalyst acidity was the strong inhibition of cracking by small amounts of nitrogen bases. In the cracking of hexadecane, an amount of quinoline sufficient to cover only 2% of the catalyst surface reduced the cracking to a small percentage of the uninhibited. Our first observation of the nitrogen base effect came from accidental contamination. We tried the effect of bicyclic aromatic compounds on the cracking of hexadecane and found strong inhibition. But the main cause of the inhibition was traced to small amounts of nitrogen compound impurities in the aromatics.

Work of Others

In 1948-50 the time was surely ripe for recognition of possible carbonium ion mechanisms for catalytic cracking. At least five groups of workers proposed carbonium ion intermediates for cracking. There were variations in the hypotheses and usually not much evidence or detail. Since there was a certain amount of intercommunication it is hard to name any one as clearly first. Carbonium ion mechanisms were proposed by Bremner (11), Hansford (12), and Ciapetta, Macuga and Leum (13). Most significant was the work of C. L. Thomas (14). At UOP he had done early work on the cracking of pure hydrocarbons. He had noted the hydrogen transfer reaction and other characteristic reactions. His paper (14) was published simultaneously with that of Greensfelder, Voge, and Good (6). His rules for the reactions of carbonium ions on the catalyst surface are very similar to ours, but he did not make numerical interpretations of products. Thomas gave an illuminating explanation of the acidity of silica-alumina. He pointed out that when tetra-valent silicon and tri-valent aluminum atoms are both tetrahedrally coordinated with oxygen atoms in a lattice, then an extra positive ion is necessary to complete the structure. By suitable preparation this positive ion is a hydrogen ion and an active catalyst results. He showed a fairly good correspondence between measured acidities and activities for a series of silica-aluminas covering a wide composition range. Maximum activity and titratable acidity were observed at Al/Si ratio of about one.

Carbonium Ions in Other Media

The idea of carbonium ions is quite old in organic chemistry. Olah has traced the early history (15). In 1902 Von Baeyer wrote of carbonium salts in explaining the deep color formed when triphenylmethyl chloride was dissolved in sulfuric acid. Carbonium ions as reaction intermediates were proposed by Meerwein in 1922, and much used by Ingold, Hughes, and others in England soon thereafter. F. C. Whitmore in the USA from 1932 on showed how carbonium ions as reaction intermediated could explain the acid-catalyzed reactions of alkylation, polymerization, and isomerization. His studies were summarized in a review article in 1948 (5). More recently, of course, there have been many spectroscopic studies of stable carbonium ions formed in highly acidic solutions at low or moderate temperatures, as, for example, in the works of N. C. Deno and G. A. Olah.

Unfortunately the transient carbonium ions presumably formed in catalytic cracking cannot be easily observed. Furthermore, there remain many questions about degrees of adsorption bonding involved with these intermediates, as well as need of information about structures and possible alternative reaction paths. In general, however, this lack of exact knowledge is characteristic of almost all reaction kinetics, catalytic or otherwise.

Later Developments

Very significant improvements are still being made in catalytic cracking, both from the engineering side and from the chemistry. Among the former are riser cracking and high-temperature regeneration. Chemically, catalyst improvements are still being made. The introduction of the zeolitic cracking catalysts, largely by Mobil workers about 1964 (16) was a major advance that brought increased catalyst activity and markedly improved selectivity. This advance could have been suggested by C. L. Thomas, who, in his 1949 paper said: "To obtain maximum activity these catalysts should be made in special ways. Silica-alumina catalysts of maximum activity probably cannot be prepared by forming a silica hydrogel and depositing alumina on it." The later method of preparation, or even less sophisticated methods of modifying natural clays, had been widley used earlier.

There is certainly need for better evidence on the details of the carbonium ion mechanism, particularly as regards the initiation of n-paraffin cracking, which has been much debated. There is always room for better catalysts that will greatly reduce coke formation, improve octane number, or permit reactions to be more specific for particular products. The future will undoubtedly bring improvements and better understanding in catalytic cracking and in the closely related reactions of hydrocracking, residue treatment, coal liquefaction, shale oil refining, and the processing of tar from sands.

Literature Cited

1. McAfee, A. M. Ind. Eng. Chem. 1915, 7, 737.
2. Houdry, E.; Joseph, A. Bull Assoc. Franc. Technicians Petrole 1956, 117, 177.
3. Murphree, E. V. Advances in Chemistry 1951, No. 5, 30.
4. Voge, H. H.; Good, G. M. J. Am. Chem. Soc. 1949, 71, 593.
5. Whitmore, F. C. J. Am. Chem. Soc. 1932, 54, 3274; Chem. Eng. News 1948, 26, 668.
6. Greensfelder, B. S.; Voge, H. H.; Good, G. M. Ind. Eng. Chem. 1949, 41, 2573.
7. Voge, H. H. "Catalysis", Emmett, P. H. Ed. Reihhold 1958; Vol. VI, 407.
8. Greensfelder, B. S. "The Chemistry of Petroleum Hydrocarbons", Brooks, B. T. Ed. Reinhold 1955; Vol. II, 137.
9. Tamele, M. W. Faraday Soc. Disc. 1950, No. 8, 270.
10. Benesi, H. A. J. Am. Chem. Soc. 1956, 78, 5490.
11. Bremner, J. G. M. Research 1948, 1, 281.
12. Hansford, R. C. Ind. Eng. Chem. 1947, 39, 849.
13. Ciapetta, F. G.; Macuga, S. J.; Leum, L. N. Ind. Eng. Chem. 1948, 40, 2091.
14. Thomas, C. L. Ind. Eng. Chem. 1949, 41, 2564.
15. Olah, G. A. Chem. Eng. News 1967, March 27, 77.
16. Plank, C. J.; Rosinski, E. J.; Hawthorne, W. P. Ind. Eng. Chem. Prod. Res. Develop. 1964, 3, 165.

RECEIVED November 17, 1982

20

A History of Early Catalytic Cracking Research at Universal Oil Products Company

CHARLES L. THOMAS
Tempe, AZ 85282

Catalytic cracking research at UOP began in the early 1930's under the supervision of Dr. Hans Tropsch (of the Fischer-Tropsch Process) with the research work being done by Dr. Julian M. Mavity. The thrust of the research was to develop a catalyst that could be produced from an economical natural clay or mineral.

It should be noted that certain clays were, at that time, well known in the refining of petroleum products. Most commonly, clays were used for decolorizing lubricating oils either alone or in conjunction with sulfuric acid treating and for stabilizing gasolines produced by thermal cracking. It was assumed that unstable hydrocarbons ("gum" formers), presumably dienes, were polymerized by the clay treating. Certainly higher boiling substances were formed so that that product had to be redistilled and the distillation curve of the gasoline (Engler or A.S.T.M) was altered. Some of the clays were "natural" clays as mined, e.g., Florida or Georgia clays. Others were natural clays that had been acid treated, e.g., the clays then supplied by the Filtrol Corporation. These clays, and others, were explored for the catalytic cracking of gas oil. Quite a few had some activity but the most active was Superfiltrol from the Filtrol Corporation. This was thought to be a sulfuric acid treated montmorillonite or halloysite.

This led to exploratory attempts to enhance the activity of Superfiltrol by adding small amonts of other substances as "promotors". Certain metal phosphates, especially acid phosphates and even phosphoric acid were mildly effective. It was known that Superfiltrol already had acidic properties but this was presumed to be residual traces of sulfuric acid remaining from the sulfuric acid treatment of the original clay.

Dr. Tropsch proposed to "open up" feldspar, an alkali aluminum silicate, to produce an alumino-silicate catalyst. He had a reference saying feldspars could be "opened" by heating the feldspar with lime and calcium chloride followed by acid leaching. This gave encouraging results and evolved into

0097-6156/83/0222-0241$06.00/0

"replaced feldspar" catalysts that were more active than Superfiltrol.

Since Superfiltrol was an aluminosilicate, it was thought that other metal silicates, natural or synthetic, might also be active cracking catalysts. I was added to the catalytic cracking research team and was assigned this part of the problem. I prepared some co-precipitated catalysts from water glass (sodium silicate) and aluminum salts and U. S. Patent 2,282,922 proposing the use of these catalysts was filed March 20, 1937.

At that time there were synthetic "zeolites" used for softening water by base exchange. These were amorphous sodium aluminosilicates. Active cracking catalysts were made from them by exchanging the sodium ions with other ions.

In 1935, when Dr. Tropsch learned he was afflicted with a terminal illness, he gave up the work and returned to his native Germany. The research was interrupted. I was assigned to work on the alkylation of isobutance under the supervision of Dr. A. von Grosse. This reaction took place in the presence of strong acid catalysts and the work later had an influence on my thinking about catalytic cracking.

A little later, a cracking research group was formed with me as supervisor and Drs. Gustav Egloff and J. C. Morrell as directors. The group included, in its early years, Drs. J. Elston Ahlberg, Herman S. Bloch, Edward C. Lee and Mr. George Tobiasson. We did research on both thermal and catalytic cracking. We started the catalytic cracking work with the "replaced feldspar" as the catalyst. The feldspars have the composition $MAlSi_3O_8$ in which M is K, Na, or Ca. It was assumed that the heating with lime and calcium chloride "opened up" the feldspar. Acid leaching then removed the lime, calcium chloride and alkali metal ions and left an alumino-silicate of some kind. Dr. Ahlberg analyzed the catalyst to put our thinking on a more quantitative basis and found that the catalyst was mostly silica with only small amounts of alumina; far less than the Al:3Si of feldspar.

From this we concluded that the "replaced feldspar" preparation was mostly a method for preparing hydrous silica containing a little alumina but substantially free of other metallic ions due to the extensive acid treatment to remove the alkali and calcium ions. We also concluded that purified silica hydrogel could be made less expensively from water glass (sodium silicate).

Starting with a large batch of undried, carefully purified silica hydrogel, highly active cracking catalysts were prepared by precipitating hydrous alumina on the silica. U. S. Patent 2,285,314 describing this method of preparation was filed October 22, 1938. Since there was no obvious stoichiometry involved, we called the products "silica-aluminas". Besides silica-alumina, we found silica-magnesia and silica-zirconia were also highly active cracking catalysts.

Similarly, we prepared a large batch of purifi- precipitated hydrous alumina and precipitated other materials on it. We found that alumina-boria was an active catalyst. Moving down the Periodic Table from silicon to titanium, we were not able to prepare an active catalyst from titania-alumina but titania-boria was active.

Besides catalyst exploration, we were also exploring other feed stocks besides our standard test gas oil and also doing some catalytic cracking of pure hydrocarbons with silica-alumina. This led to the conclusion that the silica-alumina must be a high temperature acid; a concept that came gradually rather than a "flash of genius".

For one thing, we saw that gas oils containing natural nitrogen compounds were more refractory than the standard gas oil, and deposited more coke on the catalyst during cracking even though the depth of cracking was less. We removed a few percent of the gas oil by acid treating (nitrogen bases removed) and the gas oil cracked readily. The nitrogen bases were poisoning the high temperature acid catalyst. This was confirmed by adding a little quinoline to our standard gas oil.

In the catalytic cracking of the pure hydrocarbons, the susceptibility and reaction products were compared with thermal cracking of the same hydrocarbons (1-5). The catalytic cracking of cumene was amazingly easy and selective compared to the thermal cracking. The catalytic cracking was essentially pure dealkylation to benzene and propylene. According to catalysis theory, if the catalyst dealkylates it should also alkylate. But alkylation catalysts are well known to be acids so the cracking catalyst must also be an acid.

More or less concurrently with our work, Dr. Louis Schmerling in the same UOP Research Laboratories was studying the mechanism of the alkylation of isobutane with olefins. He showed conclusively that this was a carbonium ion mechanism in the presence of strong acid catalysts, and that one of the key steps in the mechanism was the transfer of a hydride ion from a donor hydrocarbon to a tertiary carbonium ion. There were enough similarities with hydrocarbon reactions in the presence of cracking catalysts to justify concluding that these are also carbonium ion reactions. We even showed that hydride transfer reactions occurred in the presence of cracking catalysts at temperatures slightly below normal cracking temperatures, and that at such lower temperatures cracked gasoline of very low olefin content could be obtained from gas oil. Incidentally, there are catalysts that accelerate cracking that are not acidic, e.g., activated carbons. They produce reactions that are quite similar to thermal cracking and seem to be free radical reactions.

There remained the question: How can silica-alumina be a high temperature acid? Minerology described many minerals in ino-silicate complex acts as an anion with various s, e.g., Na, K, Mg, etc. It is interesting that at of the aluminum in these silicates is thought to be dral coordination with oxygen. Although our ina catalysts were amorphous, it seemed to us that a triv aluminum in a matrix of tetrahedrally coordinated silicon with oxygen would also be tetrahedrally coordinated with the oxygen consistent with one of Pauling's rules. This would leave unsaturation in the 4 oxygen atoms attached to the aluminum, i.e.

```
           |
           O
           |
  ( - O - Al - O - )⁻
           |
           O
           |
```

This could be associated with a hydrogen ion to form a protonic acid or could lead to a Lewis acid. The fact that the catalyst was a high temperature acid was the important point to us. It confirmed our specifications that, in the commercial manufacture of the catalyst, sodium ions had to be reduced to very low levels to prevent them from migrating at high temperatures and neutralizing the acidity.

There remained the suspicion that the acidity of the catalyst might be due to acid remaining from the preparation of the silica hydrogel from water glass by acid treating. This suspicion led to a series of catalysts made from ethyl silicate and aluminium isopropoxide in which no acids were used. These were also acidic and the acidity could be titrated with alkali as was the case with the silica-alumina from water glass. Interestingly, both acidity and cracking activity of the ethyl silicate catalysts came to a maximum at a Si:Al ratio of 1, i.e., as the catalyst composition approached the hypothetical $HAlSiO_4$ (6).

It may be worth noting here that the Petroleum Administrator for War during World War II had an industry group, PAW 41, coordinate their efforts on catalytic cracking. This group included Kellogg, Shell, Standard Oil Co. (Indiana), Standard Oil Co. of New Jersey (Exxon), Texaco and Universal Oil Products Co. I gave a report "What Makes the Cracking Catalyst Crack" at one of their meetings, describing the silica-alumina catalyst as a high temperature acid and giving evidence relating to the structure of the catalyst itself plus describing the cracking as carbonium ion reactions. The report was then classified. After the War and after the report was declassified, it was published (6).

This summary was written 40 to 50 years after the facts without access to the primary documents. While the facts are believed to be as stated, a faulty memory may have altered something unintentionally.

Literature Cited

1. Egloff, G.; Morrell, J. C.; Thomas, C. L.; Bloch, H. S. J. Amer. Chem. Soc., (1939), 61, 3571.
2. Thomas, C. L. J. Amer. Chem. Soc., (1944), 66, 1586.
3. Bloch, H. S., Thomas, C. L. J. Amer. Chem. Soc., (1944), 66, 1589.
4. Thomas, C. L., Hoekstra, J.; Pinkston, J. J. Amer. Chem. Soc., (1944), 66, 1694.
5. Thomas, C. L., Hoekstra, J.; Pinkston, J. J. Amer. Chem. Soc., (1944) 66, 1694.
6. Thomas, C. L. Ind. Eng. Chem., (1949), 41, 2564.

RECEIVED November 17, 1982

21

Development of the Theory of Catalytic Cracking

ROWLAND C. HANSFORD
Yorba Linda, CA 92686

The theory of catalytic cracking was developed in the 1940's almost simultaneously in the research laboratories of Socony-Vacuum Oil Company (now Mobil Oil Corporation), Shell Development Company, and Universal Oil Products Company (now UOP, Inc). This review records the history of that portion of the development which occurred at Mobil's laboratories in Paulsboro, New Jersey, beginning about 1940.

Mobil was one of the early pioneers in the commercial development and application of catalytic cracking through its collaboration with Eugene Houdry in the middle 1930's. The first commercial catalytic cracking unit employing Houdry's concept of a solid regenerable catalyst went into operation at Mobil's Paulsboro Refinery in 1936 (2). This was a 2000-barrel plant which served to demonstrate and prove out the process. Soon a larger 15,000-barrel unit was built at Sun Oil's Marcus Hook, Pennsylvania refinery across the Delaware River from Mobil's Paulsboro Refinery where a similar larger Houdry-type plant also was built beside the prototype unit. By the beginning of World War II in Europe in 1939, approximately 100,000 barrels of Houdry capacity was in operation across the United States. It has been said that this capacity to produce high-grade aviation gasoline components enabled victory in the Battle of Britian in the early 1940's. With the expansion of the war in 1941 by our entry into the conflict against both Japan and Germany, it was obvious that larger and better cracking plants were going to be necessary to support the war effort of our allies and ourselves. Already new types of catalytic cracking processes were being developed. These were the Fluid Catalytic Cracking (FCC) Process and the Thermofor Catalytic Cracking (TCC) Process. Prototypes of both process units went into commercial operation in the early years of American involvement in the war.

In addition to better and more efficient cracking plants, the exigencies of the expanding war effort required higher yields of high quality aviation gasoline. This meant improved

0097-6156/83/0222-0247$06.00/0

catalysts had to be developed quickly to fill the new cracking units. At Mobil's research laboratories in Paulsboro, Milton Marisic and coworkers soon came up with a new catalyst. This became known as "bead catalyst" since it was made in the form of opalescent spheres. The improved catalyst was used extensively both in Houdry-type plants and TCC plants during the critical latter part of the war.

Marisic's basic concept was to make a true homogeneous hydrosol of silica and alumina, which could be formed into spherical droplets in an immiscible fluid, like oil. The concentration of hydrous oxides and the pH of the hydrosol were controlled to cause the droplets to set to a firm hydrogel before reaching the bottom of the oil column. The resulting hydrogel spheres were washed, ion exchanged to remove sodium, dried, and calcined to produce hard, microporous, opalescent "beads" (3). The process developed for the manufacture of the catalyst was continuous and therefore capable of a high rate of production. The spherical shape made the catalyst very suitable for moving bed operation in the TCC Process, as well as for fixed-bed operation in the Houdry Process. But most important from the viewpoint of this discussion of catalyst mechanism development is the chemistry of the process and of its product.

Marisic recognized that the gelation of the homogeneous hydrosol of hydrous silica and alumina produced a more or less uniform crosslinked structure in which nearly every aluminum ion is tetrahedrally coordinated with silicon ions through oxygen ions. Each such aluminum ion possesses a net negative charge requiring neutralization by a cation. In short, the hydrogel is an amorphous "zeolite" having ion-exchange capacity equivalent to its aluminum ion content. The hydrogel, when washed free of soluble salts formed in the neutralization process of sol formation, still contains approximately one Na^+ for every aluminum in the hydrogel structure. This "zeolitic" sodium is easily replaced by other ions such as NH_4^+, H^+, AlO^+, etc.

Full ion-exchange capacity equivalent to the content of aluminum of the hydrogel exists before calcination. However, if the Na^+ is exchanged out with NH_4^+ and the product is heated to drive off NH_3, the ion-exchange capacity will drop to about 10% of theoretical. But this residual capacity, now present as hydrogen ions, is very important for the functioning of the calcined xerogel as a cracking catalyst. Edward Griest in Marisic's laboratory discovered that when the calcined beads were contacted with an alcoholic solution of methyl orange the acid color of the indicator developed. Unfortunately, he never published this observation, although it was made several years before similar results were published by others (4).

In 1944, the writer became head of the catalyst research group that had worked with Marisic on the development of bead catalyst. Among this group was Charles Plank, who much later was responsible for the important invention of the vastly superior crystalline aluminosilicate zeolite cracking catalysts employed in practically all commercial catalytic cracking today. Others in the group were Leonard Drake and Howard Ritter, who jointly invented the mercury porosimeter. This device has been an important research tool in determining the porous structure of all sorts of solids (5).

Plank made many fundamental studies of the properties of silica and silica-alumina hydrogels and xerogels. Drake contributed to the research on physical structure of these materials using his mercury porosimeter. It is probable that much of this background led Plank to his practical development of zeolite cracking catalysts in later years. But the group was also concerned with improving the properties of the bead catalyst, such as its pore structure and its regeneration. A practical problem associated with "after-burning" of regeneration gases after leaving the dense catalyst bed in TCC units was solved by adding small amounts of chromia to the catalyst (6). This was called the chrome bead catalyst. It is interesting that similar problems in modern catalytic cracking plants have been solved by adding traces of platinum to the catalyst.

A major effort of the group after the war was devoted to research aimed at an understanding of the origin of catalyst activity and of the mechanism of catalytic cracking. Griest's observation of the acidic nature of silica-alumina catalysts led the writer to consider the origin and nature of these acid centers. Further, the questions of whether acidity plays a role in the cracking mechanism and how it may do so became a subject for serious investigation.

In a study of the cracking of n-butane over a commercial silica-alumina catalyst, it was early observed that activity decreased as the catalyst was dehydrated at high temperature under vacuum (7). The activity could be restored simply by allowing water vapor to readsorb on the catalyst at elevated temperature. As little as 0.1% of adsorbed water significantly affected activity.

It seemed clear that since only small amounts of water are effective, and if acidity is also involved in the catalytic function, then the nature of the acidity probably is protonic. Removal of water reduces proton availability and also catalyst activity. If protons are indeed involved in the mechanism of hydrocarbon activation and reaction, then it was reasoned that an exchange of hydrogen between catalyst and hydrocarbon might occur. By hydrating the catalyst with D_2O it was shown that indeed such exchange does occur (7). The ease of exchange was found to follow more or less the expected relative ease of protonation or carbonium ion formation. Thus, the lowest temperature at which exchange was observed for a number of pure hydrocarbons was found to be as follows:

2-Butene	Below 30°C
Isobutane	35°
Benzene	70°
n-Butane	205°
n-Hexane	260°
Cyclohexane	315°

A possible explanation for the origin of catalyst acidity was proposed by the writer, based on Pauling's electrostatic valence rule:

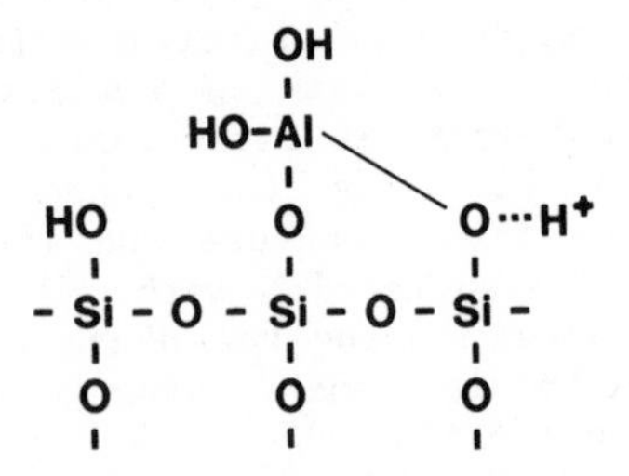

Aluminum is four-coordinated in the structure, but having a valence of 3^+ the residual bonding power for one of the hydroxyl hydrogens (in this case hydroxyl bonded to Si) is reduced enough to labilize that hydrogen as a proton. Tamele (8) later proposed a similar, but probably better, picture in which the labile proton comes from the hydroxyl attached to Al:

```
                   H+
              HO
    |    ..    ..    ..    |
  - Si : O : Al : O : Si -
    |    ..    ..    ..    |
             : O :
               ..
```

The advantage of Tamele's concept is that it more clearly shows how the acid site associated with the aluminum ion is interconvertible to Lewis acid and Brönsted acid by removing or adding a water molecule:

```
                                                         H+
                                                  HO
   |    ..    ..    ..    |    +H2O      |    ..    ..    ..    |
 - Si : O : Al : O : Si -  ------->  - Si : O : Al : O : Si -
   |    ..    ..    ..    |  <-------    |    ..    ..    ..    |
                               -H2O

        LEWIS ACID                        BRÖNSTED ACID
```

If the Lewis acid site is assumed to be inactive in cracking, this explains the effect of dehydration on activity of the catalyst.

The mechanism originally proposed by the writer for the cracking reaction invoked both carbonium ions and carbanions, depending on whether the hydrocarbon had unsaturated or saturated carbon-carbon bonds. With olefins or aromatics protonation to form a carbonium ion is relatively easy. However, at that time it was not obvious how a paraffin can be converted to a carbonium ion. So it was postulated that water might extract a proton from the paraffin to form in this case a carbanion. However, this concept was very soon abandoned because it was inconsistent with the observed rearrangements (isomerization) in reactants and products of cracking. An all-cationic mechanism was then proposed (9) in which the activation of paraffins occurs by hydrogen transfer to form a carbonium ion intermediate.

Convincing proof of the carbonium ion mechanism of activation of paraffins was obtained from a detailed study of hydrogen exchange between catalyst and isobutane (10). If the exchange occurs through a carbonium ion mechanism, the theory requires that only primary hydrogens will exchange. The tertiary hydrogen will not exchange. The results clearly showed that nine of the hydrogens of isobutane readily exchanged with hydrogen (deuterium) on the catalyst. Only traces of product containing ten deuterium atoms were found. Proof that it was the tertiary hydrogen that does not exchange was provided by an experiment with 2-methylpropane-2-d on a non-deuterated silica-alumina catalyst. No significant exchange was observed, i.e., less than 0.4% of nondeuterated isobutane was observed in the product after 16 hours of contact at 120°C. Under the same conditions, approximately 80% total exchange (disappearance of mass 58 isobutane) was observed with undeuterated isobutane on a deuterated catalyst.

The carbonium ion theory originally proposed by Frank Whitmore (11) is thus as applicable to the mechanism of catalytic cracking as it is to other acid-catalyzed reactions like isomerization or alkylation.

The question of carbonium ion formation from saturated hydrocarbons was considered in (7) by the writer when the possibility of participation by olefins from thermal cracking was mentioned. However, it was only somewhat later that this suggestion was seriously adopted (9). Then it was postulated that even traces of unsaturated hydrocarbons can activate saturated hydrocarbons by first forming a carbonium ion by proton addition. This ion can then extract a hydride ion by hydrogen transfer from the paraffin or cycloparaffin. This initiates a sort of chain reaction in which new carbonium ions are formed by hydrogen transfer with a steady-state population of ions on the catalyst surface.

The hydrogen exchange study reported by the author and co-workers (10) showed that indeed only small amounts of olefin are required to activate isobutane. The effect of traces of olefins on the exchange rate was greatest at low temperatures (60-80°C). At 120°C the addition of olefin had no effect on the exchange rate of the undoped hydrocarbon, indicating that at this temperature enough olefin was being formed by thermal cracking (or dehydrogenation) to maintain the initiation and propagation steps. The writer prefers this mechanism of carbonium ion formation from saturated hydrocarbons over the direct hydride ion extraction by the catalyst ($H^+ + H^-$ H_2). It is well known that saturation of olefins by hydrogen transfer is a major reaction occurring in commercial catalytic cracking.

The original carbonium ion theory (now called the carbenium ion theory by modern purists) was applied in a more quantitative way by Charles L. Thomas (12) of UOP and in particular by Bernard Greensfelder and Hervey Voge of Shell Development Company (13). Its success has endured few challenges now for some 35 years, and it seems unlikely that a better theory as applied to catalytic cracking will ever replace it.

Literature Cited

1. Formerly Research Associate, Mobil Research and Development Corporation; presently Retired Staff Consultant, Science and Technology Division, Union Oil Company of California.
2. Houdry, E., Burt, W. F., Pew, A. E., and Peters, Jr., W. A Petroleum Refiner (1938) 17, No. 11, 574.
3. Marisic, M. M. U. S. Patents 2,384,942 and 2,384,946, (1945); Porter, R. W. Chem. & Met. Eng. 53, 94 (1946).
4. See for example: Walling, C. J. Am. Chem. Soc. 1950, 72, 1164.
5. Ritter, H. L. and Drake, L. C. Ind. Eng. Chem., Anal. Ed., 1945, 17, 782. L. C. Drake, Ind. Eng. Chem. 1949, 41, 780.
6. Plank, C. J. and Hansford, R. C., U. S. Patent 2,647,860 (1953).
7. Hansford, R. C. Ind. Eng. Chem. 1947, 39, 849.
8. Tamele, M. W. Faraday Soc. Discussion 1950, 8, 270.
9. Hansford, R. C. Advances in Catalysis 1952, 4, 14.
10. Hansford, R. C., Waldo, P. G., Drake, L. C., and Honig, R. E. Ind. Eng. Chem. 1952, 44, 1108.
11. Whitmore, F. C. J. Am. Chem. Soc. 1932, 54, 3274.
12. Thomas, C. L. Ind. Eng. Chem. 1949, 41, 2564.
13. Greensfelder, B. S., Voge, H. H. and Good, G. M. Ind. Eng. Chem. 1949, 41, 2573.

RECEIVED December 2, 1982

22

The Invention of Zeolite Cracking Catalysts—A Personal Viewpoint

CHARLES J. PLANK[1]
Mobil Research and Development Corporation, Paulsboro, NJ 08066

The story of zeolite cracking catalysts is told from the viewpoint of Mobil inventors, C. J. Plank and E. J. Rosinski, whose patents laid the foundation for modern catalytic cracking. Their basic patent, U.S. 3,140,249, led to the induction of the inventors as the 30th and 31st members of the National Inventors Hall of Fame in 1979. It describes "Catalytic Cracking of Hydrocarbons with a Crystalline Zeolite Catalyst Composite". The catalyst consists of a finely divided crystalline alumino-silicate, having uniform pore openings between 6 and 15 Angstrom units, dispersed in an inorganic oxide matrix and has a low sodium content. Background work leading to the invention and the economic significance of zeolite catalysts are discussed. The synergistic properties of these composites are pointed out. The extraordinary activity and selectivity of the new catalysts led to a very rapid commercialization. Specifically, much higher gasoline yields were obtained at the expense of gas and coke. In addition, greatly increased gas oil conversions could be obtained without increasing coke yields. Use of the zeolite catalysts has been estimated to save the U.S. petroleum industry 200,000,000 barrels of imported crude oil a year. Thus, at the present price of crude oil the value of these catalysts to our economy is quite substantial.

[1] Current address: 522 Delaware St., Woodbury, NJ 08096

0097-6156/83/0222-0253$06.00/0

I am going to do what a retired scientist loves to do -- reminisce about his favorite subject. I retired from Mobil Oil almost three years ago after more than thirty-eight years of service. Of this all but two years were spent in the field of industrial catalysis.

When I started making and using catalysts, I saw no reason for doubting the correctness of the Ostwald definition of a catalyst. You remember that he said a catalyst is a material which changes the rate of a chemical reaction without itself appearing in the products. In other words the catalyst is unchanged by the process. Since that time I have learned the shortcomings of that definition. In fact, I now refer to "Industrial Catalysis" as "The Science of Dirty Surfaces."

My work involved making, testing and characterizing catalysts for almost all phases of petroleum processing. However, the field of catalytic cracking has occupied a major part of my efforts. Quite a few years of research in this field preceded our discovery of the zeolite catalysts. And the habits of thought and testing methods developed during those years had much to do with making that discovery.

In my work I have always felt it necessary to keep right on top of the experimental data. As a result, I decided that testing of our experimental catalysts must be carried out within our own group as much as possible. Originally we used a test developed by Alexander and Shimp of the Houdry Development Company -- called the CAT-A test. (1)

We were looking particularly for three properties of our catalysts, namely their activity, selectivity and stability. Activity is, of course, defined by the percent conversion of the charge at a given set of conditions; selectivity is the yield of desirable products at a given conversion; and stability is the ability to retain activity and selectivity during use. While all are very important, probably the most important is selectivity. In our case this means increased gasoline at the expense of dry gas (C_3's and lighter) and coke.

My Background in Catalytic Cracking

The situation which existed in catalytic cracking at the time I started working in the field has been illustrated by Dr. Oblad in his discussion of Eugene Houdry's work. This process was a fixed-bed process and it was largely replaced in the '40's by a moving-bed process, TCC, and fluid catalytic cracking, FCC. The latter processes required more rugged catalyst than did Houdry's, so the acid-treated clay was replaced by synthetic silica-alumina gels. These gave little, if any, gain in selectivity, but they were considerably more stable (especially physically) and somewhat more active.

Thus, our purpose was to improve on the silica-alumina gel catalysts. I well remember the statistic we used to quote in our early days. If we could just increase the gasoline yield by 1% at the expense of gas and coke, it would be worth a million dollars a year to our company. So the motive for increasing selectivity is obvious -- and there seemed to be plenty of opportunity to do so.

Even after several years of testing cracking catalysts it was a real eye-opener for me to read a 1953 article by Blanding (2) of Standard Oil Development Company. In it he showed the plot reproduced in Figure 1.

Here is plotted the instantaneous activity (reaction rate constant) for cracking of a clay catalyst at 850°F against time on stream. You can see that the activity at 0.01 seconds on stream is 750 times as high as it is after twenty minutes. And a static bed test with an on stream time of at least ten minutes will show an average activity no more than 5 on this scale -- hundreds of times lower than the original catalyst activity.

While I had been well aware of this effect, I had no idea of its magnitude. Quite clearly the reason for the effect is the accumulation of the undesirable product coke on the catalyst. A plot of coke accumulation would be just about the inverse of this -- very rapid at the start and quite slow at the end of the period. In fact, at the very start of the reaction, most of the charge goes to coke.

The same thing occurs in reforming and hydrocracking, just to mention two more of the most important processes in the petroleum industry. In all cases the catalyst used for most of the processing time has vastly different properties from those of the catalyst at the start.

As one further example, in the middle '60's we developed a dehydrogenation activity (DA) test which measured the rate constants of reforming or hydrocracking catalysts for converting cyclohexane to benzene. We found that fresh platinum/alumina catalysts had a DA of about 1000. After only one day in a unit the DA dropped to about 1. And, of course, such catalysts are used for months without regeneration in a reforming unit. Now you know why I refer to industrial catalysis as the science of dirty surfaces.

To go back to Blanding's data, they led me to the idea of searching for a catalyst whose original activity could be better preserved than existing catalysts. The plot also shows how the newer synthetic gels compared to clay catalysts. They start at about the same point, but level off with the synthetic about twice as active as clay. This was a quantitative measure of what we already knew. And, it left room for a large measure of improvement.

At about this time we decided that the CAT-A test simply did not recognize catalyst differences which we thought it should. Nor did we think the conditions of the test -- especially the use of a very light gas oil -- properly represented industrial conditions. Therefore, I put one man to work for about a year to develop a new test.

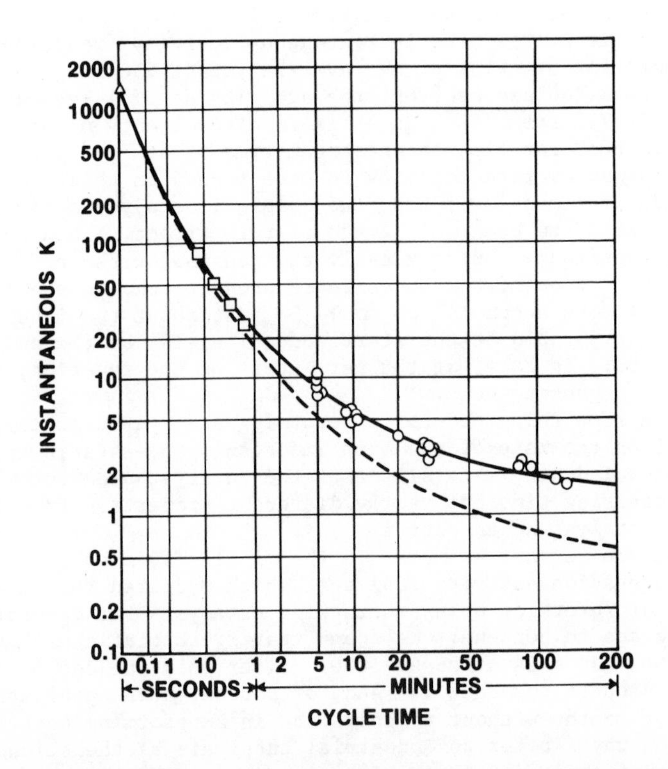

Figure 1. Cracking light east Texas gas oil at 850 °F and atmospheric pressure. Key: ---, clay catalyst; —, silica-alumina gel catalyst. (Reproduced with permission from Ref. 2.)

The resulting test proved able to recognize catalyst differences in much the same way they were shown by TCC pilot units. It was labeled the CAT-C test and its test conditions were: Gas Oil - a full range Mid-Continent Gas Oil; Temperature - 900°F; Liquid Hourly Space Velocity - 2; Catalyst-Oil-Ratio (Vol.) - 3; On-Stream Time - 10 Minutes. The unit is illustrated in Figure 2.

The first thing we did after we were satisfied with test conditions was to set up a series of standard curves. These are shown in Figure 3. We have plotted the C_5+ gasoline, total C_4's, dry gas and coke yields against the volume percent conversion of the gas oil. The conversion was controlled by varying the activity of the standard silica-alumina gel catalyst. This was done by controlled steam treatment.

The dot on each graph represents the results for a hypothetical new catalyst, and shows our method of making comparative evaluations. Comparison is made at the activity level of the new catalyst. And here the hypothetical catalyst is a very good one. It gives much more gasoline than the standard silica-alumina at the expense of C_4's, dry gas and coke. The differences which we found were referred to as Δ (delta) values.

At this point it is important to note that with both CAT-A and CAT-C test procedures we examined both fresh and partially deactivated catalysts. This was necessary to evaluate catalyst stability -- both as to activity and selectivity. The major cause of catalyst deactivation is its exposure to high temperature steam in the regenerator. Therefore, our deactivation procedure involved a high temperature steam treatment.

Early Zeolite Work with E. J. Rosinski

In the latter part of 1956, I was extremely fortunate to have a young fellow named Ed Rosinski join my group. Ed already had a reputation for prodigious energy. He not only earned his BS, ChE degree while working a full shift at Mobil, he also built a house for himself and family during his spare time in the same period. On receiving his degree I was very happy that he requested a transfer into my group. Even before he reported to the group we discussed what we were looking for in improved cracking catalysts. I found that we thought extraordinarily alike and that our discussions led to a real stimulation of each other's ideas. As a result, he fitted perfectly into my style of research and we remained a close-knit team until my retirement.

Very early we decided that what we needed was a catalyst whose active sites were located in controlled pores not much larger than the molecules to be cracked. This we reasoned should have two beneficial results -- more selective cracking and, as a result, less coking. This in turn should lead to maintenance of

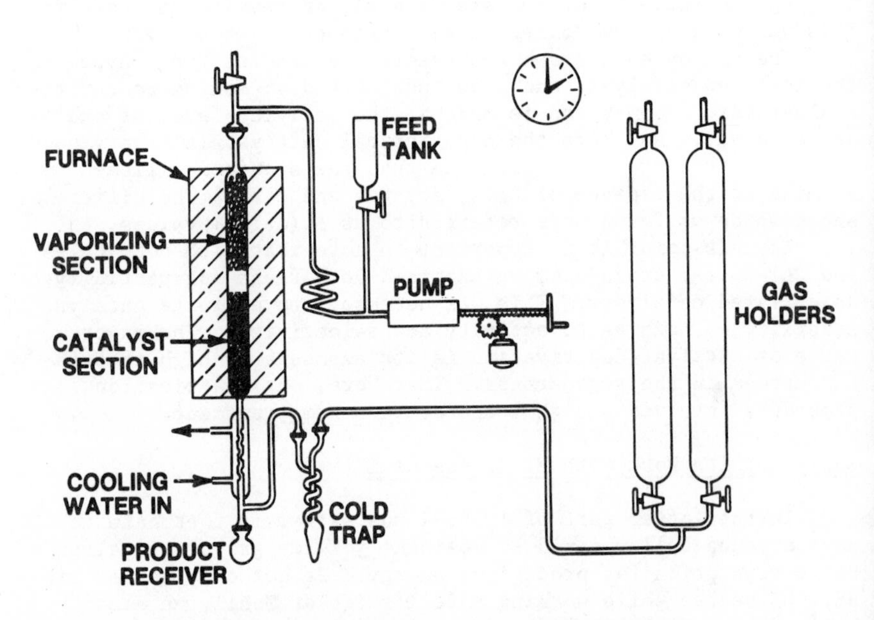

Figure 2. Diagram of the CAT-C unit.

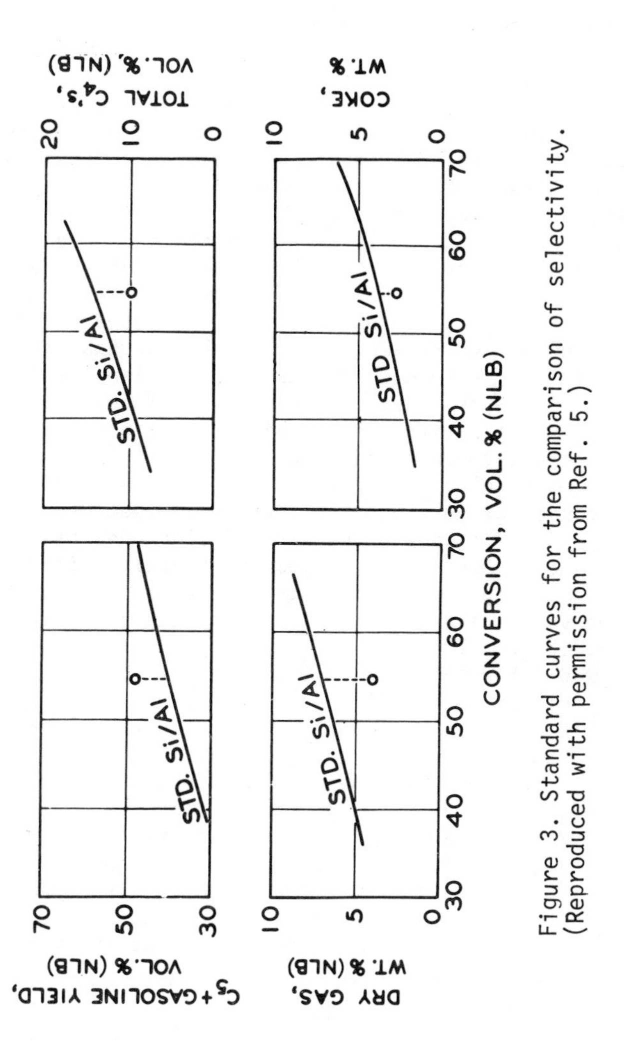

Figure 3. Standard curves for the comparison of selectivity. (Reproduced with permission from Ref. 5.)

a higher activity level. Two types of material which we picked for early study were molecular sieve zeolites and silica-alumina gels formed around molecular templates.

The molecular template idea was based on a paper by Dickey (3a) working with Pauling (3b) in which they described selective silica gel adsorbents. These were prepared by gelling in the presence of indicator molecules such as methyl orange, ethyl orange, etc. While I had never been able to confirm the results given in the paper, the idea stayed with me.

Tests of the molecular template idea were originally very encouraging. The fresh catalysts consistently showed improved selectivity. However, the improved selectivity disappeared after the catalysts were steamed. As a result we concentrated our efforts on the zeolite catalyst.

A zeolite, of course, is a crystalline aluminosilicate mineral with a very open pore structure and a very large water adsorption capacity. It possesses two well-defined chemical properties. First, it can be reversibly hydrated and dehydrated. Second, it has a high reversible ion exchange capacity. This capacity corresponds to one equivalent of cation per Al atom in the anionic framework. The anionic nature of the framework derives from the substitution of trivalent Al atoms for Si atoms in the tetracoordinated structure.

The stick model of the zeolite faujasite which is very familiar to most of us gives the impression that the zeolite is all open space. Figure 4 shows a Courtauld model of the same structure made at Mobil which gives an entirely different impression. The holes are still there but the structure is shown to be very solid indeed. The 12-member ring of faujasite is shown at the center.

This model has the advantage that Fisher-Hirschfelder models of hydrocarbons can be used to show what kinds of molecules can enter the zeolite pores. While much more realistic, the model is hard to make and very heavy. In fact, this particular model fell apart due to its own weight in a few years.

The X-form of faujasite is the one with which we began to do our experiments. In our very first experiment we put 25% NaX into a silica gel matrix and base-exchanged with NH_4Cl to obtain a low sodium product. The unsteamed product gave us a surprisingly high activity and a selectivity much superior to silica-alumina. After steam treatment both activity and selectivity decreased. However, the selectivity advantage was enough to encourage us to look for improvements.

It should be noted that this very first catalyst possessed the three essential characteristics of our final patented catalyst. It consisted of a finely divided crystalline aluminosilicate having uniform pore openings between 6 and 15 Angstrom units, dispersed in an inorganic oxide matrix and had a low sodium content.

Weisz and Frilette (4) at our laboratories investigated

Figure 4. Courtauld model of faujasite.

commercial NaX and CaX, called 10X and 13X (their designated pore sizes), for cracking n-decane. They found both to be active, but very different in selectivity. The CaX gave higher activity with a product distribution like silica-alumina's, while NaX had about the same activity as silica-alumina with a product distribution like thermal cracking.

We obtained the same type of results when we tested these two materials for gas oil cracking in the CAT-C test. From the data we decided that we needed to enhance the acidic nature of the zeolite catalysts to improve their properties.

Results Leading Our Early Patents

A thorough discussion of the results which we achieved and on which our early patents are based is given in a paper by Plank, Rosinski and Hawthorne (5) in 1961. At this point, I will immodestly mention that this paper was chosen for republication in Chem Tech during the Bicentennial Year, 1976. It was chosen as one of the twelve most significant papers published in Industrial and Engineering Chemistry during its sixty-four year history.

I will summarize the data briefly. Since we thought we needed to introduce acidic sites into NaX, we decided to study the pure zeolites -- replacing the sodium with a combination of metal and ammonium ions. This was done by base-exchange with a number of different metal/NH_4^+ combinations.

One group of these products is illustrated in Table I.

Table I

Wt. % Composition of Faujasites

	Ca-Acid	Mn-Acid	RE-Acid
Na^+	0.46	0.52	0.22
Metal Oxide	10.7 CaO	15.4 MnO	26.9 $(RE)_2O_3$
Al_2O_3	34.9	33.6	29.6
SiO_2	52.1	48.6	41.4

Note first of all that we obtained a very low sodium content in each case. All contain much less than 5% of the original sodium. Furthermore, the metal ion equivalence per Al atom is 0.7 for the Ca, 0.67 for the Mn, and 0.84 for the rare earth (RE) zeolite. The remainder of the cationic sites were satisfied by protons giving us our metal-acid faujasites. In order to achieve the degree of ion exchange obtained here it was necessary to carry out the ion exchange at elevated temperature (180°F) for many contacts.

The catalytic data obtained with these materials are shown in Table II.

Table II

Cracking Over Metal-Acid Faujasites

	Ca-Acid	Mn-Acid	RE-Acid
Conversion, vol %	54.7	53.1	65.4
C_5+ Gasoline, vol %	48.0	48.0	52.3
Total C_4's, vol %	9.9	8.4	14.7
Dry Gas, wt %	4.9	4.2	6.5
Coke, wt %	2.6	1.5	3.2
Delta Advantage (over Si/Al)			
C_5+ Gasoline, vol %	+8.1	+9.3	+7.1
Total C_4's, vol %	-4.4	-5.3	-3.5
Dry Gas, wt %	-3.1	-2.6	-2.1
Coke, wt %	-1.0	-1.9	-2.1
Run Conditions	(a)	(a)	(b)
LHSV	10	16	16
Cat./Oil	0.6	0.38	0.38

(a) 100% steam for 20 hr. at 1225°F and atm. pressure.
(b) 100% steam for 24 hr. at 1200°F and 15 psig.

First of all, it was necessary to run the catalysts at much higher space velocities than the standard catalyst to obtain the indicated conversions because of their very high activities. The first was run at 10 LHSV and the other two forms at 16 LHSV. Furthermore, all had been steamed to simulate aged catalysts. The rare earth catalyst was the most stable and was steamed quite a bit more severely than the other two.

If we took a silica-alumina catalyst and steam-treated it at the conditions used for this rare earth catalyst we would only get about 50% conversion at 1 LHSV and 6 catalyst/oil ratio. That is, the space velocity is only about 1/16 as high and the catalyst/oil is 16X as great. These two factors are considered (6) equally important in catalytic cracking. So we have an activity advantage of more than 16 x 16 or 256 times for the rare earth-acid catalyst. Conservatively, we said the rare earth form was at least 100 times and the other forms at least 30 to 50 times as active as standard silica-alumina catalysts. We believed that the factors were really much greater.

This is true in spite of the fact that the coke level on the rare earth zeolite is more than five times that of the silica-alumina at the much lower conversion level. This is one of the most important and surprising features of the zeolites. They continue to function well at coke levels that would completely deactivate silica-alumina.

It is possible that in silica-alumina many of the active sites are located in dead-end pores. Thus, a little coke at one end may kill many active sites, while coke in the three-dimensional pore structure of a zeolite would have little effect on access to the interior.

I will point out again that this comparison is on the steam-treated catalysts. The rare earth catalyst run fresh has such a high activity that it forms coke like mad and the conversion level drops extremely rapidly. As a result the selectivity is horrible and the apparent overall activity is quite low. Thus, it is most fortunate that our practice was to test steamed as well as fresh catalysts. Had we only looked at the fresh catalysts, we would probably have discarded them.

Now consider the cracking data in Table III. I can only describe the results as fantastic. The most important numbers are those under "Delta advantage." The numbers for gasoline and C_4's are volume %, the others weight %. Comparing these numbers as a percentage change over the base catalyst yields at the same conversion level, the calcium-acid zeolite gives a 20% increase in gasoline and a 28% reduction in coke. The manganese-acid zeolite gives a 24% increase in gasoline with 56% less coke and the rare earth-acid zeolite gives 16% more gasoline and 40% less coke.

Table III

Durabead Type Catalysts - 10% Na Faujasite

	Ca-acid(a)	Mn-Acid(b)	RE-Acid(b)
Conversion, vol %	67.9	54.6	61.5
C_5+ Gasoline, vol %	54.8	45.5	49.3
Total C_4's, vol %	16.4	10.8	15.2
Dry Gas, wt %	6.5	5.3	6.4
Coke, wt %	4.5	2.2	2.6
Delta Advantage (over Si/Al)			
C_5+ Gasoline, vol %	+9.4	+6.7	+5.9
Total C_4's, vol %	-6.2	-4.0	-1.3
Dry Gas, wt %	-1.8	-1.7	-1.5
Coke, wt %	-2.0	-1.4	-2.2

Run Conditions: LHSV = 4; Cat./Oil = 1.5

(a) 100% steam for 20 hr. at 1225°F and atm. pressure.
(b) 100% steam for 24 hr. at 1200°F and 15 psig.

Having converted NaX into extraordinarily active, selective and stable materials we returned to our original idea of suspending these little crystallites in gel matrixes. All three of these types of metal-acid faujasites were incorporated into silica-alumina matrixes in bead form to the extent of 3 to 15%. Quantitative comparisons for all three catalysts at the 10% zeolite level are shown in Table III. The manganese and rare earth forms were steamed at more severe conditions than the calcium catalyst. But all showed extraordinary selectivity advantages. Compared to the gasoline yield given by the standard, the calcium catalyst gave 20% more gasoline, the manganese form 17% more, and the rare earth form 14% more. And coke decreased by 31%, 39% and 40%, respectively.

Economic Considerations

Qualitatively, the economic consequences of such data are quite clear. At the time of this work, dry gas and coke had a value equal only to fuel value. And even butanes were valued at about 2/3 that of gasoline. In 1967, we presented (7) some simplified calculations as to the value of incremental gasoline yields to the whole U.S. petroleum industry. These are shown in Figure 5.

The first column shows the additional product value resulting from a given amount of incremental gasoline. The last shows how much crude oil could be saved in making a given amount of gasoline for domestic consumption.

Even 1% more gasoline is worth $20,000,000 to the industry and is equivalent to saving 27,000,000 barrels of crude oil a year. It is clear from our data that we should expect at least a 7% increment in gasoline as indicated in the cross-hatched plot. This corresponds to an added product value of $140,000,000 and to a potential saving of 190,000,000 barrels of crude oil per year at the 1963 rate of demand. Since the Arab oil embargo of 1973, these numbers have escalated tremendously. Even by 1968, the refining editor of the Oil and Gas Journal, Stormont, (8) estimated refinery savings as $300,000,000 in capital costs and more than a million dollars a day in reduced crude needs and increased product value.

The economic incentive to commercialize these catalysts was so great that Mobil expended a tremendous effort to that end. The rare earth in silica-alumina gel form was the most stable and in a very short time became our first commercial catalyst. Results of our first commercial test were described by Elliott and Eastwood in 1962. (9)

The first patents covering our zeolite cracking catalysts were granted in 1964. (10-12) Of these, the basic patent, U.S. 3,140,249, led to the induction of the inventors, C. J. Plank and E. J. Rosinski, as the thirtieth and thirty-first members of the National Inventors Hall of Fame in 1979.

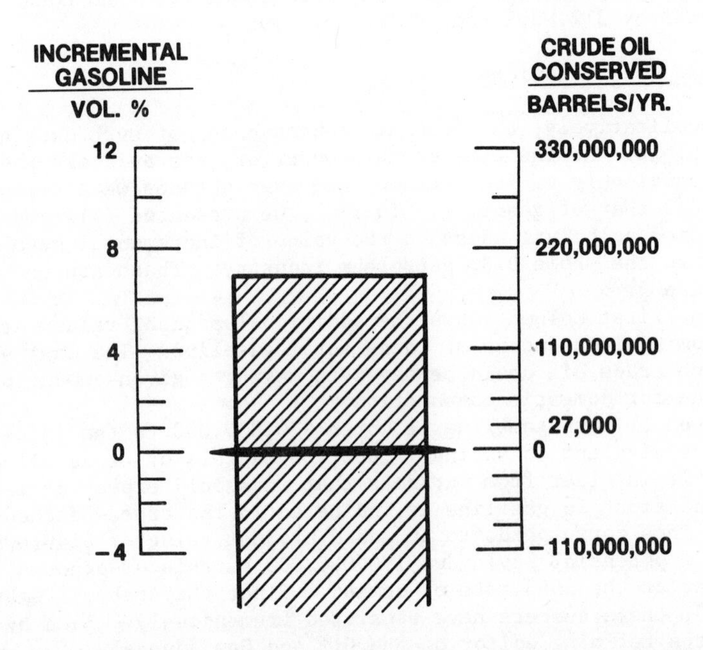

Figure 5. Economic value of incremental gasoline. (Reproduced with permission from Ref. 7.)

Catalyst Improvements

Since 1961, zeolite cracking catalysts have quite substantially improved. Much has been done by ourselves and others. Figure 6 summarizes the results of recent Mobil improvements over the original catalyst. These data, unlike the earlier data, were developed in a fluid-bed pilot unit rather than in the fixed-bed CAT-C unit. All, of course, involve zeolite/matrix catalysts.

The first point to be made is that I want to make gasoline yield comparisons at the 65% conversion level. At this level, the standard catalyst gives 45% gasoline. The "early zeolite" (REHX) gives 51% gasoline, while the "improved zeolite" (REHY) gives 54% gasoline, and the recent experimental catalyst containing a CO promoter gives 57% gasoline.

The second point refers to the black dots on each curve. This is the level at which the catalyst produces 4% coke based on the charge. This occurs at a conversion level of 56% on silica-alumina, 68% conversion on REHX, 75% on REHY, and almost 85% on the experimental catalyst.

If the unit should be limited by regeneration capacity to 4% coke on charge, these conversion increases could be attainable with the attendant increase in gasoline yield. Those gasoline numbers are 40.5% for silica-alumina, 52.5% for REHX, 59.5% for REHY, and almost 65% for the experimental catalyst. Of course, such results have only been obtained in pilot units and have not yet been realized commercially. But they indicate an amazing yield potential.

Zeolite-Matrix Synergism

I have emphasized the selectivity advantages of these low sodium/zeolite/matrix catalysts without, so far, expanding on the functions of the matrix. Many people have made the assumption that its only function is the physical one of transforming tiny crystallites into particles in the size range useful for commercial operation. While this function is important, it is certainly not the only one provided.

When we first thought about using zeolites for cracking, we theorized that the matrix would stabilize the zeolite. This has definitely proved to be true. In an article written with Eastwood and Weisz (13) for the 1970 World Petroleum Congress we presented data showing a very great stability advantage for REHY/silica-alumina over REHY along toward high temperature steam. These are shown in Table IV. Here we compared the catalytic activity of pure REHY, the same REHY dispersed as approximately 4μ crystallites in a silica-alumina gel matrix and that of the same silica-alumina gel alone after two different deactivation treatments. The test conditions were chosen so

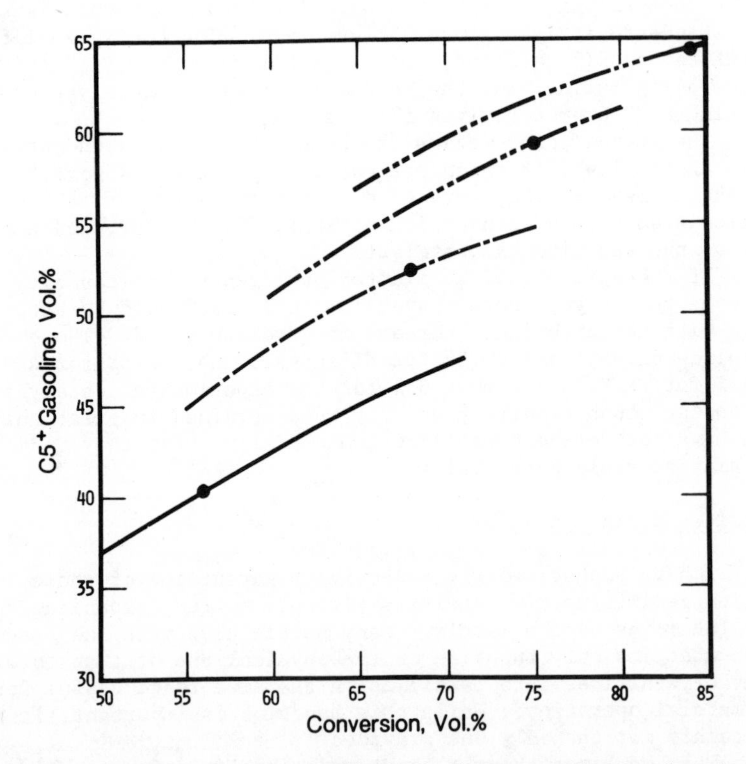

Figure 6. Improvements in zeolite cracking catalyst selectivity. Key: ——, standard silica-alumina gel; —·—, early zeolite catalysts (REHX); —··—, improved zeolite catalyst (REHY and copromoter); and ●, point where 4% coke (on charge) occurs.

that the pure REHY and the REHY in silica-alumina gave about the same conversion after the standard steaming (Note 4). The silica-alumina was tested at the same conditions as the composite catalyst. The synergistic effect of the matrix in stabilizing the REHY is obvious.

Table IV

Effect of Matrix on Stability of REHY

Catalyst	REHY[2]		REHY (10%)[3] in Silica-Alumina		Silica-Alumina[3]	
	4	5	4	5	4	5
Conversion[1] (Vol %)	67.8	6.3	67.5	58.2	34.4	35.2
Gasoline (Vol %)	58.3	5.8	56.7	49.7	28.8	30.6

[1]Conversion of standard wide range Mid-Continent gas oil.
[2]Run conditions -- 900°F, liquid hourly space velocity = 16, time on stream = 10 min.
[3]Run conditions -- 900°F, liquid hourly space velocity = 4, time on stream = 10 min.
[4]Calcined 10 hrs at 1000°F in air, then treated 24 hrs at 1200°F and 15 psig in 100% steam.
[5]Treatment of [4] followed by 48 hrs at 1575°F in air containing 5% steam.

The synergistic effect of the matrix on the activity of a zeolite is illustrated by the data from the same article summarized in Table V. Here we have compared the conversions given by a pair of catalysts containing 4% and 10% REHY in silica-alumina with the amount of REHY required to give about the same conversions as the two composite catalysts. It takes 9.5 gm of pure REHY to give a 21.0% increase in conversion, while 3.8 gm of added REHY gives a 19.0% increase in conversion with the zeolite/matrix combination. Thus, it is quite clear that the matrix in our composite catalyst is far from a simple diluent and physical stabilizer.

Table V

Effect of Matrix on Activity of REHY

66.7 ± 0.3 ml of Gas Oil Charge in 10 Minutes (900°F)

	REHY[1]		REHY in Silica-Alumina[2,3]	
Conversion (Vol %)	43.3	64.3	46.9	65.9
Wt REHY Present in Reactor (g)	7.5	17.0	2.9	6.7
% Conversion Change / g REHY Change	$\frac{21.0}{9.5} = 2.21$		$\frac{19.0}{3.8} = 5.0$	

[1]These catalysts are diluted with quartz to 200 cc.
[2]100 cc in bead form (varying REHY content -- 4% vs 10%) diluted with quartz to 200 cc.
[3]REHY dispersed in silica-alumina hydrogel. Base exchanged with $(NH_4)_2SO_4$ + $RECl_3$.

I have not attempted to make this a "History of Zeolite Cracking Catalysts." To do so would involve a discussion of hundreds of patents. Since our original patents, Rosinski and I, alone, have received at least twenty-eight more in the field while others at Mobil have probably been responsible for scores. The tremendous interest of the whole petroleum industry in the field attests to the importance of these catalysts. This interest has further spread to the development of new zeolites and their use as catalysts for many reactions in addition to cracking. Thus, the success of the zeolite cracking catalysts has spawned a whole new field of catalyst research. We and many others at Mobil have been in the forefront of this work as well. I am looking with great interest at what will be accomplished in the future with zeolite catalysts.

Acknowledgment

It is quite clear that my greatest thanks are due to my partner in catalyst research and invention -- Edward J. Rosinski. Second, we both owe much to Wendell P. Hawthorne, our supervisor in the early years of our research on zeolite catalysts. His early recognition and untiring efforts towards promoting our work were extremely helpful. And, finally, a tremendous Mobil team in research, management, and patent counsel were indispensable in the development and commercialization of zeolite cracking catalysts.

Literature Cited

1. Alexander, J. and Shimp, H. G., Nat'l. Petroleum News, 1944, 36, (31), R-537.
2. Blanding, F. H., I.E.C., 1953, 45, 1186.
3. (a) Dickey, F. H., Proc. Nat. Acad. Sci. U.S., 1949, 35, 227.
   (b) Pauling, L., Chem. Eng. News, 1949 27, 913.
4. Weisz, P. B. and Frilette, V. J., J. Phys. Chem., 1960, 64, 382.
5. Plank, C. J., Rosinski, E. J., and Hawthorne, W. P., IEC Prod. Res. & Dev., 1964, 3, 165.
6. Moorman, J. W., Oil Gas J., 1954, 52, 106.
7. Plank, C. J. and Rosinski, E. J., Chem. Eng. Progress Symposium Series, 1967, 63, (73), 26.
8. Stormont, D. H., Oil Gas J., 1968, 66, (14), 104.
9. Elliott, K. M. and Eastwood, S. C., Oil Gas J., 1962, 60, (June 4), 142.
10. Plank, C. J. and Rosinski, E. J., U.S. Patent 3,140,249, 7/7/64.
11. Plank, C. J. and Rosinski, E. J., U.S. Patent 3,140,251, 7/7/64.
12. Plank, C. J. and Rosinski, E. J., U.S. Patent 3,140,253, 7/7/64.
13. Eastwood, S. C., Plank, C. J. and Weisz, P. B., Eighth World Petroleum Congress Proceedings, 1971, 4, 245.

RECEIVED March 14, 1983

The Development of Fluid Catalytic Cracking

C. E. JAHNIG, H. Z. MARTIN, and D. L. CAMPBELL
Exxon Research and Engineering Company, Linden, NJ 07036

This paper reviews the history of developing the Fluidized Solids Technique and its commercial applications, particularly the original application to catalytic cracking of heavy oil to make high octane gasoline. For many years about half of the U.S. motor gasoline has come from fluid catalytic cracking. The development provided a new simple system to carry out catalytic operations using powdered catalyst. Early plants circulated catalyst between reactor and regenerator at high flow rates of more than a railroad car full (60 tons) every minute.

The development took only three years from the initial concept to the first commercial operation. This paper reviews the development of the Fluidized Solids Technique, which revolutionized petroleum refining and created a new field of chemical engineering.

An important aspect of history is to provide a record that makes it possible to benefit from the experience of others. This is especially valuable today in view of the difficult economic situation confronting R&D projects as a result of inflation and high interest rates. The history of developing Fluid Catalytic Cracking is a most useful example in that the development was completed in only 5 years, from initial concept to extensive commercial application (1939-1944). The Fluidized Solids Technique was a basically new chemical engineering system having broad potential application to many processes. Although the development was expedited due to extraordinary wartime conditions, it does show how rapidly an R&D project can be carried through successfully when there is a high incentive and the situation is favorable. History of the development is reviewed in this paper with emphasis on the first application, namely, Fluid Catalytic Cracking. A previous paper gave a somewhat broader review of the development (1). It should be pointed out that in later stages of commercializing fluid catalytic cracking, other companies began to play a significant

0097-6156/83/0222-0273$06.00/0

role. This larger effort, which began just before World War II, as part of a group called Catalytic Research Associates, was expanded further after the start of the war under the aegis of the Petroleum Administration for War.

It is interesting that the development did not start from the well known Winkler technology on coarse particles and extend it to use fine powder with solids circulation between vessels. Rather, the development started with pneumatic transport technology used for handling powders, and extended it to include circulation between fluidized dense bed reaction and regeneration vessels for catalytic cracking. A key element was the discovery that a dense phase fluidized bed could be maintained at gas velocities far exceeding the Stokes' Law free fall velocity of the particles.

A most important invention was the use of a standpipe of fluidized powder to repressure the solids, allowing continuous rather than cyclic operation of cracking and regeneration while circulating catalyst to transfer a large amount of surplus heat from regeneration to the reactor where it was needed. The standpipe allowed circulation rates an order of magnitude higher than with technology known at the time, such as the compression screw pump or lock hoppers. Moreover, it did not require mechanical equipment or moving parts subject to severe erosion since circulation was based on the "air-lift" principle. This unique combination of standpipe plus dense fluid bed reaction zones became the basis for a whole new chemical engineering unit operation. The first major application was in Fluid Catalytic Cracking.

This paper covers the history at Exxon where substantially all of the development took place, resulting in the basic patents on the fluid bed (2) fluidized standpipe (3) the integrated system (4) and the downflow design (5). The work at Exxon produced the first designs for commercial plants that were built in 1940-1945 of both the upflow type (4) and the downflow type (5). Some of the business aspects and contributions by others to the development have been reviewed elsewhere (6), as well as some technical aspects (7).

Overview of Development

Fluid Catalytic Cracking revolutionized the oil industry and provides about half of the motor gasoline used today. This spectacular growth is shown in Figure 1. It is therefore of interest to examine the history of inception and development of the processs as recorded by some of the primary participants.

Catalytic cracking decreases the boiling point of heavy oils to form naphtha that can be included in motor gasoline. Before 1940, this was accomplished mainly by thermal cracking, but in the 1930's Houdry showed that use of a catalyst resulted in higher yield and much higher octane. Early catalysts were clay or a synthetic silica gel impregnated with alumina. Cracking took place at 800 - 950°F; however, the catalyst deactivated severely in only 10 - 20 minutes due to carbon deposits. This required periodic regen-

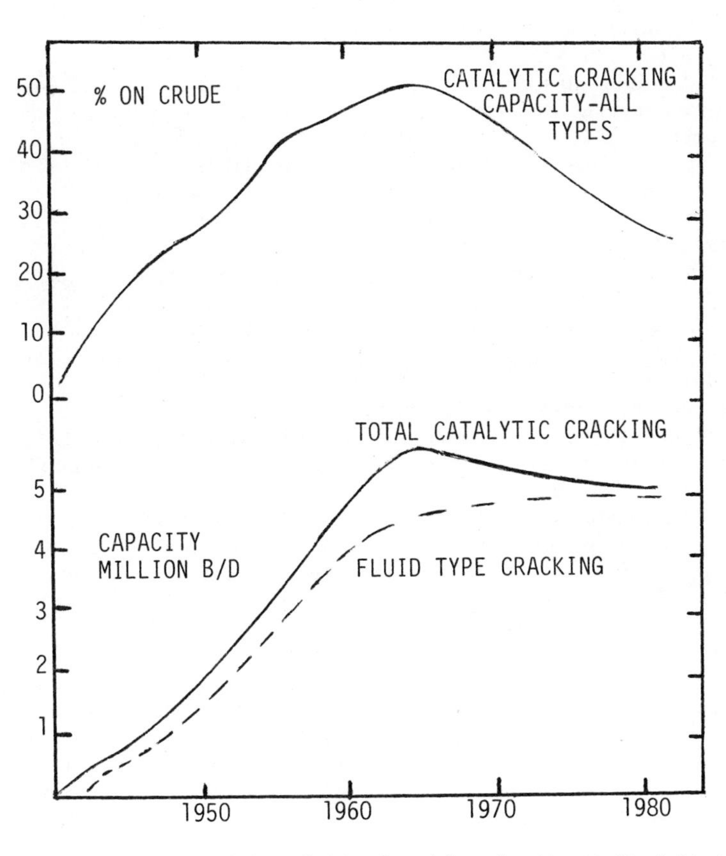

Figure 1. Growth of Catalytic Cracking in the United States.

eration which meant that oil flow had to be stopped so that the catalyst could be stripped, purged of combustible vapors, and then regenerated by burning at 1000 - 1100°F in a gas containing only a few percent oxygen to prevent excessive temperature. Multiple reactors had to be used to allow continuous flow of oil and regeneration gas.

The fixed bed process was commercialized by Houdry and associates in the late 1930's. Exxon (Standard Oil of New Jersey at that time) applied for a license but the licensing cost was very high and consequently Exxon decided to develop their own process. Small scale tests were started in a number of fixed bed laboratory units and a 100 B/D pilot plant was commissioned at Baton Rouge, LA. Mr. E. V. Murphree was the vice-president in charge of the overall development. Extensive engineering studies were made on various fixed bed designs, leading to the conclusion that the inefficiencies and problems inherent in a cyclic fixed-bed design could not be overcome. Therefore all efforts were shifted to circulating catalyst systems that would allow steady-state operations. Although early studies included systems using moving beds of catalyst pellets, Mr. Murphree soon decided that powdered catalyst was the way to go. Efforts were concentrated on powder rather than pellets, and the 100 B/D pilot plant was rebuilt to use pneumatic transport of powder in pipes at high velocity, 50 - 100 ft/sec. A Fuller-Kinyon compression screw was used to pump the catalyst to higher pressure and thereby allow continuous circulation, but this pump caused agglomeration of the catalyst into "poppy seeds" with serious deactivation.

In the pneumatic transport designs, the reactor and regenerator pipes were so long that they had to be "folded" into upflow and downflow sections. It was realized that catalyst concentration would be higher in the upflow sections due to settling. The only guide for estimating this settling was Stokes' Law but it was not thought to be strictly applicable; therefore, Professor W. K. Lewis at M.I.T. was asked to define the slippage over a wide range of conditions. The work at M.I.T. unexpectedly showed that a stable dense bed could be maintained at velocities far exceeding the Stokes' Law free fall velocities of the individual particles. For example, gas velocities of 1-3 ft/sec could be used, compared to the Stokes' free fall velocity of 0.1 ft/sec or less for the smallest particles in the powder mixture.

At about the same time the concept of a standpipe to build up pressure was conceived in the work at Exxon. A dense column of aerated solids would build up a fluostatic pressure, similar to a true liquid. It was pictured that aeration gas had to be added to offset the effects of compression, the amount of aeration being proportional to catalyst flow rate down the standpipe. Without the standpipe it would not be practical to circulate catalyst at the high rate needed to transfer all of the heat release in the regenerator over to the reactor.

The combination of the standpipe together with dense bed reaction zones was a "break-through" and the 100 B/D pilot plant was rebuilt again at top priority to demonstrate this fluidized solids system. At the same time, construction of three 12000 B/D commercial units was begun. The pilot plant worked perfectly from the start. It used upflow reactor and regenerator vessels in which all the catalyst fed into a vessel was removed from the top and passed to cyclones for recovery.

While the 100 B/D demonstration was proceeding the idea of bottom-drawoff or "downflow operation" was conceived; that is, most of the catalyst that entered a vessel was drawn off directly from the dense bed and fed to the other vessel. Only a small fraction of the circulating catalyst was carried overhead; therefore, a single stage of cyclones was adequate in place of the previous three stages. Also, any desired bed level could be selected and maintained since the catalyst holdup was no longer dependent on catalyst circulation rate and gas velocity. Again, the 100 B/D pilot plant was rebuilt to demonstrate this new design. At the same time, construction of five large (15,000 B/D each) commercial downflow plants was started before the first upflow plant ever operated.

Another important innovation was the injection of oil as a liquid rather than vapor (8). This was very desirable in order to make best use of the large amount of heat available from regeneration, but it required a high rate of catalyst circulation. There was considerable concern about possibly forming mud and rendering the system inoperable. When the first commercial upflow plant started up at the Baton Rouge refinery a critical test was made to see if liquid feed injection was operable. Fortunately, the test was completely successful.

Possible effect of scaleup on entrainment in the down-flow design was also of great concern since the theoretical basis was nebulous at best. Therefore when the upflow commercial plant at the Bayway NJ refinery was being started up, a 10 hour test period was allotted to measure entrainment at 1.5 ft/sec gas velocity with simulated downflow conditions. It confirmed the pilot plant results.

Many other contributions were essential to the success of the development, such as innovations in the design of expansion joints, insulation, slide valves, erosion protection, cyclones, electrostatic precipitators, instrumentation to control flow rates, pressures, temperature, etc. In addition, before the large scale plants were started up detailed studies were made of emergency procedures - that could go wrong, how to prevent it, proper startup and shutdown procedures, operator training, etc. In a very short time a large number of plants were constructed throughout the country, with capacities totalling 1M B/D by 1948. The outstanding success in these operations attests to the high level of competence and the extra efforts of all those involved in the extensive application of this new technology.

It may be useful to point out the following aspects of the Fluid Catalytic Cracking development that made it possible to develop such a radically new technology in such a short time:

- The country and the company needed the new process in view of the limitations of thermal cracking, the need for aviation gasoline and for raw materials to use in making synthetic rubber.
- Exxon encouraged innovation and invention and new ideas were welcomed and information was exchanged freely and promptly.
- Commercial application was expedited without waiting for complete knowledge or understanding of the new technology.

Successful application of the new technology required effective coordination of many areas, including research and development activites, as well as design, construction and startup of commercial plants. These activities were coordinated by a task force headed by E. J. Gohr, reporting directly to top management who gave full support and were receptive to innovations and willing to take risks. Ideas were exchanged freely - both up and down the line. In addition, there were major impacts on refinery operations to accommodate the new process and to redirect the operations to maximize production of aviation gasoline and synthetic rubber.

The 1938 Environment

During the 1930's, the petroleum industry was growing rapidly in response to the need for more motor gasoline of higher octane. Until about 1915 the main product had been kerosene for lighting and heating, but by 1938 the major market was to supply fuel for automobiles. As shown in Table 1 there was some industry experience on catalytic operations, including catalytic cracking, but the major refining process was thermal cracking which provided increased yield of olefinic gasoline having a high octane relative to that available from crude petroleum by simple distillation.

TABLE I

STATE OF THE ART - 1938

- Some catalytic cracking, experimental and commercial.
- Catalytic hydrocracking of residuum, plant built.
- Thermal cracking, gas oil or naphtha, 1000°F, 100 psi.
- Continuous fractionation, bubble caps.
- Winkler coal gasification.
- Clay treating, catalytic polymerization.
- Exxon expanding its small R&D effort.
- Chemical Engineering developing as special field.

Chemical Engineering was just emerging as a special field and continuous fractionation had been in general use for only about 15 years, distillation having been previously carried out in "shell stills" (heated drums). Professor W. K. Lewis of M.I.T. was a leader in the new field of Chemical Engineering and was a highly valued consultant to Exxon Research and Engineering (Standard Oil Development Co. at that time), where he effectively promoted applying chemical engineering principles to the petroleum industry. A photograph of Professor Lewis is shown in Figure 2. Exxon was vigorously expanding its research activities at that time, with emphasis on process developments.

Discussion

An overall summary of the development is given in Table 2 showing the timing of key elements, and the remarkably short time span of 3 years to commercial operation. The initial experiments on catalytic cracking used a fixed bed of catalyst in a cyclic operation, with periodic regeneration by burning off the carbon deposits in a gas of low oxygen content. This fixed bed type of system had been used by Houdry and associates in a small commercial plant for a few years by 1938 (6). However, it appeared to be expensive and relatively inefficient. It required valves that operated at high temperature, roughly 900°F or higher, to seal between oil vapors and the combustion gases. Heat released during regeneration was transferred to molten salt, but it was difficult to use the heat effectively to supply heat required for the cracking reaction.

TABLE II

EARLY DEVELOPMENTS IN FLUID CATALYTIC CRACKING

Year	Development
to 1938	Some catalytic operation, fixed bed cracking.
1939	100 B/D dilute phase unit, screw pump.
1939	Fluid Bed concept, Professor, W. K. Lewis.
1940	PECLA modified to upflow Fluid Beds.
1940	Design 3 large plants, 12,000 B/D each.
1941	Model II Bottom drawoff, PECLA.
1941	Design 5 Model II plants, 15,000 B/D each.
1942	May 1st, Startup first large plant at Baton Rouge.
1943	January, Startup second large plant, Bayway.
1943	June 21, Startup first Model II plant, Baton Route.

It appeared that a continuous process could be simpler and less costly, and that one might be developed for considerably less than the cost of licensing the fixed bed process. This is a good example of how the patent system can act to stimulate innovation. A large number of concepts for a continuous process were screened and evaluated, including the use of moving beds with bucket elevators, or having the catalyst in "box-cars" on wheels so that they could move between cracking and regeneration zones. Outside of Exxon most of the efforts on developing a continuous process were focused on moving bed systems, several of which were later commercialized. An early selection at Exxon decided on the use of powdered catalyst

Figure 2. Professor W. K. Lewis of M.I.T.

in a pneumatic transport system somewhat similar to technology used in the cement industry at the time. The catalyst was mixed with oil vapors and flowed through pipes forming the reactor. After being separated in cyclones and stripped with steam, the catalyst was regenerated by mixing it with air and flowing it through a regeneration zone. The pipes had to be so long that they were "folded" into upflow and downflow sections.

Since there was considerable pressure drop through the circuit it was essential to have a method for pumping the catalyst to a higher pressure. In the first $100^B/D$ powdered catalyst experimental cracking unit a screw pump was used to build up pressure, again similar to technology of the cement industry. The screw pump was limited to a low rate of circulation, and as might be expected, erosion on the screw pump was serious. A more serious problem soon showed up - the catalyst became agglomerated by the pumps compression action, to the size of poppyseeds, and soon lost most of its activity. Although the system was operable and did demonstrate the basic operations of cracking and regeneration, it left much to be desired. Fortunately, several other avenues were being studied actively at the same time. Indeed, the $100^B/D$ experimental unit was rebuilt to successfully demonstrate four very different process systems during a period of only three years.

Of the other avenues being explored one turned out to be a breakthrough. As mentioned, the initial design using finely divided catalyst had a "folded pipe" system with upflow and downflow pipes. It was realized that catalyst powder would tend to settle, resulting in a higher concentration in the upflow pipes. The magnitude of this effect could not be estimated from available theory since Stokes' Law applies only to a discrete single particle and not to a group or cluster of particles. The gas velocities used were orders of magnitude greater than the Stokes' Law settling velocity for individual particles. Following consultation with Professor W. K. Lewis in 1938 a bench scale program was started at M.I.T. directed by Professors Lewis and Gilliland to define the magnitude of catalyst slippage in upflow pipes over a wide range of gas and catalyst flow rates. It was found that a turbulent dense bed of fine powder could be maintained indefinitely under certain conditions provided catalyst was fed in at the proper rate. This result was not predictable from knowledge available at the time. Later it was realized that most of the gas passed through the bed in the form of bubbles, rather than having roughly uniform spacing between particles as might have been expected.

The type of turbulent dense bed used by Winkler was well known at the time; however, the Winkler type bed was predictable from Stokes' Law since it used only large particles of fairly uniform size such that the free fall velocity of each particle was more than the existing gas velocity. Consequently, the particles could not blow overhead, although they could be supported by the upflowing gas. Small particles would be rapidly blown overhead, and today this is still a limitation of the coarse bed system in that it cannot retain small particles in the bed for processing. On the contrary, in a bed of fluidized fine particles (at lower velocity) the entrainment rate is unexpectedly very low and it becomes quite practical to operate the bed at velocities far exceeding the Stokes' Law free fall velocity of individual particles. Moreover, the original Winkler technology has not been adapted to catalytic operations nor does it provide for solids circulation between two reaction zones. In fluid bed combustion the technology has been extended by using cyclones to recover the coarse particles.

A dense phase of fine particles deaerates at a slow rate, making it possible to use the "hydrostatic" type of pressure buildup in a standpipe to allow continuous circulation of powder at high rates between different zones in a process, almost as though it were a liquid. This was a very important advantage that allowed regenerating the catalyst frequently, while transferring all of the heat of regeneration over to the reactor where it was used effectively to permit feeding liquid oil instead of vapor. The combination of the standpipe with a fluidized bed was the key to the new technology. Rate of catalyst circulation in the early plants was about 60 tons/min, an order of magnitude larger than in the moving bed plants or through other previous solids handling technology.

It was pointed out earlier that a compression screw pump was used to provide catalyst circulation in the initial 100 $^{B}/D$ plant and that it caused serious operating problems. Fortunately the concept of the standpipe to build up pressure was conceived at this opportune moment and was soon put to use. It may seem surprising, but the standpipe concept was only accepted after considerable persuasion and discussion. Confirmation was needed, so a standpipe 100 feet high was set up on one of the refinery fractionation columns at Baton Rouge, filled with catalyst, and aerated. Pressure gauges confirmed the calculated buildup in pressure, and when a valve at the bottom was opened the catalyst ran out as though it were a liquid.

The essential elements of the new system were now assembled, namely, the standpipe to build up pressure, combined with a low-density riser to give an "air-lift" effect for circulation, together with a fluidized dense bed to provide ample catalyst holdup in vessels of reasonable size. This was the contribution that created a new type of Fluidized Solids Technique that resulted in such extensive application to catalytic cracking and other processes (9). Thus, Fluidized Solids Processing was originated as a result of a program whose objective was to develop a continuous process for catalytic cracking based on using finely divided powdered catalyst (3,4). The inventors of this process became known in Exxon circles as "the four horsemen" and are shown in the photograph of Figure 3. Figures 4 and 5 show the arrangement of equipment visualized at that time for carrying out catalytic cracking.

Figure 3. The "Four-Horsemen" of Exxon. Left to Right: H. Z. Martin, D. L. Campbell, C. W. Tyson, and E. V. Murphree.

Oct. 19, 1948. D. L. CAMPBELL ET AL 2,451,804

METHOD OF AND APPARATUS FOR CONTACTING SOLIDS AND GASES

Filed Dec. 27, 1940 2 Sheets-Sheet 2

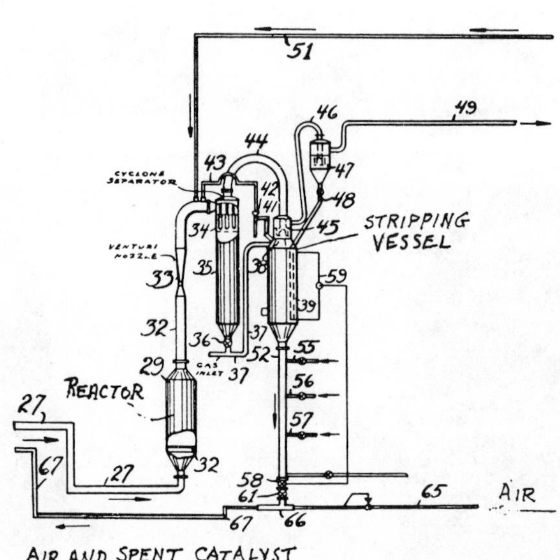

Donald L. Campbell
Homer Z. Martin
Eger V. Murphree Inventors
Charles W. Tyson
By P. L. Young Attorney

Figure 4. Design of Upflow Reactor Section.

Oct. 19, 1948. D. L. CAMPBELL ET AL **2,451,804**

METHOD OF AND APPARATUS FOR CONTACTING SOLIDS AND GASES

Filed Dec. 27, 1940 2 Sheets–Sheet 1

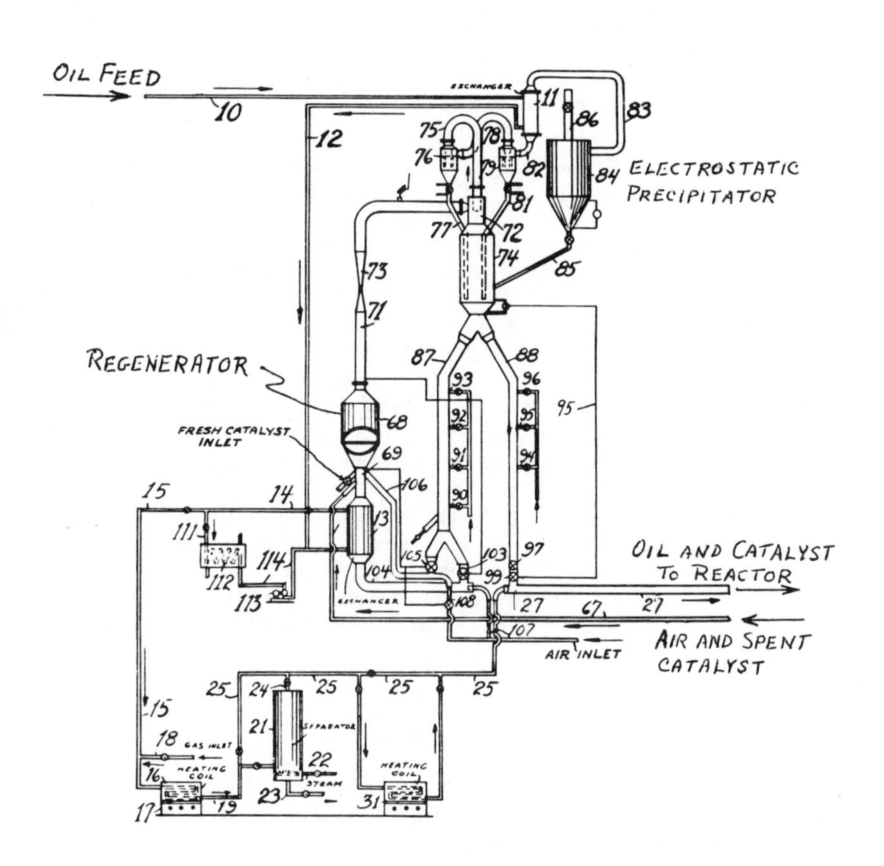

Donald L. Campbell
Homer Z. Martin
Eger V. Murphree
Charles W. Tyson
Inventors

By P. L. Young Attorney

Figure 5. Design of Upflow Regeneration Section.

To demonstrate this new concept the 100 B/D pilot plant was completely rebuilt for the second time. The demonstration was successful in 1940, meeting all expectations. Construction of three commercial plants was underway each with a capacity of 12,000 B/D. The first of these was put on-stream in May 1942, in a remarkably short time from the initial concept of about 3 years. This was by far the largest construction project carried out in the petroleum industry up to that time. These first plants started up easily and with only minor problems which were mostly mechanical, such as the expected erosion, and problems with expansion joints of which there were about two dozen. All problems were soon overcome, extending run lengths to a year or more.

In the first commercial plants the reactor and regenerator were operated upflow, that is, all the circulating catalyst was entrained and taken out from the top of the vessel. Catalyst was circulated at a high rate to maintain a dense bed and all of this catalyst passed through cyclones for recovery. Developments were moving very fast during this time period. In retrospect it is amazing how great was the flexibility and receptivity to innovations on such a large project undergoing initial commercial application.

Early pilot plant results were analyzed to relate entrainment/circulation rate to gas velocity and catalyst holdup in the vessels. Such an analysis suggested that it was not necessary to take all the circulating catalyst overhead and through the cyclones. Thus, by purposely operating with a greater height of dilute phase above the bed it should be possible to draw circulating catalyst directly from the dense bed so that the circulating catalyst bypassed all cyclones and thereby unloaded the catalyst recovery system (10). Moreover, it became possible for the first time to control inventory of catalyst in the vessels independent of catalyst circulation rate or gas velocity. This was known as the "downflow" or Model 2 design. The original concept of this system is shown in Figure 6.

The 100 B/D plant was again rebuilt to demonstrate the downflow operation in 1941, and simultaneously five more large scale plants of this type were under construction by Exxon, the first of which started operating in mid 1943. Figure 7 shows the Fluid Cracking units at the Baton Rouge refinery.

In a period of less than five years a radical new technology was developed from the initial concept to the point of having eight large plants in operation. Before the first upflow plant was ever operated, five plants of the downflow design were constructed, giving a combined capacity of over 100,000 B/D. Included in the development was a 100 B/D pilot plant demonstration of three entirely different systems for using powdered catalyst. Figure 1 illustrates the extensive application of Fluid Catalytic Cracking since its inception. Although conditions were extraordinary for this development, the implications regarding today's R&D problems are apparent: innovations need to be developed and applied fast in order to generate favorable economics.

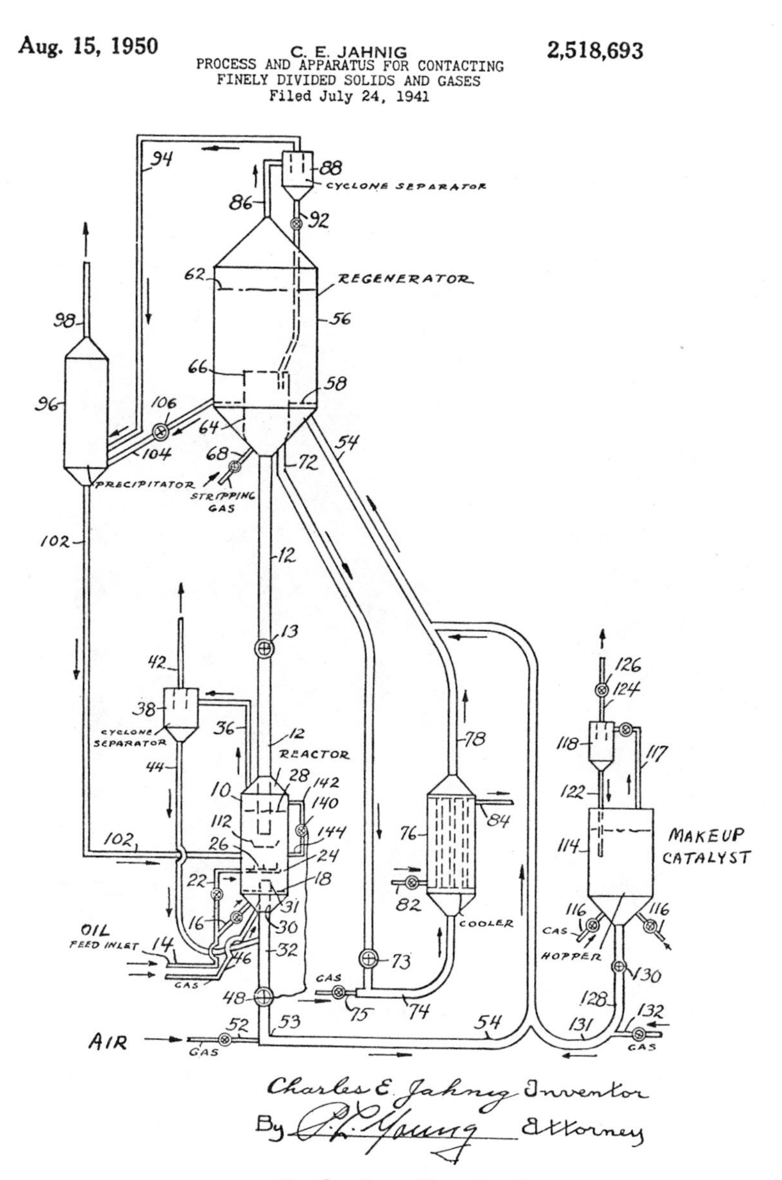

Figure 6. Early Downflow Design.

Figure 7. The First Commercial Fluid Catalytic Cracking Plants at Baton Rouge, LA. PCLA 1 at left and at right are two of downflow designs.

This discussion has focused on the basic developments leading to successful commercialization of the Fluidized Solids Technique; however, many areas were important to success of the project and were pursued vigorously at the same time. Some of these are listed in Table 3. For example, the availability of special equipment had to be assured, including cyclones, slide valves and expansion joints. Also, metals and refractories had to be tested and methods of fabrication developed. A large supply of catalyst was needed. Fortunately, the natural clay type catalyst used initially was readily available, having been used for clay treating of lubricating oils, etc. It was soon found that a synthetic silica-alumina catalyst was much better, with the result that a whole new industry was started to supply it. Twenty years later the silica alumina catalyst was displaced by the more active zeolites.

TABLE III

COMPANION DEVELOPMENTS

Design Methods	Metals
Scale-up	Insulation
Cyclones	Instrumentation
Electrostatic ppt.	Analytical
Slide Valves	Slurry System
Expansion Joints	Afterburning
Exchangers	Synthetic catalyst
	Patents and Licensing

Catalytic cracking to make high-octane gasoline was the first application of the new Fluidized Solids technique, and provided the impetus for commercial development. Since then, a great many other applications have been explored and many of them carried through to commercial use. Table 4 lists some of these. Fluid coking, as one example, fluidizes particles that are coated with a sticky liquid.

A number of new applications look very promising, such as the use of Fluid Bed Combustion of coal in a bed of limestone or dolomite to control emissions of SO_x and NO_x. In the area of synthetic fuels, the technology has led to several processes for gasification to make fuel gas or synthesis gas, and in South Africa it is used to make synthetic hydrocarbons from CO plus H_2. Interesting modifications of the technology include using two different solids of different sizes that can be separated for separate processing, as well as controlled agglomeration in a fluid bed to make large particles, e.g., from coal ash. A recent innovation is the magnetically stabilized fluid bed which makes it possible to operate a reactor in a manner similar to a moving bed of pellets, while operating the rest of the system in a conventional fluidized manner (11).

TABLE IV

OTHER APPLICATIONS OF FLUIDIZED SOLIDS TECHNIQUE

1940's	Fluid Hydroforming Fluid Coking Hydrocarbon Synthesis Coal Gasification Shale Retorting
1950's	Commercial Fluid Coking and Hydroforming
1960's	Iron Ore Reduction Hydrogen Manufacture Coke Calcining
1970's	Flexicoking CO Boiler, CO Combustion Catalyst Expander on Flue Gas Zeolite Catalysts Flue Gas Scrubbing of SO_x and particulates Pressurized Fluid Bed Combustion of Coal Magnetically Stabilized Fluid Bed

Literature Cited

1. Jahnig, C. E., Campbell, D. L. and Martin, H. Z., History of Fluidized Solids Development at Exxon in "Fluidization" (1980) edited by T. R. Grace and J. M. Matsen, Plenum Pub. Co., NY 10013
2. Lewis, W. K., and Gilliland, E. R., US Patent 2,498,088 "Conversion of Hydrocarbons with Suspended Catalyst" Filed January 3, 1940, issued February 21, 1950.
3. Campbell, D. L., Martin, H. Z. and Tyson, C. W., US Patent 2,451,803 "Method of Contacting Solids and Gases", Filed October 5, 1940, issued October 19, 1948.
4. Campbell, D. L., Martin, H. Z., Tyson, C. W. and Murphree, E. V., US Patent 2,451,804, "Method and Apparatus for Contacting Gases and Solids" Filed October 5, 1940, issued October 19, 1948.
5. Jahnig, C. E., US Patent 2,518,693, "Process and Apparatus for Contacting Finely Divided Solids and Gases", Filed July 24, 1941, issued August 15, 1950.
6. Enos, J. L. "Petroleum Progress and Profits", M.I.T. Press 1962.
7. Shankland, R. V., "Industrial Catalytic Cracking" in Advances in Catalysis Vol. VI (1954) p. 272-466, Academic Press.
8. Belchetz, A., US Patent 2,253,486, "Catalytic Conversion of Hydrocarbons" Filed May 20, 1939, issued August 19, 1941.
9. Campbell, D. L., Martin, H. Z., and Tyson, C. W., US Patent 2,735,822, "Method and Apparatus for Contacting Solids and Gases", Filed October 5, 1940, issued February 21, 1956.
10. Murphree, E.V., et al. AIChE Transactions 41, 19 (1945), p. 19-33, "Improved Fluid Process for Catalytic Cracking".
11. Lucchesi, P. J., et al, 10th World Petrol Congress Bucharest Rumania, September 1979, "Magnetically Stabilized Beds - New Gas Solids Contacting Technology".

RECEIVED October 29, 1982

24

The Development of Hydrocracking

RICHARD F. SULLIVAN and JOHN W. SCOTT
Chevron Research Company, Richmond, CA 94802

In 1959 Chevron Research announced the "world's first commercially proven low temperature hydrocracking process" called Isocracking. Among its attractive features was the ability to efficiently crack aromatics and to produce paraffins that were far on the "iso" side of thermodynamic equilibrium. Although hydrocracking had been practiced previously at high temperatures and pressures, the invention of superior catalysts permitted operation at moderate temperatures (200-400°C) and lower pressures (35-140 atm.). Isocracking was developed as a response to major changes in the domestic petroleum market during the 1950's. The trend toward automobile engines with high compression ratios resulted in an increased demand for high octane gasoline. The shift by the railroads from steam to diesel locomotives caused corresponding downward shifts in fuel oil demand. The overall consequence was a need to convert excess refractory cutter stocks to high octane gasoline, and Isocracking addressed this problem. In the intervening years, further catalyst improvements made both at Chevron and in the petroleum industry generally have extended the range of modern hydrocracking feedstocks to high boiling distillates and residua; and hydrocracking has become one of the most useful and flexible refining processes. Catalysts can be tailored to fit specific needs of a refiner by careful control of

0097-6156/83/0222-0293$06.00/0

> their chemical and physical properties. Depending upon the catalyst and processing conditions, the major product can be liquefied petroleum gas (LPG), gasoline, jet fuel, middle distillates, lubricants, petrochemicals, or a combination of products.

In a broad sense, hydrocracking can be defined as any cracking of molecules in the presence of hydrogen, whether it takes place in the presence or absence of a catalyst. Because some cracking occurs in many hydrorefining processes, the Oil and Gas Journal arbitrarily defines "hydrocracking" as a conversion process utilizing hydrogen in which at least 50% of the reactant molecules are reduced in molecular size (1). This rough definition is suitable for the purposes of this paper.

Most modern hydrocracking processes are catalytic, and the catalyst employed is usually dual functional with both a hydrogenation component and an acidic component. Typical acidic components include amorphous silica-alumina, alumina, and a large family of zeolites. Typical hydrogenation components are noble metals such as palladium and platinum and nonnoble metals such as nickel, cobalt, tungsten, and molybdenum. The latter metals are usually in sulfided form.

Modern catalytic hydrocracking is probably the most versatile and certainly one of the most important conversion processes in modern refining technology. Research in the 1950's led to the large commercial development of hydrocracking in the 1960's, and modern commercial hydrocracking processes are continuing to evolve. However, hydrocracking in an earlier form is one of the oldest hydrocarbon conversion processes. It was the first catalytic cracking process to attain appreciable commercial importance. An extensive hydrocracking technology for coal conversion was built up in Germany between 1915 and 1945 (2-5). The driving force that led to this productive effort was strategic rather than economic. Germany needed a secure supply of liquid fuels derived from a domestic energy source; namely, coal. In 1943, 12 plants were operating which provided Germany with 98% of the aviation gasoline and 47% of the total hydrocarbon products consumed in Germany during the latter years of World War II (6). Similar, though less extensive, efforts took place in Great Britain, France, Manchuria, and Korea (6, 7, 8). A parallel development in the United States was directed toward the conversion of heavier petroleum fractions (9, 10). In general, the conversion of coal was accomplished in two or three separate catalytic steps. Reaction conditions were typically 200-700 atm. (3,000-10,000 psig) and 375-525°C (700-975°F). Although somewhat less severe conditions were appropriate for petroleum hydrocracking, design pressures were 200-300 atm. (3,000-4,500 psig); and temperatures generally exceeded 375°C (700°F).

Both technical and economic changes reduced the importance of hydrocracking after the end of World War II. The general availability of Middle Eastern crude oils removed the incentive for conversion of coal to liquid fuels. New catalytic cracking processes, which subtract carbon rather than add hydrogen, proved more economic for converting heavy petroleum gas oils to gasoline. Construction of new hydrocracking plants stopped. Although a few of the existing plants were adapted to petroleum hydrocracking (11, 12), most were shut down or converted to other service (13, 14). A modest coal conversion industry was continued in East Germany, Czechoslovakia, and the U.S.S.R., countries in which the government controlled the industry and no competitive market existed (6).

The field lay dormant until 1959 when Chevron Research Company, then known as California Research Corporation, announced that a new hydrocracking process, "Isocracking," was in commercial operation in the Richmond Refinery of the Standard Oil Company of California (15). The following year, the Universal Oil Products Company announced a hydrocracking process called "Lomax" (16) and Union Oil Company announced the "Unicracking" process (17). Publications in the early 1960's showed that most of the other major petroleum companies also had a significant research effort in hydrocracking. The rapid acceptance of hydrocracking in the 1960's as a major refining process indicated the timeliness of the development. By 1966, seven different hydrocracking processes were offered for license (18).

In this paper, we will review the considerations that led to the development of modern hydrocracking and some aspects of the chemistry of hydrocracking. Also, we will briefly discuss some of the advances in hydrocracking during the 23 years since the Isocracking process was announced.

Need of a New Conversion Process

A series of related events in the late 1940's and early 1950's led to the need for a new hydrocarbon conversion process. Reduced overseas shipments as a result of the end of

World War II, the conversion of railroads from steam to diesel power, and the availability of a large supply of cheap natural gas all contributed to reduced fuel oil demand. The rapid growth of catalytic cracking led to an excess of refractory cycle stocks. The trend in the automobile industry was to make higher compression ratio, high performance cars with high octane requirements. Relatively inexpensive gasoline was available and there was little emphasis on fuel economy. There was a pressing need to convert refractory stocks to gasoline.

A goal-oriented research program was started by Chevron in 1952 with the exploratory research of Scott et al. (19). The object was to find a processing route to convert excess gas oils to high octane gasoline components. The papers of M. Pier describing the German hydrocracking process had given an indication that hydrocracking had the potential to fill this need (3, 4, 11). It was clear, however, that both selectivity and product character needed to be modified in order to satisfy gasoline quality requirements economically.

New pilot plant capability and multiple screening units permitted a wide variety of test conditions, and direct comparisons of catalysts could be made rapidly. What was needed was a catalyst that could: (1) produce paraffins on the iso side of equilibrium, (2) crack aromatics and cycloparaffins without loss of ring structure, (3) control demethanation reactions, (4) minimize hydrogen consumption, (5) operate at lower pressures than the earlier hydrocracking processes, and (6) operate with a variety of feedstocks. Furthermore, such a process should be sufficiently flexible that product distributions could be changed as product demand shifted. For example, when commercial jet aircrafts were introduced in the 1950's, low freeze point kerosene jet fuel became another important petroleum product.

Branched paraffins (collectively referred to as "isoparaffins") generally have high octane numbers; normal paraffins have low octane numbers. Normal paraffins of carbon numbers of seven or fewer are particularly hard to reform to aromatics and relatively hard to isomerize. Therefore, the target selectivity was a maximum production of isoparaffins and a minimum production of normal paraffins, particularly in the lower carbon number range (C_4-C_7).

The thermodynamic equilibrium for isoparaffins compared to normal paraffins improves as the temperature decreases; however, a relatively large amount of normal paraffins are present at equilibrium even at low temperatures. Therefore, the desired isoparaffin to normal paraffin selectivity was on the "iso" side of equilibrium.

Aromatic compounds generally have very high octane numbers. With the introduction of catalytic reforming in the 1950's, paraffins and cycloparaffins could be catalytically reformed to high octane aromatic components. However, dehydrocyclization of paraffins is much harder to accomplish than dehydrogenation of ring-containing compounds and is accompanied by the unwanted side reaction of cracking to light gases. Therefore, a maximum conservation of cyclic structures was a target selectivity.

Liquid products are usually more valuable than methane and ethane. At the high temperatures of the earlier hydrocracking processes, a considerable amount of demethanation took place. Therefore, the goal was to find catalysts that operated at low temperatures where demethanation reactions could be avoided.

Then, as now, hydrogen was relatively expensive. By minimizing aromatics saturation and cracking to light gases, hydrogen consumption could be minimized.

The early high pressure hydrocracking processes were very costly. For a hydrocracking process to be cost effective, lower pressures were necessary.

Hydrocracking Reactions

As promising new catalysts were developed, a research program was designed to study the chemistry of hydrocracking by testing pure compounds and simple mixtures to determine the mechanism of hydrocracking reactions. The introduction of gas chromotography in the middle 1950's, in combination with mass spectrometry, provided a powerful new analytical tool for identifying individual compounds in complex mixtures of product hydrocarbons.

A fortunate combination of theoretical considerations and experimental circumstances directed attention to some unusual reaction paths. This led to detailed studies of reactions of typical hydrocarbon classes. New and highly specific nonequilibrium reactions were identified. Techniques and catalysts were discovered which permitted desirable reactions of the individual hydrocarbon classes to dominate the conversion of mixtures. These studies became the basis for the commercial hydrocracking process called Isocracking and the catalysts developed for this service. The technical objectives of a low pressure, low temperature process were achieved as demonstrated in many long pilot plant tests prior to the first commercial test.

The chemistry of hydrocracking is discussed in detail in other review papers (20 21). In this paper, we will briefly discuss some of the reactions of the unusual reactions of alkylbenzenes, alkylcyclohexanes, and paraffins.

Aromatics. One of the earliest reactions to gain attention in screening tests was an unexpected type of hydrodealkylation observed when hydrocracking polysubstituted alkylbenzenes. This reaction produced lower boiling aromatics as desired. The concentration of aromatics in the product was, however, higher than would be explained by any known reaction mechanism, and the missing methyl substituents appeared in the product mainly as isoparaffins rather than as methane. The importance of alkylbenzene disproportionation to the course of this reaction was readily apparent; but the fate of one important intermediate, hexamethylbenzene, was obscure. Our research (22) showed hexamethylbenzene cracked over such simple catalysts as NiS-silica-alumina to give, mainly, lower boiling aromatics, isobutane, and isopentane. Figure 1 illustrates the unusual product distribution observed. This reaction, in its apparent effect, peels or pares methyl groups from the ring and, therefore, was named the paring reaction. A reaction mechanism was proposed which involves repeated contraction and expansion of aromatic rings adsorbed on acid sites on the catalytic surface. This probably proceeds by way of an isomerization between an aromatic C_6 ring and a relatively stable cyclopentadienyl cationic intermediate. Isomerization proceeds until a branched side chain is formed that can crack off to form an isoparrafin. The remainder of the molecule desorbs as a lower molecular weight aromatic. The plausibility of the cyclopentadienyl cations postulated as intermediates is strongly supported by the work on DeVries (23) and Winstein and co-workers (24, 25). These workers propose that such intermediates have a nonclassical form.

Further research showed that the paring reaction occurs on silica-alumina in the absence of hydrogen or the hydrogenation component. Cycloparaffins are not formed at the reaction conditions; therefore, they are not essential intermediates in the reaction. Under these conditions, the silica-alumina is deactivated rapidly, and the observed reactions rates are much lower. An important function of the metal sulfide and hydrogen

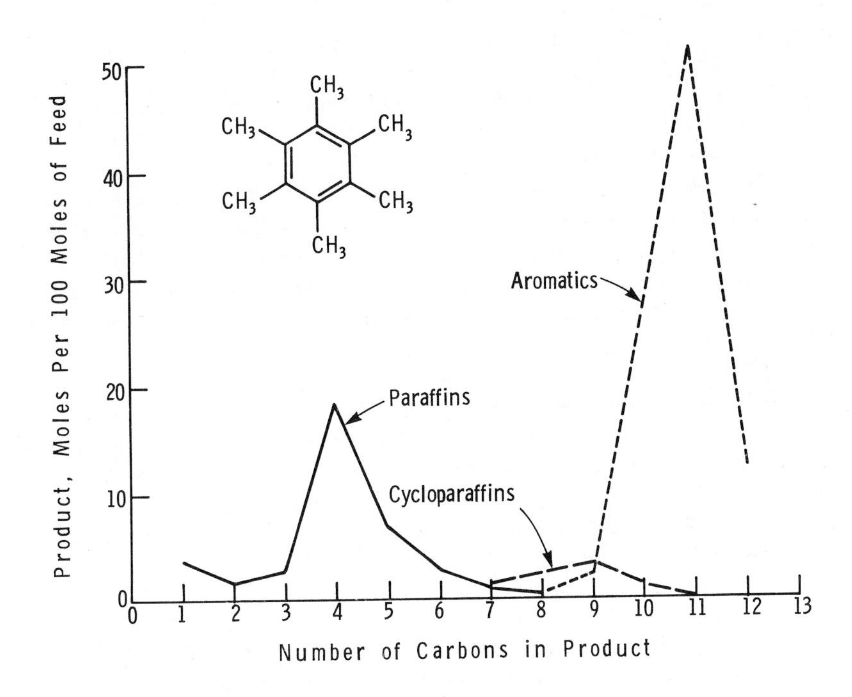

Figure 1 - Hydrocracking of hexamethylbenzene at 349°C and 14 atm.

is to maintain catalyst activity by preventing buildup of carbonaceous deposits on the catalyst.

Cycloparaffins. A similar, even more rapid paring reaction occurs with cycloparaffins (26). For example, Figure 2 shows that hexamethylcyclohexane reacts to form isobutane and a mixture of C_8 cycloparaffins (mainly cyclopentanes) as the most important products. Similarly, diisopropylcyclohexane reacts to form isobutane and C_8 cycloparaffins instead of forming (as one might expect) large quantities of propane. As with aromatics, essentially all of the ring structures are preserved in the paring reaction of cycloparaffins.

The cycloparaffins tend to form isobutane and a cycloparaffin of four carbon numbers lower than the reactant molecule. Therefore, the dominant products from C_{10} cycloparaffins are isobutane and methylcyclopentane. A sequence of reactions to produce these compounds from tetramethylcyclohexane is given in Figure 3. The product distributions from different cycloparaffins of any given carbon number are very similar to each other--a strong indication that similar intermediates are involved in each case.

Paraffins. The reactions of paraffins, while somewhat less unexpected than the paring reaction of cyclic compounds, gave us an important key as to how to tailor catalysts to fit specific refining needs and to yield different product slates. We found a profound difference between the behavior of normal paraffins in hydrocracking with a catalyst containing a strong hydrogenation component such as nickel metal or a noble metal and a relatively weak hydrogenation component such as nickel sulfide.

Mechanisms of hydrocracking of paraffins have been studied extensively (27-33). A carbonium ion mechanism is usually proposed similar to the mechanisms previously proposed for catalytic cracking except that hydrogenation and hydroisomerization are superimposed. The paraffins are first dehydrogenated to an olefin, then are adsorbed as a cation on an acidic site, isomerized to the preferred tertiary configuration, and undergoes beta scission. Virtually no methane and ethane are formed. The reaction becomes more selective for isoparaffin production as the temperature is decreased.

Figure 4 shows the iso-to-normal ratio of the combined pentanes and hexanes for the reaction of pure normal decane over catalysts with a strongly acidic component and illustrates the advantage of operating at low temperatures for maximum

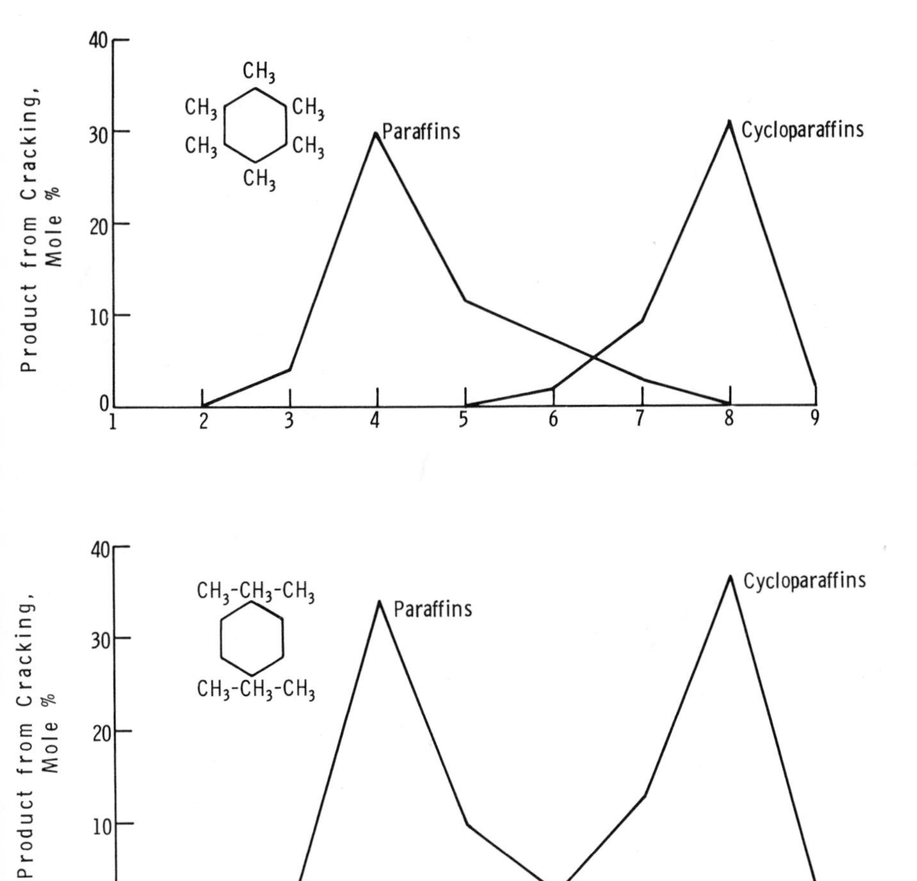

Figure 2 - Hydrocracking of C_{12} cyclohexanes at 233°C and 82 atm.

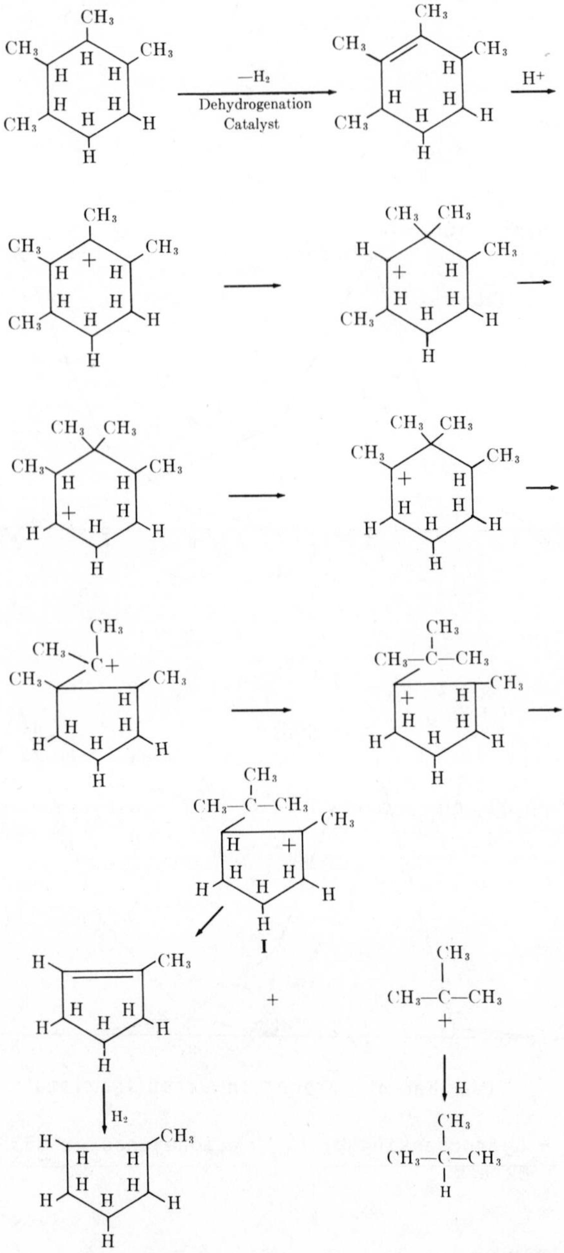

$-H_2$
Dehydrogenation
Catalyst

H^+

I

Figure 3 - Proposed mechanism for the hydrocracking of tetramethylcyclohexane.

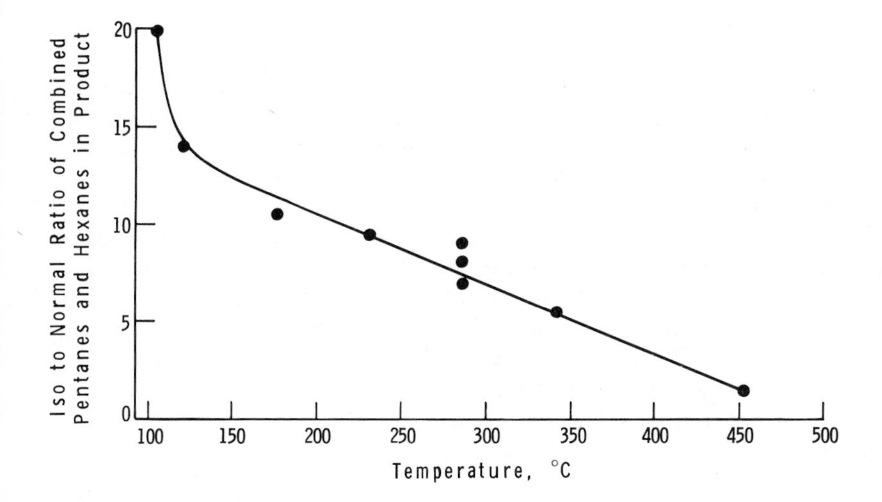

Figure 4 - Isoparaffin to n-paraffin ratios in the product from hydrocracking of n-decane using strongly acidic catalysts.

isoparaffin yields. The high iso-to-normal ratios in the product gave Isocracking its name.

By increasing the hydrogenation activity of the catalyst relative to its acidity, the product distribution from normal paraffin reactants can be dramatically changed. For example, for the hydrocracking of normal decane at comparable conditions, the following iso-to-normal ratios for the combined pentanes and hexanes were obtained with three different metal compositions on a single, acidic silica-alumina support:

	Iso/Normal
Nickel Metal	0.1
Platinum	1
Nickel Sulfide	8

The catalysts with the high hydrogenation activity give product with low iso-to-normal ratios. Apparently, olefinic intermediates are quenched and less hydroisomerization occurs. With nickel metal, appreciable hydrogenolysis occurs without isomerization.

Figure 5 compares the reactions of n-hexadecane on a catalyst with strong hydrogenation activity (platinum on silica-alumina) to one with strong acidity and weaker hydrogenation activity (nickel sulfide on the same silica-alumina support). Isoparaffins of carbon numbers of four, five, and six are the preferred products with the strongly acidic catalyst. The catalyst with the stronger hydrogenation activity makes much less isobutane, and the product distribution is spread more evenly over a wide molecular weight range. Therefore, the C_5+ liquid yield is higher for the catalyst with strong hydrogenation activity, although iso-to-normal ratios are much lower.

Commercial Feeds. The same effect shown for pure compounds is here illustrated in Figure 6 with a commercial feed, a California vacuum gas oil. All of the feed boiling above the recycle cut point of 288°C (550°F) was recycled to extinction. The catalyst with the higher hydrogenation-to-acidity ratio gives the higher liquid yield. The increased liquid yield is largely due to more paraffins in the higher boiling product and less isobutane formation (34).

Figure 7 shows results for the same feed with a variety of catalysts with varying hydrogenation-to-acidity ratio. The C_5+

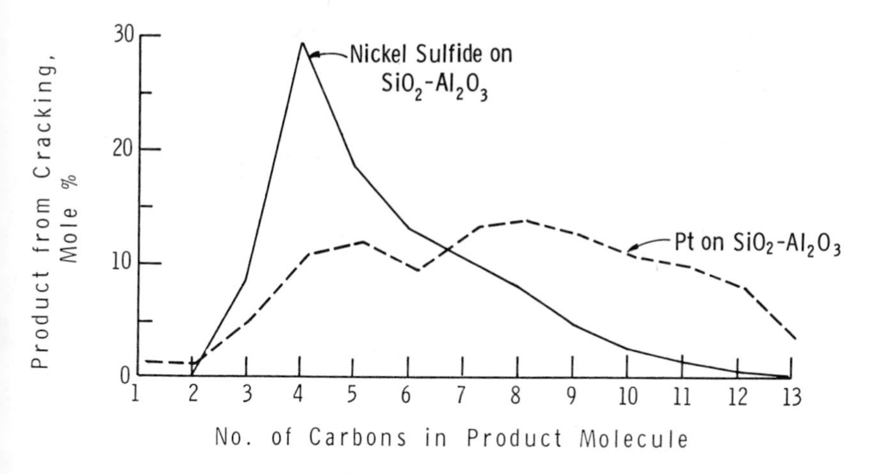

Figure 5 - Products from hydrocracking of n-hexadecane with two different catalysts.

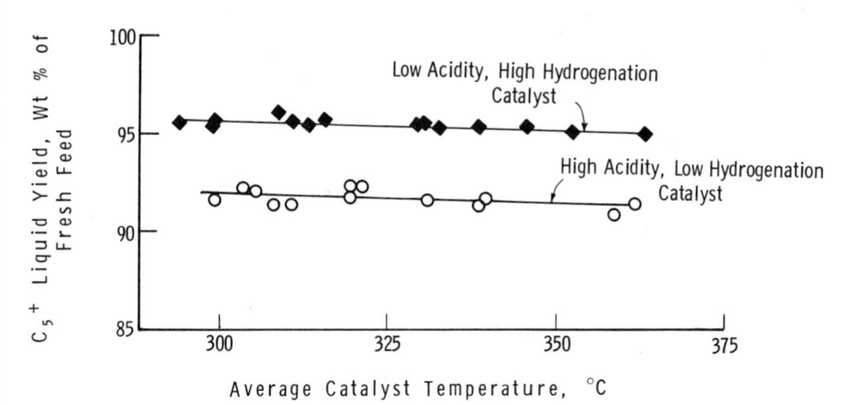

Figure 6 - Effect of temperature on C_5^+ liquid yield in the hydrocracking of California gas oil.

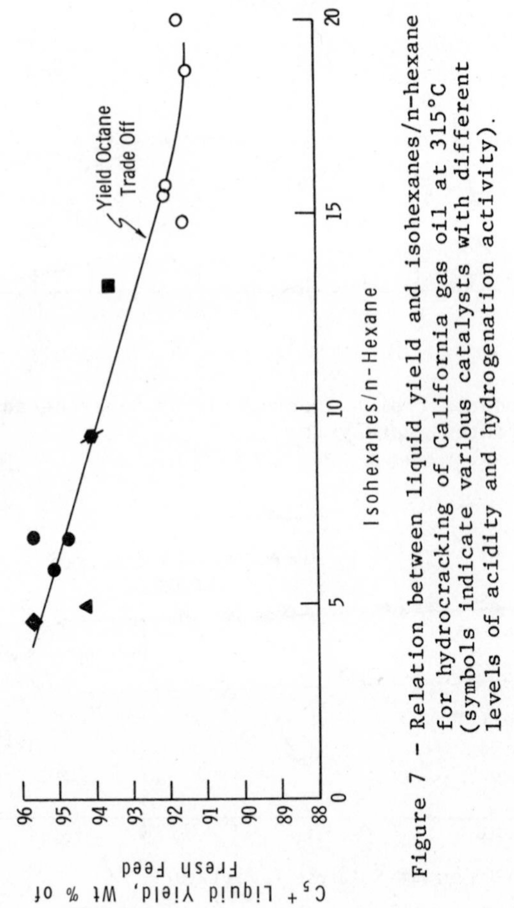

Figure 7 - Relation between liquid yield and isohexanes/n-hexane for hydrocracking of California gas oil at 315°C (symbols indicate various catalysts with different levels of acidity and hydrogenation activity).

yield varies as the ratio changes, and this can be correlated with the iso-to-normal hexane ratio in the product. As the iso-to-normal ratio increases, the octane number of the naphtha product increases. This is particularly important for the light naphtha (C_5-C_6) which is an important light gasoline blending component.

In the previous paragraphs, we have shown how to change product distribution and iso-to-normal ratios by altering the catalyst hydrogenation-to-acidity ratio. Another way to do this is by preferentially poisoning one or the other catalytic site. Nitrogen (as ammonia) is a typical poison for an acid site; sulfur (as hydrogen sulfide) is a typical poison for a metal hydrogenation site. Table I shows that the same effects just discussed can be achieved in the presence of heteroatoms.

Table I

Product from Pd on SiO_2-Al_2O_3 with California Gas Oil - Effects of Sulfur and Nitrogen

	Isohexanes/ n-Hexane	C_5+ Yield, Wt %	138-288°C (280-550°F) Yield, LV %	C_5-82°C (C_5-180°F) Octane, F-1 Clear
Before Addition of Sulfur or Nitrogen	5	95	65	80
8 ppm Nitrogen Added to Feed (Catalyst Equilibrated)	3	96	71	78
100 ppm Sulfur Added to Feed (After 170 Volumes of Feed Containing Sulfur)	13	93	58	84

With these principles, we found it possible to make specific catalysts to achieve desired product distributions. In

1959, at the time Isocracking was introduced, the strong acid catalysts were the preferred hydrocracking catalyst to produce high octane gasoline. For maximum production of kerosene jet fuel, catalysts with more hydrogenation activity could be employed.

The first Chevron Isocrackers operated as two-stage processes. The first stage involved hydrotreating to remove heteroatoms; the cracking reactions occurred in the second stage over catalysts with strong acidity and moderate hydrogenation activity.

The 1960's - Growth of Hydrocracking

The 1960's were years of rapid growth of hydrocracking. By the end of the decade, nine different processes were operating commercially or had plants under construction. New catalysts were developed, both more active and more stable than the earlier catalysts. Of particular note, molecular sieve catalysts were introduced (35, 36). Because of their high surface area and large number of catalytic sites, they were more tolerant of heteroatom impurities, such as nitrogen, than the previous catalysts. In general, the reaction mechanisms for sieve catalysts were believed to be similar to those for amorphous catalysts. However, in some specialized cases, the shape selective properties of zeolites could be used to permit limited access to the catalyst sites, thus allowing certain molecules to react while excluding others. (The more recent catalytic hydrodewaxing developments are an example of such an application. However, because typically fewer than 50% of the molecules are cracked, they fall outside of our working definition of hydrocracking.)

Another development of the 1960's was that of stable, large-pored catalysts which could crack very heavy feeds (37). For example, in Richmond, Chevron has had a hydrocracker processing deasphalted oil (DAO) since the middle 1960's. The H-oil process was commercialized by Hydrocarbon Research, Inc., also in the 1960's as a residuum hydrocracking process (38).

The versatility of hydrocracking was demonstrated in the 1960's as the demands for a variety of products increased. In addition to gasoline and jet fuel, the product range included diesel fuel, lubricating oils, low sulfur fuel oils, LPG, and chemicals.

A wide variety of flow schemes, both single-stage and two-stage, were practiced commercially. In particular, single-stage processes were used advantageously to produce middle distillates (37). The tolerance of sieve containing catalysts for hetoatoms permitted operation of two-catalyst systems in series without intermediate removal of ammonia and hydrogen sulfide in the Unicracking process (39). Most fixed bed processes operated in a downflow configuration. In contrast, the H-Oil process employed an ebullated catalyst bed and operated in an upflow mode (38).

The 1970's

Figure 8 shows the rapid growth of hydrocracking in the United States during the 1960's and early 1970's. However, in the later 1970's, the rate of growth of the hydrocracking was more moderate. Among the reasons for this leveling off were major improvements in catalytic cracking due to the widespread use of zeolite catalysts. The high cost of hydrogen generally made hydrocracking a more expensive process than catalytic cracking for gasoline production.

By the 1970's, hydrocracking was a mature process. Although there was limited growth of hydrocracking itself, there was a large growth in related hydrotreating processes such as hydrodesulfurization.

The 1970's were years in which additional use of new catalysts permitted better utilization of existing facilities. The trends toward heavier feeds continued.

Development of superior analytical tools permitted more detailed studies of the mechanisms of hydrocracking reactions. In particular, the work of Weitkamp and coworkers should be noted (40, 41).

The Future

At present, the United States' hydrocracking capacity is over 900,000 barrels per stream day (BPSD); worldwide capacity is approaching 1.5 million BPSD (42).

Despite all of the uncertainties of the present economic climate, some trends can be predicted for the 1980's. Because of their versatility, hydrocrackers offer the refiner the ability to meet these changing demands. In the early 1980's, the demand for gasoline decreased due to more energy conservation measures, smaller and more efficient automobile engines, higher prices, and reduced economic growth. It is generally believed that the demand for gasoline will continue to decrease. However, we expect an increased demand for middle distillates. Hydrocracking is a particularly effective route for production

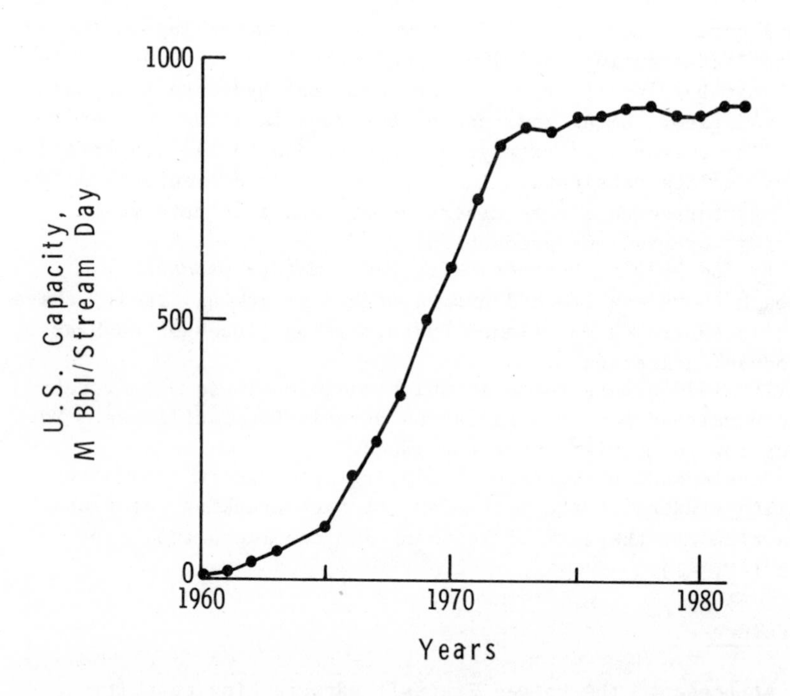

Figure 8 - Growth of hydrocracking in the United States.

of incremental jet and diesel fuels. With modern hydrocracking catalysts, yields of at least 95 LV % of specification diesel fuels (based on feed) can be obtained from heavy gas oils by hydrocracking.

The demand for fuel oil is expected to decrease. As it decreases, refiners will consider using hydrotreaters that were originally built to make low sulfur fuel oil for possible conversion to hydrocracking units.

Bottom of the barrel conversion continues to be a top refining priority. Hydrocracking is expected to find an important place in residuum conversion technology. For example, the German Veba processes are modern versions of the early German hydrocracking processes adapted to residuum (43).

Because some of the pressures on world petroleum supplies have been relieved, it is now expected that, at most, synthetic fuels will have only a minor impact in the 1980's. However, at least some synthetic crudes from oil shale are expected to be available in the latter part of the decade. Hydrocracking remains the logical choice for conversion of shale oil to jet fuel (44). Similarly, when coal liquids become available, they too will be likely candidates for hydrocracking (45). We are continuing our research on the upgrading of synthetic crudes by hydrotreating and hydrocracking.

In conclusion, we expect hydrocracking to play an important role in the refinery of the future.

Literature Cited

1. Oil and Gas Journal, 1982, 80 (12), 128-154.
2. Bergius, F. Proceedings of World Petroleum Congress, London, 1933, 2, 282-289.
3. Pier, M. Proceedings of World Petroleum Congress, London, 1933, 2, 290-294.
4. Ministry of Fuel and Power, "Report on the Petroleum and Synthetic Oil Industry of Germany," B.I.O.S. Overall Report No. 1., Section C, Hydrogenation Processes, 1947, 46-73.
5. Pier, M. Zeitschrift fur Elektrochemie, 1949, 53, (5), 291-301.
6. Wu, W. R. K.; Storch, H. H., Bureau of Mines Bulletin 633, "Hydrogenation of Coal and Tar," United States Department of Interior, Washington D.C., 1968, 3-10, 193.
7. Gordon, K. Journal of the Institute of Fuel, 1935, 9, (44), 69-89.
8. Gordon, K. Journal of the Institute of Fuel, 1946, 20, (110), 42-58.

9. Haslam, R. T.; Russell, R. P. Industrial and Engineering Chemistry, 1930, 22, (10), 1030-1037.
10. Murphree, E. V.; Brown, C. L.; Gohr, E. J. Industrial and Engineering Chemistry, 1940, 32, (9), 1203-1212.
11. Pier, M.; von Funer, W.; Horing, M.; Nonnnmacher, H.; Oettinger, W.; Reitz, O. Proceedings of Third World Petroleum Congress, The Hague, 1951, Section IV, 81-90.
12. Pier M. Proceedings of Fourth World Petroleum Congress, Rome, 1955, Section III, 517-530.
13. Clough, H. Industrial Engineering Chemistry, 1957, 49, (4), 673-678.
14. Chemical and Engineering News, 1948, 26, (50), 3694.
15. Stormont, D. H. Oil and Gas Journal, 1959, 57, (44), 48-49.
16. Sterba, M. J.; Watkins, C. H. Oil and Gas Journal, 1960, 58, (21) 102-106.
17. Oil and Gas Journal, 1960, 58, (16), 104-106.
18. Scott, J. W.; Paterson, N. J. Proceedings of the Seventh Worth Petroleum Congress, Mexico City, 1967, 4, 97-111.
19. Scott, J. W.; Robbers, J. A.; Mason, H. F.; Paterson, N. J.; Kozlowsk, R. H. Proceedings of the Sixth World Petroleum Congress, Frankfurt, 1963, Section III, 201-218.
20. Langlois, G. E.; Sullivan, R. F. Advances in Chemistry Series 97, "Refining Petroleum for Chemicals," American Chem. Soc.: Washington, D.C., 1970; 38-67.
21. Choudhary, N.; Saraf, D. N. Ind. Eng. Chem. Prod. Res. Dev., 1975, 14, (2), 74-83.
22. Sullivan, R. F.; Egan, C. J.; Langlois, G. E.; Sieg, R. P. J. Am. Chem. Soc., 1961, 83, (5), 1156-1160.
23. de Vries, L. J. Am. Chem. Soc., 1960, 82, 5242.
24. Winstein, S.; Battiste. J Am. Chem. Soc., 1960, 82, 5244.
25. Childs, R. F.; Sakai, M.; Winstein S. J. Am. Chem. Soc., 1968, 90, 7144.
26. Egan, C. J.; Langlois, G. E.; White, R. J. Am. Chem. Soc., 1962, 84, 1204.
27. Flinn, R. A.; Larson, O. A.; Beuther, H. Ind. Eng. Chem., 1960, 52, 153-156.
28. Archibald, R. C.; Greensfelder, B. S.; Holzman, G.; Rowe, D. H. Ind. Eng. Chem., 1960, 52, 745-750.
29. Beuther, H.; Larson, O. A.; Ind. Eng. Chem., Process Design Develop., 1965, 4, 177-181.
30. Beuther, H.; McKinley, J. B.; Flinn, R. A. Preprints, Div. Petrol. Chem., Am. Chem. Soc., 1961, 6, (3), A-75 - A-91.

31. Langlois, G. E.; Sullivan, R. F.; Egan, C. J. Phys. Chem., 1966, 70, 3666-3671.
32. Coonradt, H. L.; Garwood, W. E. Preprints, Division Petrol. Chem., Am. Chem, Soc., 1967, 12, (4), B-47.
33. Coonradt, H. L.; Garwood, W. E. Ind. Eng. Chem., Process Design Develop., 1964, 3, 38-45.
34. Sullivan, R. F.; Meyer, J. A. ACS Symposium Series 20, "Hydrotreating and Hydrocracking," Am. Chem. Soc.: Washington, D.C., 1975; 28-51.
35. Baral, W. J.; Huffman, H. C. Eighth World Petroleum Congress, Moscow, 1971, 4, 119-127.
36. Bolten, A. P. ACS Monograph Series, 171, "Zeolite Chemistry and Catalysis," Am. Chem. Soc.: Washington, D.C., 1976; 714-779.
37. Scott, J. W.; Bridge, A. G. Adv. in Chem. Series 103, "Origin and Refining of Petroleum," Am. Chem. Society: Washington, D.C., 1971; 113-129.
38. Johnson, A. R.; Papso, J. E.; Happel, R.; Wolk, R. NPRA Annual Meeting, San Francisco, March 21-23, 1971 (Preprint AM 71-17).
39. Duir, J. H. Hydrocarbon Processing, 1967, 46 (9), 127-134.
40. Weitkamp, J. ACS Symposium Series 20, "Hydrocracking and Hydrotreating," Am. Chem. Soc.: Washington, D.C., 1975; 1-27.
41. Weitkamp, J.; Jacobs, P. A. Preprints, Div. Petrol. Chem., Am. Chem. Soc., 1981, 26, (1), 9-13.
42. Oil and Gas Journal, 1982, 79, (52), 148-193.
43. Graeser, U.; Niemann, K. Oil and Gas Journal, 80 (12), 1982, 121-127.
44. Lander, H. R. "Jet Fuel from Shale Oil - 1981 Technology Review," Aero Propulsion Laboratory, Wright Patterson Air Force Base, Ohio, Technical Paper AFWAL-TR-81-2135, December 1981.
45. Sullivan, R. F.; O'Rear, D. J.; Stangeland, B. E. Petro. Div. Reprints, Am. Chem. Soc., 1980, 25, (3), 583-607.

RECEIVED October 29, 1982

RESEARCH DEVELOPMENTS

25

Selective Oxidation by Heterogeneous Catalysis

ROBERT K. GRASSELLI
The Standard Oil Company, Research Center, Cleveland, OH 44128

Selective oxidation of hydrocarbons by heterogeneous catalysis is a versatile approach to commercial production of many important monomers such as acrylonitrile, acrylic acid, acrylates, ethylene oxide, maleic anhydride, and phthalic anhydride.

Over the past twenty-five years the development of efficient catalysts for selective oxidation resulted in a new generation of commercial processes which utilize inexpensive olefinic and paraffinic feeds, replacing more reactive and costly raw materials. The catalysts are complex solid metal oxide systems which selectively activate hydrocarbons. Olefins, in particular, are activated via an allylic intermediate formation. The catalysts contain facile solid state redox couples which allow for efficient electron and lattice oxygen transport between reactant, adsorption and surface active site, and the surface reoxidation site which is then reconstituted by gaseous oxygen.

Historically, the key discoveries were based on an understanding of the important features of oxidation catalysis in terms of oxygen bond strength, site isolation, and redox mechanisms. Further advancement of fundamental catalyst science will continue to serve as a basis for the design of even more efficient catalyst systems of the future.

The relationship between the petroleum and chemical industries is one of direct interdependence. About 85% of the primary organic chemicals produced today are derived from petroleum and natural gas sources. Thus, rapidly changing supplies of petroleum have a double impact by affecting both energy and chemical industries. This is reflected, for example, in the automotive industry, by the trend toward production of more

0097–6156/83/0222–0317$08.25/0

energy-efficient engines and lightweight bodies, to extend the dwindling petroleum supplies.

In the chemical sector, it becomes necessary to discover more efficient processes which produce, with higher selectivity, the desired useful products to the exclusion of waste in order to maintain petroleum as an attractive primary source. This will become increasingly important, since the percentage of the available petroleum which is used for chemicals is expected to increase as petroleum supplies decrease.

A comparison of petroleum utilization by the energy and chemical industries shows that the major portion (90-92%) of every barrel of oil is used for fuels, i.e., energy production via total combustion to carbon dioxide and water. In contrast to this unselective use of petroleum, oxidation reactions which are carried out on the much smaller fraction devoted to the chemicals industry result in conversion to useful products via selective processes which stop short of total combustion. The selective nature of nearly all of the oxidation reactions of industrial significance is made possible by the use of a catalyst, which lowers the activation energy for the selected process and provides a facile path by which useful products can form. Thus, the key to both the discovery of new routes to useful chemicals and improvements in existing industrial processes lies in catalysis.

The majority of today's important industrial organic chemicals is produced by catalytic reactions. Based on a comparison of the weight of end-product produced, five major reaction types account for about 91% of the top twenty organic chemicals produced in the United States by catalytic processes (Figure 1). These are, aromatic alkylation (24%), heterogeneous oxidation (23%), dehydrogenation (15%), methanol synthesis (16%), and homogeneous oxidation (13%). Styrene, an important monomer for the manufacture of thermoplastics, is produced *via* a two-step route which utilizes both aromatic alkylation (propylene + ethylene → ethylbenzene) and dehydrogenation (ethylbenzene → styrene + hydrogen). Methanol, a product of the catalytic conversion of synthesis gas (a mixture of carbon monoxide and hydrogen), and phenol, which is produced from the homogeneous catalytic oxidation of cumene, are both used extensively in the production of important industrial resins. Adipic acid, used in the manufacture of nylon 6,6, is also produced from a homogeneous catalytic oxidation, which involves the two-step oxidation of cyclohexane, first to a mixture of cyclohexanol and cyclohexanone. Catalytic reduction and hydration account for the remaining 9% of the total which is not produced by one of these five major reaction classes. Cyclohexane production from catalytic hydrogenation of benzene, and the acid catalyzed hydration of ethylene to ethanol, which is used extensively as a paint solvent in the form of ethyl acetate, are important examples of these reactions.

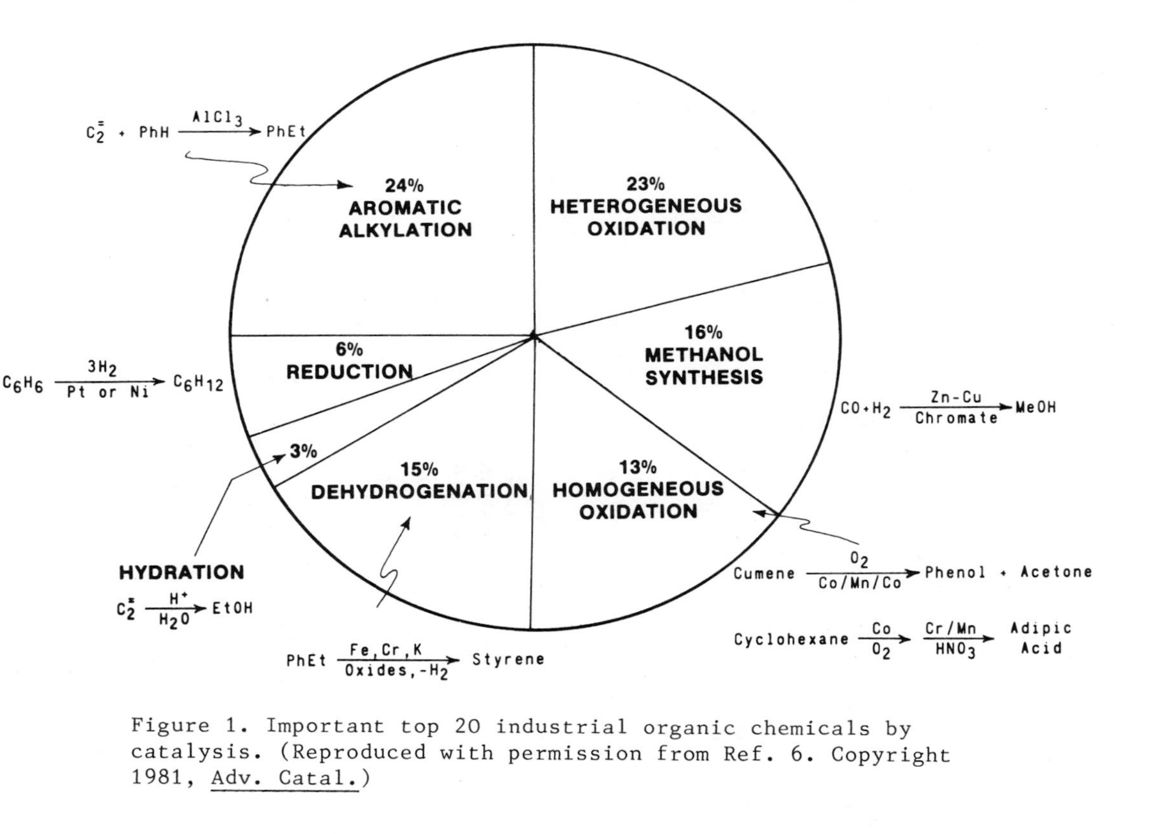

Figure 1. Important top 20 industrial organic chemicals by catalysis. (Reproduced with permission from Ref. 6. Copyright 1981, Adv. Catal.)

Selective oxidation by heterogeneous catalysis (heterogeneous oxidation in Figure 1) refers to those processes where organic feeds are converted in the vapor phase to useful products containing the same number of carbon atoms using solid phase catalysts. Of the oxidation processes by which important (top twenty) organic chemicals are produced by heterogeneous catalysis in the U.S., selective oxidation of hydrocarbons accounts for 79% of the total supply (Figure 2). This type of oxidation may be divided into four classes: Allylic oxidation, epoxidation, aromatic oxidation and paraffin oxidation (Table 1). Allylic oxidation of olefins gives, α,β-unsaturated nitriles from olefins in the presence of NH_3 and O_2, but can also produce, in the absence of ammonia, aldehydes, acids, dienes, and allyl acetates. The epoxidation of olefins gives epoxides, while aromatic side chain oxidation and paraffin oxidation yield carboxylic acids and anhydrides. The catalytic oxidation of methanol to formaldehyde, an important chemical in the manufacture of resins, accounts for the remaining 24% of the important organic chemicals produced by heterogeneously catalyzed processes.

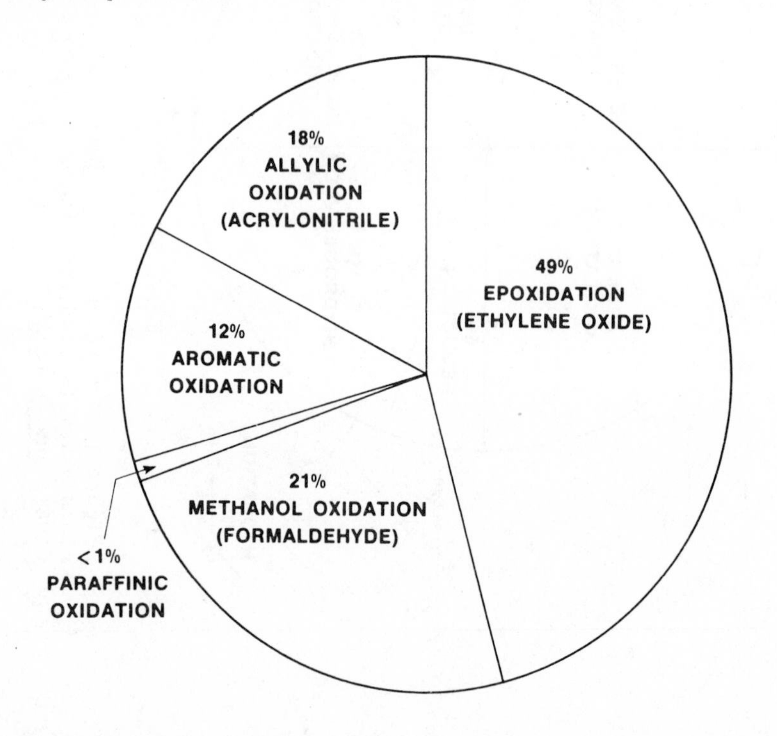

Figure 2. Important organic chemicals by selective heterogeneous oxidation. (Reproduced with permission from Ref. 6. Copyright 1981, _Adv. Catal._)

Table I Examples of Selective Heterogeneous Catalytic Oxidation Reactions.

OXIDATION CLASS	STARTING MATERIAL	PRODUCT	END USE	MILLIONS TON (WORLD)
1. ALLYLIC	PROPYLENE C_3H_6	ACRYLONITRILE C_3H_3N	ACRYLIC FIBERS, RESINS, RUBBERS ADIPONITRILE	4.3
2. EPOXIDATION	ETHYLENE C_2H_4	ETHYLENE OXIDE C_2H_4O	ETHYLENE GLYCOL, ANTIFREEZE POLYESTERS, SURFACTANTS	5.8
3. AROMATIC	o-XYLENE $C_6H_4(CH_3)_2$	PHTHALIC ANHYDRIDE $C_6H_4(C_2O_3)$	POLYESTERS, PLASTICIZERS, FINE CHEMICALS	1.7
4. PARAFFINIC	n-BUTANE C_4H_{10}	MALEIC ANHYDRIDE $C_4H_2O_3$	UNSATURATED POLYESTER RESINS FUMARIC ACID, INSECTICIDES AND FUNGICIDES	0.5

Epoxidation

The most important example of this reaction is the formation of ethylene oxide (Eqn. 1), over Ag-catalysts which displaced the two-step chlorohydrine route (Eqn. 2). Ethylene oxide is used in the production of ethylene glycol, antifreeze, polyesters and surfactants, and accounts for 18% of U.S. ethylene consumption (Figure 3).[2]

Silver is unique in its ability to catalyze the reaction forming a molecular O_2 adsorbed species which reacts with ethylene to form ethylene oxide (Scheme 1). Absorbed atomic oxygen [Ag(O)ads], a by-product of this process, is responsible for waste formation. From the stoichiometry of the mechanism, the maximum yield of selective product (ethylene oxide) is 80%.[3]

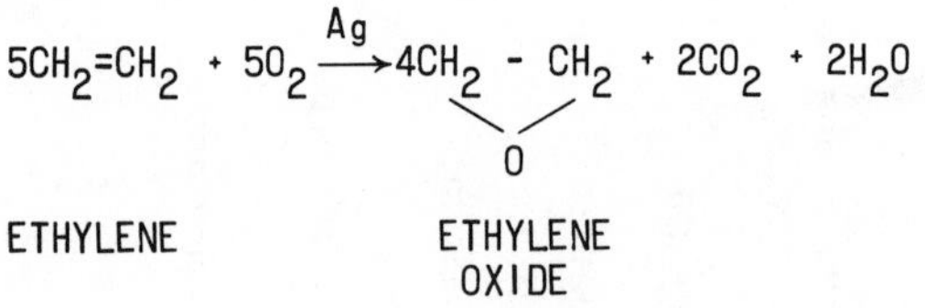

$$5CH_2{=}CH_2 + 5O_2 \xrightarrow{Ag} 4\underset{\diagdown O \diagup}{CH_2 - CH_2} + 2CO_2 + 2H_2O$$

ETHYLENE ETHYLENE OXIDE

Equation 1.

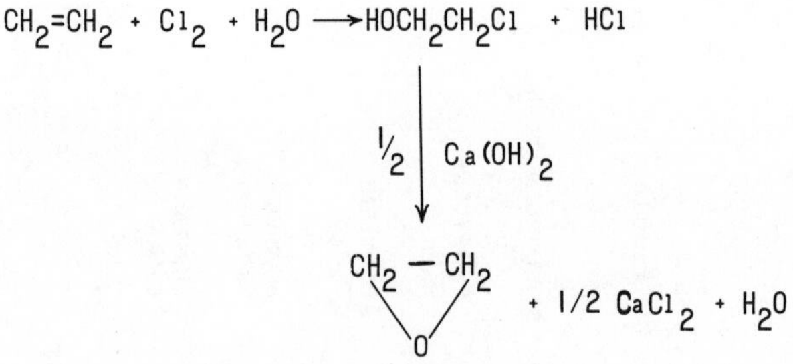

$$CH_2{=}CH_2 + Cl_2 + H_2O \longrightarrow HOCH_2CH_2Cl + HCl$$

$$\downarrow \tfrac{1}{2}\ Ca(OH)_2$$

$$\underset{\diagdown O \diagup}{CH_2 - CH_2} + 1/2\ CaCl_2 + H_2O$$

Equation 2.

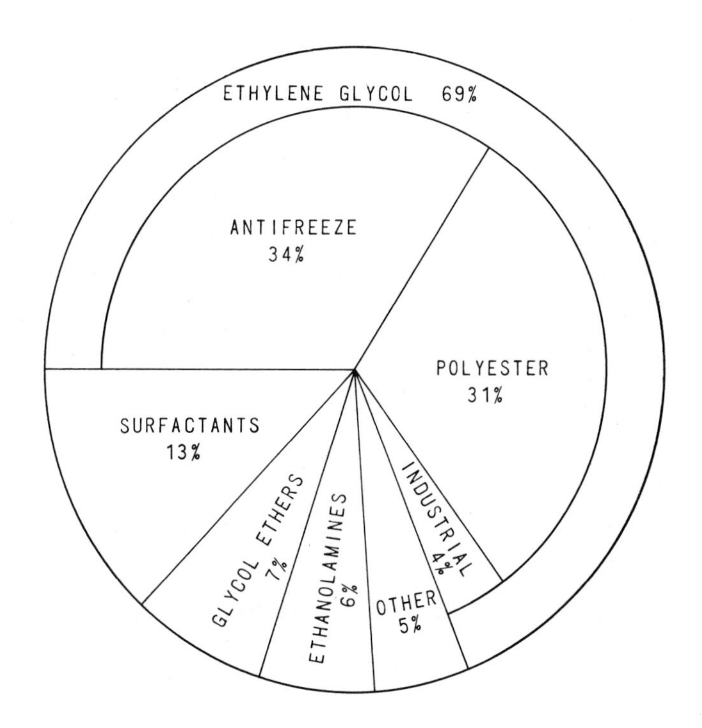

Figure 3. Ethylene oxide utilization.

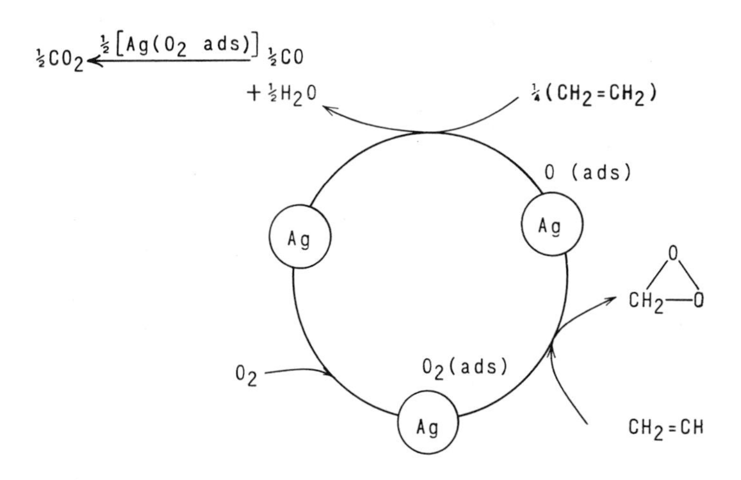

Scheme 1. Ethylene epoxidation mechanism.

Aromatic Oxidation

A process akin to the allylic oxidation in activation is aromatic side chain oxidation to produce acids or anhydrides. Phthalic anhydride, an important intermediate in production of polyesters, plasticizers, and fine chemicals synthesis, can be produced via selective oxidation of o-xylenes using vanadium oxide catalysts (Eqn. 3). This process today accounts for over 85% of the phthalic anhydride produced worldwide, and has largely displaced the partially wasteful and more expensive naphthalene-based route (Eqn. 4), by which nearly all PA was produced in 1960 (Figure 4).[3] Nearly all of the phthalic anhydride produced today is used for manufacturing vinyl plasticizers, with a much smaller application in the fine chemicals industry.

$$\text{o-XYLENE} + 3\,O_2 \xrightarrow[375\text{-}410^\circ C]{V_2O_5} \text{PHTHALIC ANHYDRIDE} + 3H_2O$$

Equation 3.

$$\text{NAPHTHALENE} + 9/2\,O_2 \xrightarrow[350\text{-}550^\circ C]{V_2O_5} \text{PHTHALIC ANHYDRIDE} + 2H_2O + 2CO_2$$

Equation 4.

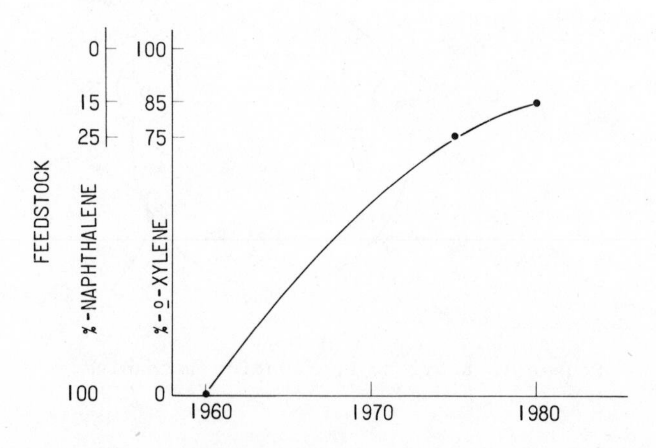

Figure 4. Change in feedstock for phthalic anhydride production.

Paraffin Oxidation

Catalysts which can selectively activate the normally unreactive paraffins have been developed in recent years. The production of maleic anhydride from butane over vanadium-phosphorous-oxide catalysts has received much attention (Eqn. 5), and is beginning to replace the more wasteful production of maleic anhydride from benzene (Eqn. 6) which is still the major feedstock. Maleic anhydride production from butene or butadiene is also possible (Eqn. 7), but cannot compete with the cheaper butane feed. Maleic anhydride is mainly used in the manufacture of unsaturated polyester resins, fumaric acid manufacture, insecticides, and fungicides (Figure 5).[3]

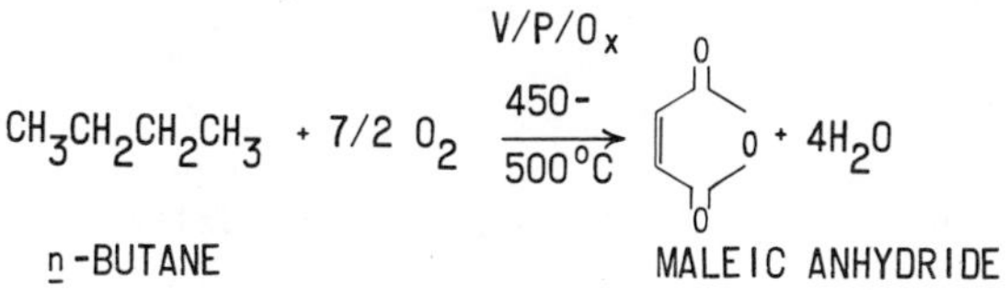

$$CH_3CH_2CH_2CH_3 + 7/2\ O_2 \xrightarrow[450\text{-}500^\circ C]{V/P/O_x} \text{maleic anhydride} + 4H_2O$$

n-BUTANE — MALEIC ANHYDRIDE

Equation 5.

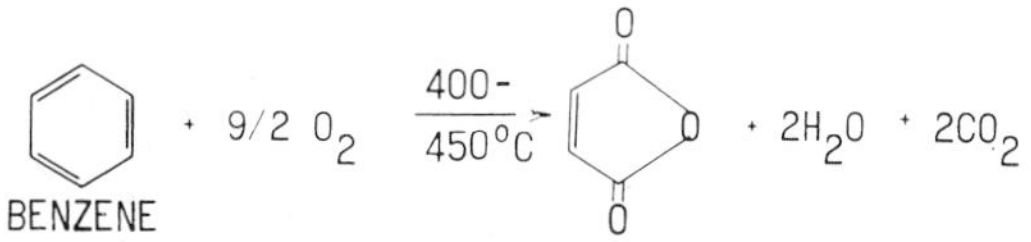

$$\text{benzene} + 9/2\ O_2 \xrightarrow{400\text{-}450^\circ C} \text{maleic anhydride} + 2H_2O + 2CO_2$$

BENZENE

Equation 6. Catalysts: $V_2O_5/MoO_3, V_2O_5/Sb_2O_3, V_2O_5/P_2O_5$

$$\text{1,2-butenes, or butadiene} + O_2 \xrightarrow{350\text{-}450^\circ C} \text{maleic anhydride} + H_2O$$

1,2-BUTENES, BUTADIENE

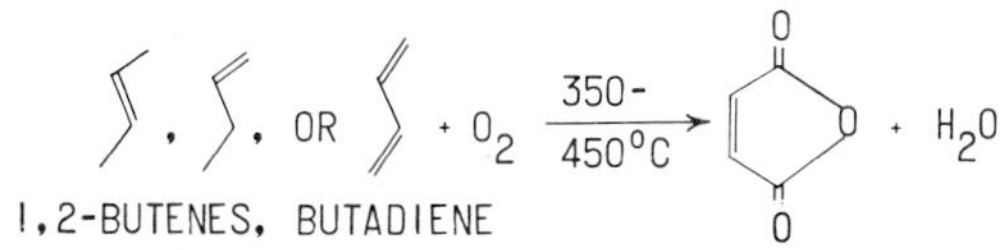

Equation 7. Catalysts: $V_2O_5/MoO_3, V_2O_5/Sb_2O_3, V_2O_5/P_2O_5$

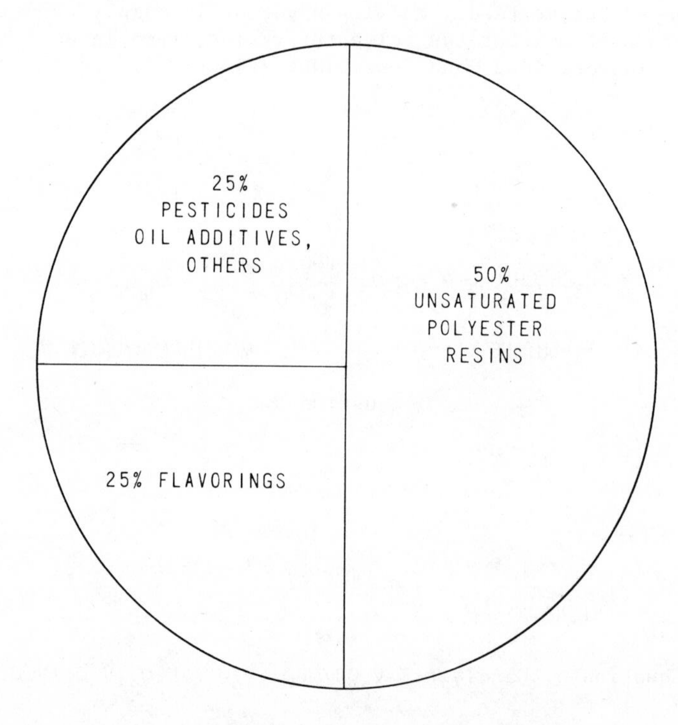

Figure 5. Maleic anhydride utilization.

Allylic Oxidation

In allylic oxidation reactions (Eqns. 8-13), an olefin (usually propylene) is activated by the abstraction of a hydrogen α to the double bond to produce an allylic intermediate. This intermediate can be intercepted by catalyst lattice oxygen to form acrolein or acrylic acid (oxidation), lattice oxygen in the presence of ammonia to form acrylonitrile (ammoxidation), HX to form an allyl-substituted olefin (e.g. acetoxylation), or it can dimerize to form 1,5-hexadiene. If an olefin containing β-hydrogens is used, loss of H from the allylic intermediate occurs faster than lattice oxygen insertion, to form a diene with the same number of carbons, e.g. butadiene from butene (oxydehydrogenation). In all of these processes a common allylic intermediate is formed.

$$CH_2{=}CH{-}CH_3 + NH_3 + 3/2\ O_2 \longrightarrow CH_2{=}CHCN + 3H_2O$$

Equation 8. Ammoxidation

$$CH_2{=}CH{-}CH_3 + O_2 \longrightarrow CH_2{=}CHCHO + H_2O$$

Equation 9. Oxidation

$$CH_2{=}CH{-}CH_3 + 3/2\ O_2 \longrightarrow CH_2{=}CHCO_2H + H_2O$$

Equation 10. Oxidation

$$2(CH_2{=}CH{-}CH_3) + 1/2\ O_2 \longrightarrow CH_2{=}CH{-}CH_2CH_2{-}CH{=}CH_2 + H_2O$$

Equation 11. Dimerization

$$CH_2{=}CH{-}CH_2CH_3 + 1/2\ O_2 \longrightarrow CH_2{=}CH{-}CH{=}CH_2 + H_2O$$

Equation 12. Oxydehydrogenation

$$CH_2{=}CH{-}CH_3 + 1/2\ O_2 + HOAc \longrightarrow CH_2{=}CH{-}CH_2OAc + H_2O$$

Equation 13. Acetoxylation

Allylic olefins of higher molecular weight than propylene can also be converted to the corresponding α-β unsaturated nitriles, aldehydes, and dienes by catalytic vapor phase oxidation and ammoxidation. Examples include the conversion of isobuthylene to methacrylonitrile (eq. 14) or methacrolein (eq. 16), α-methyl styrene to atroponitrile (eq. 15) or atropoldehyde (eq. 17), and 2-methylbutene to isoprene (eq. 18).

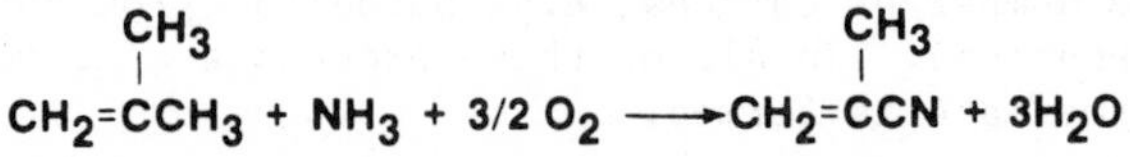

$$CH_2{=}C(CH_3)CH_3 + NH_3 + 3/2\ O_2 \longrightarrow CH_2{=}C(CH_3)CN + 3H_2O$$

Equation 14. Ammoxidation

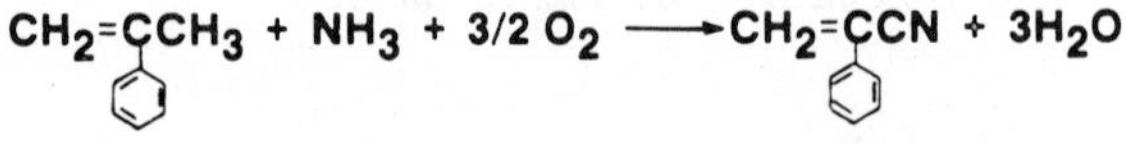

$$CH_2{=}C(C_6H_5)CH_3 + NH_3 + 3/2\ O_2 \longrightarrow CH_2{=}C(C_6H_5)CN + 3H_2O$$

Equation 15. Ammoxidation

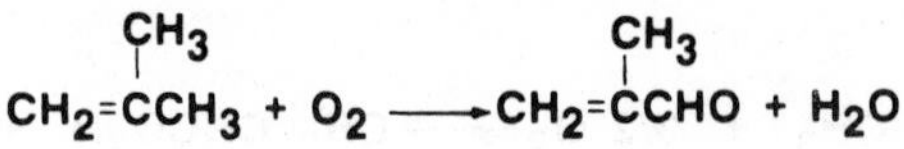

$$CH_2{=}C(CH_3)CH_3 + O_2 \longrightarrow CH_2{=}C(CH_3)CHO + H_2O$$

Equation 16. Oxidation

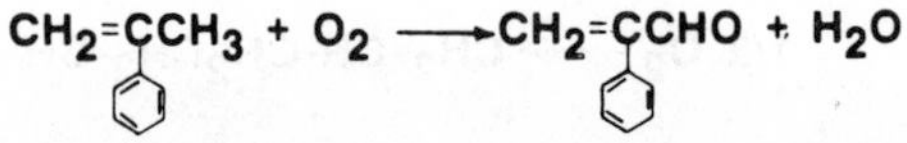

$$CH_2{=}C(C_6H_5)CH_3 + O_2 \longrightarrow CH_2{=}C(C_6H_5)CHO + H_2O$$

Equation 17. Oxidation

$$CH_2{=}C(CH_3)CH_2CH_3 + 1/2O_2 \longrightarrow CH_2{=}C(CH_3)CH{=}CH_2 + H_2O$$

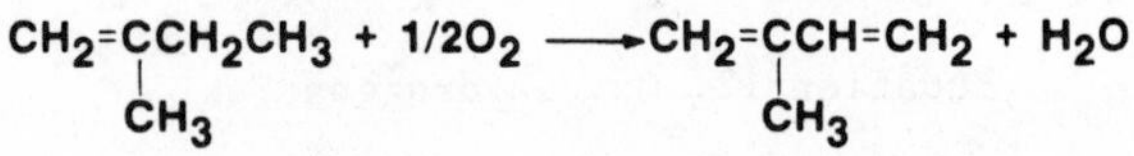

Equation 18. Oxydehydrogenation

Ammoxidation of Propylene to Acrylonitrile

The most industrially significant and well-studied allylic oxidation reaction is the ammoxidation of propylene (eq. 8) which accounts for virtually all of the 8 billion pounds of acrylonitrile produced annually world-wide. The related oxidation reaction produces acrolein (eq. 9), another important monomer. Although ammoxidation requires high temperatures, the catalysts are, in general the same for both processes and include bismuth molybdates, uranium antimonates (USb_3O_{10}), iron antimonates, and bismuth molybdate based multicomponent systems. The latter category includes many of todays highly selective and active commercial catalyst systems.

The ammoxidation process (eq. 8) displaced the more expensive acetylene-HCN-based route in the early 1960's (eq. 20). Other obsolete processes also involve more expensive reagents (e.g. ethylene oxide, eq. 19, and acetaldehyde, eq. 21) and oxidants (e.g. NO, eq. 22). The impact of the introduction of the ammoxidation process in 1960 was an immediate drastic reduction in acrylonitrile price and greatly increased production which made possible many of today's high-volume applications of acrylonitrile (Figure 6A).[4] The production of acrylonitrile, which accounts for 17% of the total U.S. propylene consumption, is used extensively in fibers, plastics and resins (ARS/SA) and rubber industries, with a growing number of miscellaneous applications, including the electro-hydrodimerization process for adiponitrile production (Figure 6B).

$$\underset{\text{(ethylene oxide: } CH_2-CH_2 \text{ bridged by O)}}{CH_2-CH_2} + HCN \xrightarrow[\text{BASE CAT.}]{} HO\text{-}CH_2CH_2CN \xrightarrow[200^\circ]{-H_2O} CH_2{=}CHCN$$

Equation 19.

$$H\text{-}C{\equiv}C\text{-}H + HCN \xrightarrow[\text{CuCl-}NH_4Cl]{80\text{-}90^\circ C} CH_2{=}CHCN$$

Equation 20.

$$CH_3CHO + HCN \longrightarrow CH_3\underset{\large OH}{\underset{|}{C}}HCN \xrightarrow[H_3PO_4,\ 600\text{-}700^\circ C]{-H_2O} CH_2{=}CHCN$$

Equation 21.

$$CH_2{=}CHCH_3 + 3/2NO \xrightarrow{Ag_2O/SiO_2} CH_2{=}CHCN + 3/2H_2O + 1/4\,N_2$$

Equation 22.

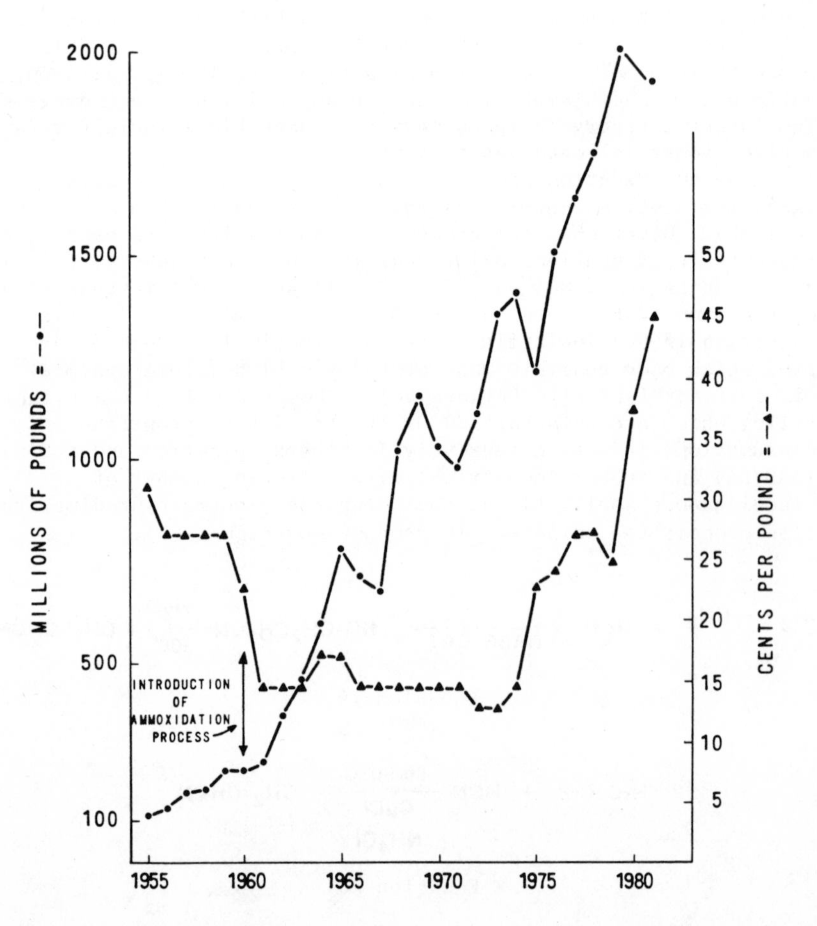

Figure 6A. U.S. production and price of acrylonitrile. (Reproduced with permission from Ref. 6. Copyright 1981, Adv. Catal.)

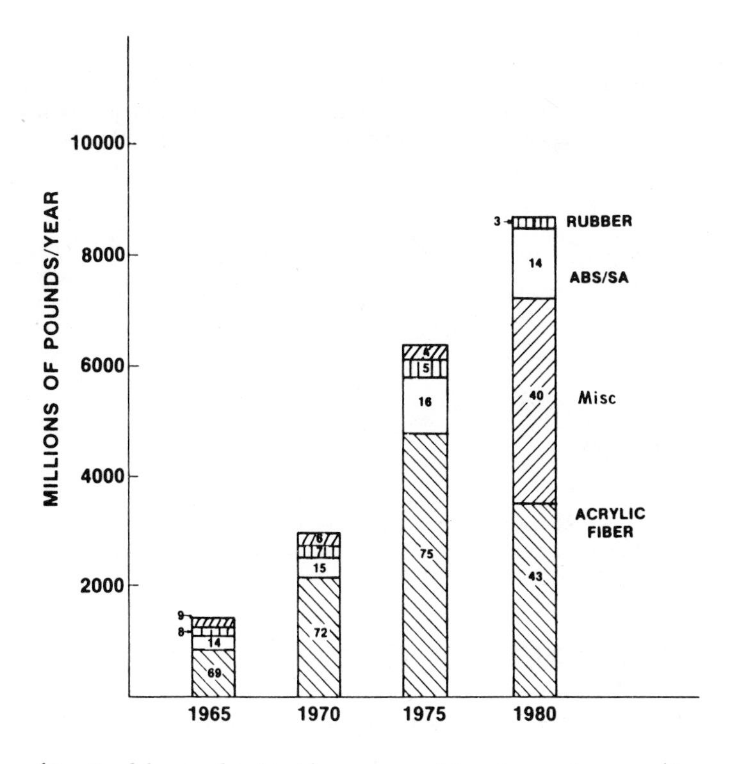

Figure 6B. World acrylonitrile consumption by end use. (Reproduced with permission from Ref. 6. Copyright 1981, Adv. Catal.)

Development of Selective Ammoxidation and Oxidation Catalysts

The development of the Sohio ammoxidation process was based on the theory that lattice oxygen from a solid metal oxide would serve as a more selective and versatile oxidant than would molecular oxygen[5], and has recently been reviewed.[6,7] When applied to the ammoxidation of propylene (eq. 8), acrylonitrile is formed in the catalyst reduction step, while molecular oxygen serves to reconstitute lattice oxygen vacancies (Fig. 7). For the oxidation of propylene to acrolein, lattice oxygen is used to form acrolein in catalyst reduction step (Fig. 8, eq. 23), and reoxidation of metal oxide by gaseous O_2 reconstitutes the active oxidant (eq. 24). The metal oxide becomes a catalyst when the process can be carried out in the presence of molecular oxygen, the stoichiometric oxidant (eq. 9).

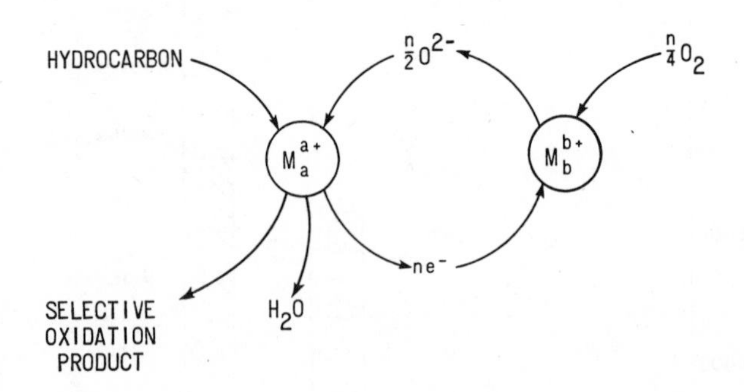

Figure 7. Catalytic selective oxidation-reduction cycle. (Reproduced with permission from Ref. 6. Copyright 1981, *Adv. Catal.*)

$$2M^{(n)+}O_x + CH_2{=}CHCH_3 \longrightarrow 2M^{(n-2)+}O_{x-1} + CH_2{=}CHCHC{=}O + H_2O$$

Figure 8. Catalytic redox cycle. Reduction of metal oxide.

$$\underset{\text{Propylene}}{CH_2{=}CHCH_3} + \underset{\text{Oxygen}}{O_{2(g)}} \xrightarrow{CAT} \underset{\text{Acrolein}}{CH_2{=}CHC{=}O} + \underset{\text{Water}}{H_2O}$$

Equation 23. Catalysis.

$$2M(n-2)+O_{x-1} + O_{2(g)} \longrightarrow 2M^{(n)+}O_x$$

Equation 24. Reoxidation of metal oxide.

In an early embodiment of this reaction, known as the oxidant process (Fig. 9)[8], hydrocarbon is fed to a reactor filled with metal oxide at high temperature to produce the desired selective oxidation products. The resulting reduced metal oxide is then lifted to a separate regeneration vessel where air is also fed, which reoxidizes the catalyst before it is

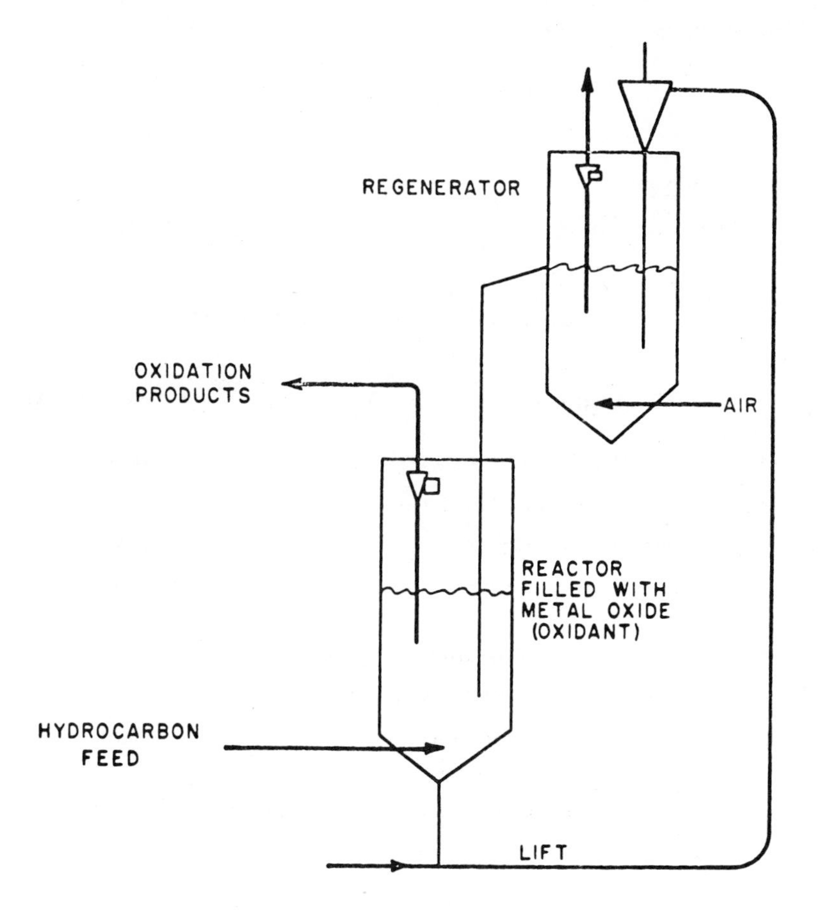

Figure 9. Oxidant process.

returned to the reactor. This process, while theoretically possible, is not practical due to the large quantity of catalyst which must be cycled (400 pounds per pound of useful product produced).

Since non-selective, deep oxidation processes are even more thermodynamically favorable than the selective ones (Table 2), it is necessary to intercept the desired products kinetically. Catalysts, therefore, must be designed which lower the activation energy of the desired reactions and thus allow the process to operate at lower temperatures than for non-catalytic equilibrium limited processes. In this manner undesirable waste formation (deep oxidation) is minimized.

Table II

Thermodynamics of Oxidation Reactions.

REACTIONS	$\Delta G°$ 427° C (KCAL/MOLE)
(A) $C_3H_6 + O_2 \longrightarrow CH_2{=}CHCHO + H_2O$	-80.92
(B) $C_3H_6 + 3/2\ O_2 \longrightarrow CH_2{=}CHCOOH + H_2O$	-131.42
(C) $C_3H_6 + 3O_2 \longrightarrow 3CO + 3H_2O$	-304.95
(D) $C_3H_6 + 9/2\ O_2 \longrightarrow 3CO_2 + 3H_2O$	-463.86
(E) $C_3H_6 + NH_3 + 3/2\ O_2 \longrightarrow CH_2{=}CHCN + 3H_2O$	-136.09
(F) $C_3H_6 + 3/2\ NH_3 + 3/2\ O_2 \longrightarrow 3/2\ CH_3CN + 3H_2O$	-142.31
(G) $C_3H_6 + 3NH_3 + 3O_2 \longrightarrow 3HCN + 6H_2O$	-273.48

From these fundamental concepts of lattice oxygen utilization, evolved two basic requirements of a heterogeneous selective oxidative catalyst:[5] (1) Oxygen atoms must be distributed on the surface of a selective oxidation catalyst in an arrangement which provides for limitation of the number of active oxygen atoms in various isolated groups (site isolation); (2) Metal-oxygen bond energy of the active oxygen atoms at the conditions of reaction, must be in a range where rapid removal (hydrocarbon oxidation) and addition (regeneration by oxygen) is assured (characteristic, unique M↔O bond strength).

The isolation of active sites can be accomplished by the partial removal of lattice oxygen, i.e. reduction, of a metal oxide. Using statistical (Mote Carlo) methods, the distribution of oxygen atoms and vacancies on a surface at various stages of reduction, represented by a surface oxide acid (Fig. 10)[5], can be calculated, as well as the corresponding density of oxygen clusters containing 1-5 oxygen atoms and > 5 oxygen atoms (Fig. 11)[5]. Based on the assumption that clusters of 1-5 oxygen atoms will result in selective product, the relative propylene conversion to selective product (acrolein) should be maximized at about 65% reduction (Fig. 12)[5]. In fact, the experimentally determined dependence of selective product yield on % reduction for CuO (Fig. 13) is very similar to the calculated curve (Fig. 12), but with the maximum shifted to ∿25% reduction.

While CuO provides an excellent illustration of the importance of active site isolation in selective oxidation over heterogeneous catalysts, it is not a practical catalyst since the small range of partially reduced states, required for selectivity, is difficult to maintain. A metal oxide catalyst which would operate selectively in a fully or nearly fully oxidized state would be a much more practical system.

One alternative to partial reduction for selectivity is active site isolation chemically by modification of solid state structure. The zig-zag chain structure of V_2O_5 (Fig. 14) provides inherent structural isolation, which can be enhanced by the addition of alkali metal oxide. The dependence of conversion to selective product on K_2O/V_2O_5 ratio is similar to the effect of reduction (Fig. 15), and gives a maximum at ∿0.25 K/V. Bifunctional systems, in which the components individually have very low activity, are among the most efficient catalysts. Among the most well studied and commercially successful in this class are the bismuth molybdates, $Bi_2O_3 \cdot nMoO_3$, where n=3 (α), 2(β) and 1(γ).

The $Bi_2Mo_3O_{12}$ (α-phase) has a Scheelite -derived structure (Fig. 16) in which the Mo atoms are tetrahedral-like, and one cation vacancy is present per 2 Bi atoms. The Bi_2MoO_6 (γ-phase) however, has a layered structure of alternating corner-shared octahedral MoO_2 layers and Bi_2O_2 layers (Fig. 17).

The mechanism of bismuth molybdate catalyzed oxidation has been extensively studied. In the catalytic cycle, the function

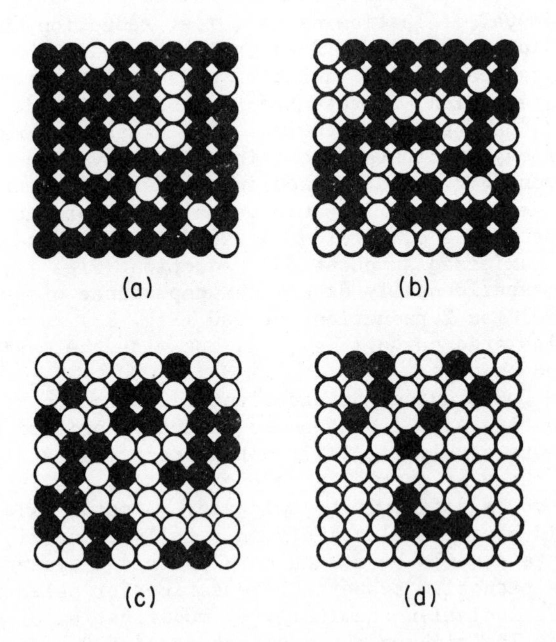

Figure 10. Reduction of an oxidized surface grid. (a) 80%, (b) 60%, (c) 20% oxidized.(Reproduced with permission from Ref. 5. Copyright 1963, A. I. Ch. E. Journal.)

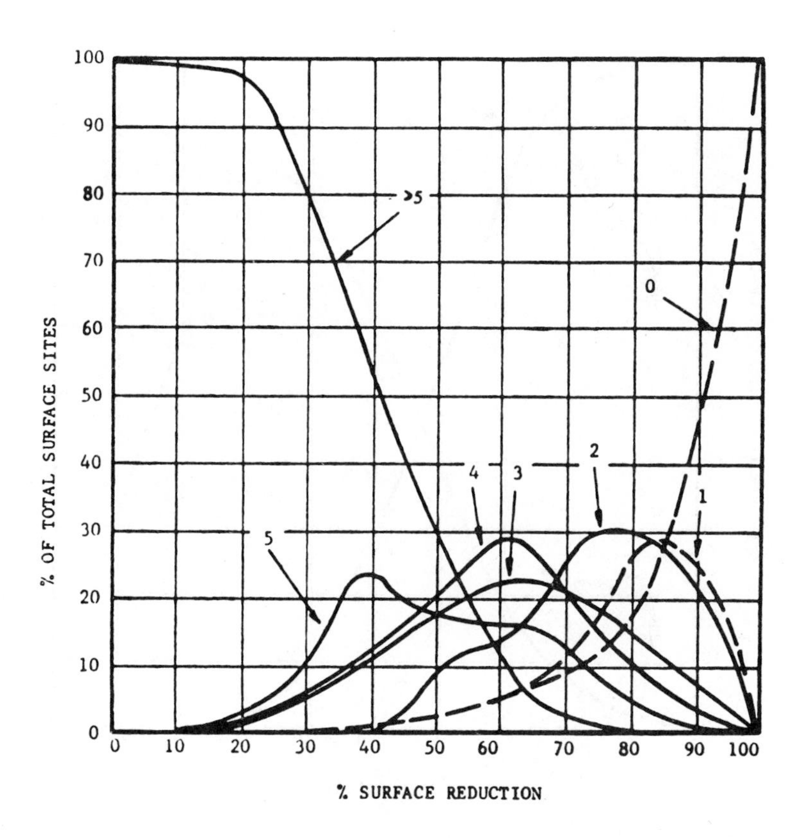

Figure 11. Site population as a function of surface coverage - Oxygen regeneration of reduced grid. (Reproduced with permission from Ref. 5. Copyright 1963, A. I. CH. E. Journal.)

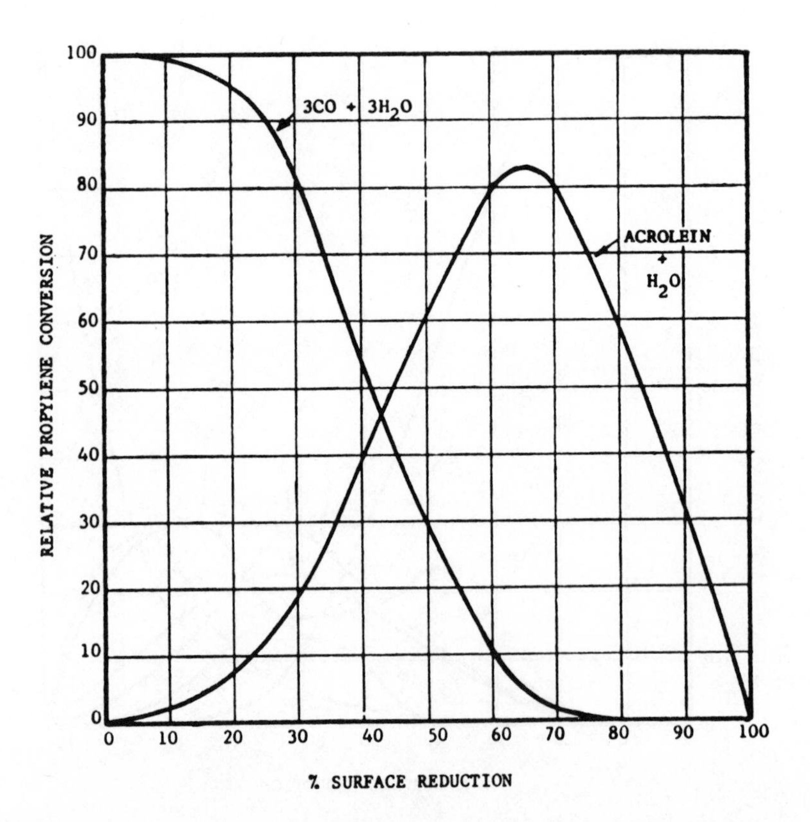

Figure 12. Relative propylene conversion as a function of oxidation state - Oxygen regeneration of reduced grid. (Reproduced with permission from Ref. 5. Copyright 1963, A. I. Ch. E. Journal.)

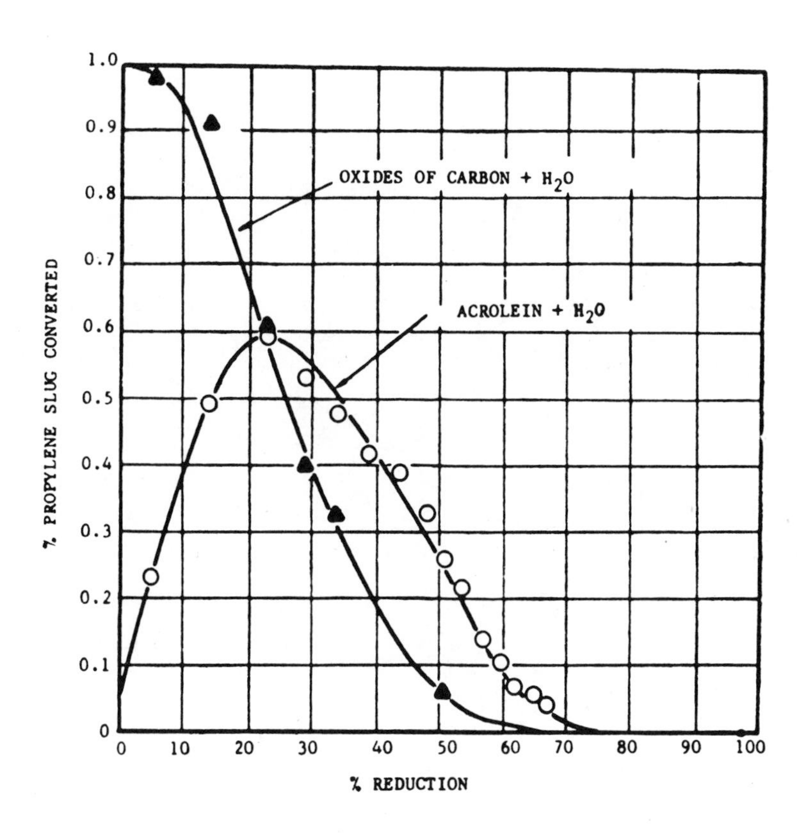

Figure 13. Experimental propylene oxidation activity vs. catalyst oxidation state - Copper oxide catalyst, 300 °C reaction temperature.(Reproduced with permission from Ref. 5. Copyright 1963, A. I. Ch. E. Journal.)

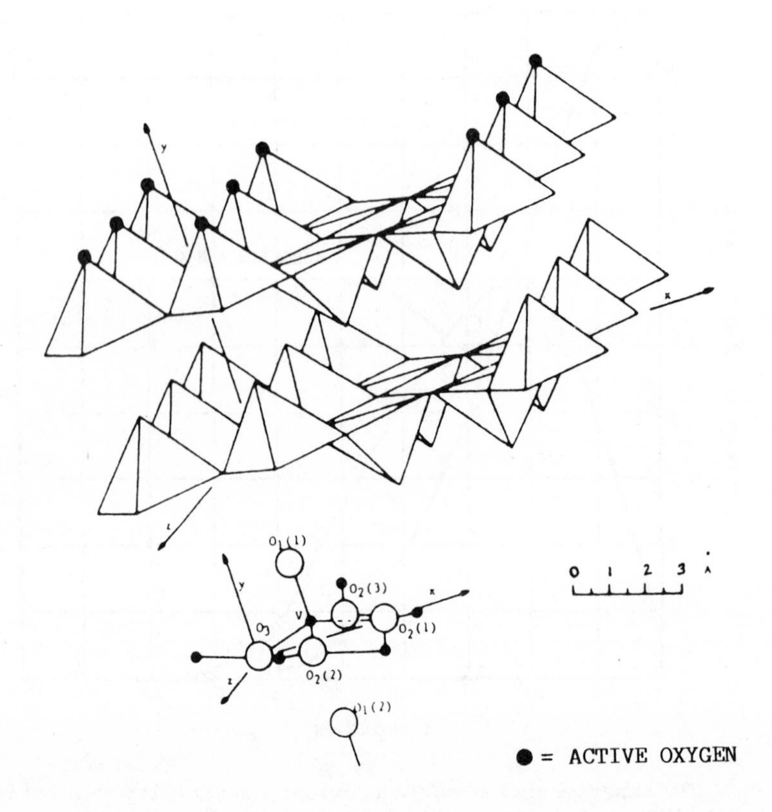

Figure 14. Crystal structure of V_2O_5 - side view. (Reproduced with permission from Ref. 9. Copyright 1961, Zeit Kristall.)

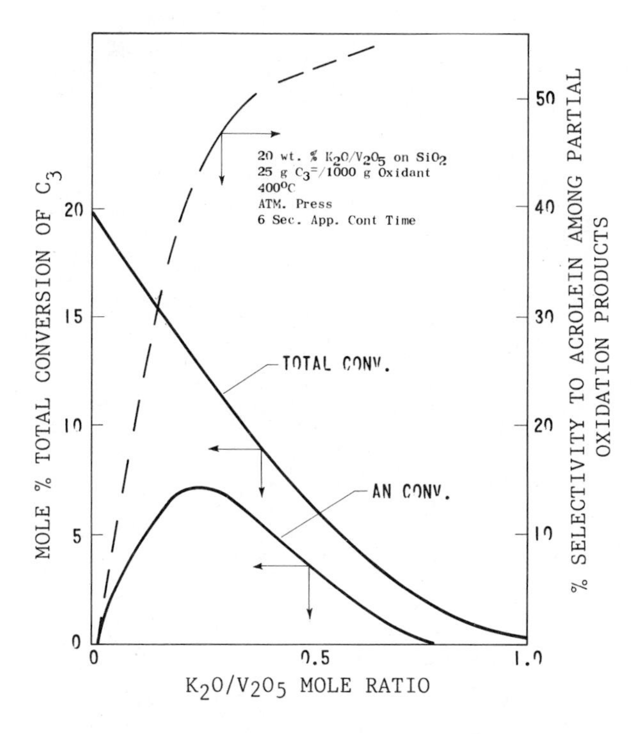

Figure 15. Effect of K_2O/V_2O_5 ratio on $C_3^=$ conversion and acrolein selectivity.

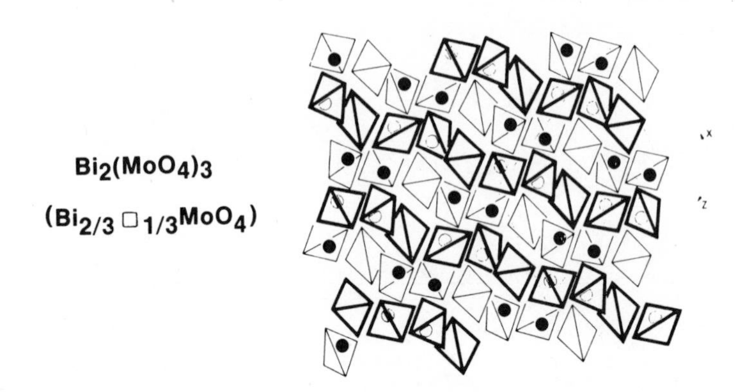

Figure 16. Scheelite derived structure. (Reproduced with permission from Ref. 10. Copyright 1973, Acta. Cryst.)

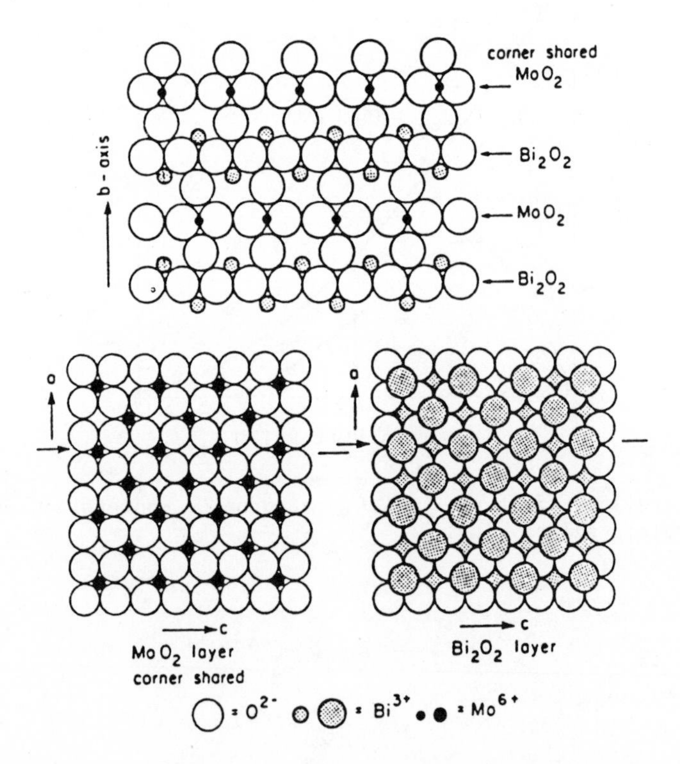

Figure 17. Layered structure of Bi_2MoO_6. (Reproduced with permission from Ref. 11.)

of the oxygens associated with Bi is to perform the rate-determining α-hydrogen abstraction step, while those in Mo polyhedra are sites for olefin chemisorption and O-insertion (F.18). The active and selective site is composed of Bi:Mo pairs, represented by structure 1 (Fig. 19) in which the α-hydrogen abstracting bismuthyl (Bi=O) is bonded through a bridging oxygen to the olefin chemisorption/oxygen-insertion Mo-dioxo functionality. Initial chemisorption, and α-H abstraction produces O-π-allyl 2 and subsequently O-σ-complex 3, and then acrolein after a 2nd H-abstraction. Formation of the analogous N-bonded species (4 and 5) followed by a 3rd H-abstraction produces acrylonitrile when ammonia is present, after initial NH_3 activation by formation of a Mo-di imido species 6.

Ammoxidation of Propylene in Practice

In a typical bench scale experiment a first generation $Bi_9PMo_{12}O_{52}$ catalyst produces 65.2% per pass propylene conversion to acrylonitrile with 4.1% HCN, 4.0% acetonitrile, 0.1% acrolein, and 16.8% CO_2 (Table 3). Yields of useful products have been greatly improved with newer generation catalyst systems. Advancement of new catalyst systems progresses in several stages, eventually resulting in a 30 million-fold scale-up from a 5-gram laboratory microreactor to a 26 ft. diameter commercial reactor (Table 4).

Future Trends

The trend in chemical feedstocks is towards less expensive, more available ones and away from the expensive, more reactive feeds. For example, the extensive use of acetylene as a feedstock in the 1930-1940's has been replaced in the 1960-1980's by olefins and diolefins. The future trend appears to be towards paraffins and synthesis gas (Figure 20). Continued developments in fundamental catalyst science will serve as the basis for the design of the catalytic single-step processes of the future which will efficiently utilize inexpensive, readily available feeds.

Acknowledgments

I should like to acknowledge my early mentor, Dr. Franklin Veatch, my long time friend and collaborator, Dr. James L. Callahan, and my many co-workers, in particular, Dr. Dev D. Suresh, Dr. James D. Burrington, and Dr. James F. Brazdil.

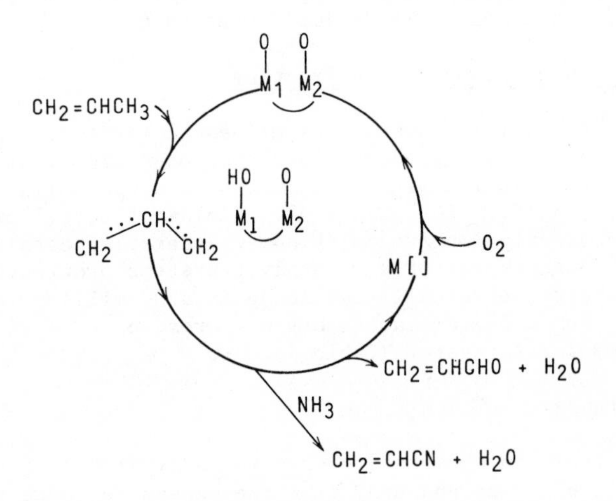

Figure 18. Allylic oxidation mechanism. (Reproduced with permission from Ref. 6. Copyright 1981, Adv. Catal.)

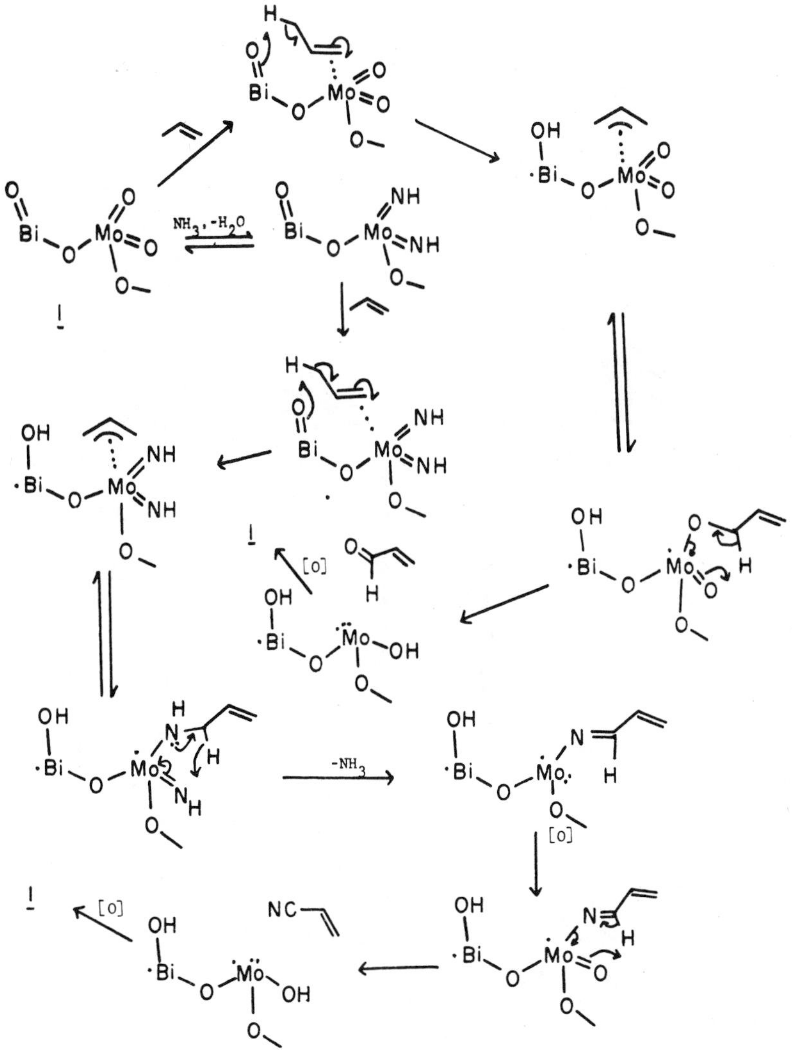

Figure 19. Mechanism for selective oxidation and ammoxidation of propylene. (Reproduced with permission from Ref. 12. Copyright 1980, J. Catal.)

Table III Bench-Scale Acrylonitrile Run

CONDITIONS		
REACTORS 1-1/2 INCH PIPE FLUID BED		
CATALYST CHARGE	566 G	
TEMPERATURE	470° C	
PRESSURE, PSIG	ATMOSPHERIC	
CONTACT TIME	9.1 SEC.	
MOLES FEED/HOUR		RATIO
PROPYLENE	0.255	1
AIR	3.355	13.1
NH_3	0.282	1.1
PROPYLENE WT. HOURLY SPACE VELOCITY	0.0253	
RESULTS		
MOLE % EXCESS O_2	2.2	
% PER PASS CONVERSION (C-ATOM BASIS) TO:		
ACRYLONITRILE	65.2	
ACETONITRILE	4.0	
ACROLEIN	0.09	
HCN	4.1	
CO	5.5	
CO_2	16.8	
TOTAL	95.69	
% CARBON BALANCE	95.6	
% UNREACTED NH_3	8.2	

(50% $Bi_9PMo_{12}O_{52}$ - 50% SiO_2 CATALYST)

Table IV Scale up of Catalysts.

MICROREACTOR A	1.5 GMS	(0.003 LBS)
MICROREACTOR B	5.0 GMS	(0.01 LBS)
1-1/2 INCH REACTOR	550 GMS	(1.2 LBS)
3 INCH REACTOR	2,000 GMS	(4.4 LBS)
18 INCH REACTOR	249,700 GMS	(550 LBS)
24 INCH REACTOR	681,000 GMS	(1,500 LBS)
26 FT. COMMERCIAL REACTOR	136,200,000 GMS	(300,000 LBS)

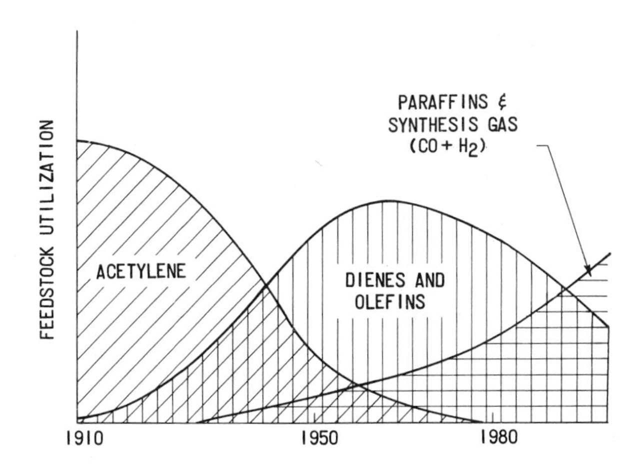

Figure 20. Trends in chemical feedstock. (Reproduced with permission from Ref. 6. Copyright 1981. Adv. Catal.)

Literature Cited

1. Chem. and Eng. News, June 8, 1981.
2. Hydrocarbon Processing, October 1979, 59, 123.
3. Weissermel, K., and Arpe, H.J., "Industrial Organic Chemistry," trans. Mullen, A., Verlag Chemie, New York, 1978.
4. Synthetic Organic Chemicals, U.S. Production and Sales, U.S. Fundamental Trade Commission; Chemical Marketing Reputes.
5. Callahan, J.L., and Grasselli, R.K., A.I.Ch. E. Journal 1963 9, 755
6. Grasselli, R.K., and Burrington, J.D., Adv. Catal. 1981, 30, 133.
7. Grasselli, R.K., Burrington, J.D., and Brazdil, J.F., J. Chem. Soc. Faraday Soc. Disc., 1981, 72, 203.
8. Callahan, J.L., and Grasselli, R.K., Milberger, E.C., and Strecker, H.A., I & E.C. Prod. Res. Dev., 1970, 9, 134
9. Bachman, H.G., et.al., Zeit. Kristall, 1961, 115, 110.
10. Jeitschko, W., Acta. Cryst. 1973, B29, 2074.
11. Gates, B.C., Katzer, J.A., and Schuit, G.C.A., "Chemistry of Catalytic Processes," McGraw-Hill, New York, 1979.
12. Burrington, J. D., Kartisek, C. T., and Grasselli, R. K., J. Catal., 63, 235 (1980) (Figure 19).

RECEIVED November 29, 1982

26

Methanol: Bright Past–Brilliant Future?

ALVIN B. STILES

University of Delaware, Department of Chemical Engineering, Center for Catalytic Science and Technology, Newark, DE 19711

Methanol presently is a large volume chemical used primarily for chemical purposes. The possibility is increasing that methanol itself, as a blend or after further processing, will become a major liquid fuel in the future.

This chapter surveys the history over the past 75 years of methanol synthesis from CO, H_2 and CO_2. Factors considered are first the catalyst developments but also describing the improvements in gas purification, reactor design, catalyst life, operating pressure and productivity.

The very first methanol was probably produced by nature herself when lightning struck the first plant which contained lignin and which emerged from the primordial sea. The rapid pyrolysis of the lignin content of this primitive plant very likely generated methyl or methoxy radicals which combined with hydroxyl radicals or protons to form methanol (Figure 1).

In medieval times the alchemist also pyrolized wood in his retort and in so doing generated gases some of which were condensible to liquids and some of the liquid was methanol or, in those days, termed wood alcohol. It may seem anachronistic but wood pyrolysis was the method used in World War I to make wood alcohol for methyl ester solvents. The esters were the solvents for the pyroxylin used to strengthen and tighten the fabrics on the wings of those early planes.

Germany, however, had two much more serious problems - one relating to their isolation from sources of petroleum and the other from nitrates. As a

0097–6156/83/0222–0349$07.25/0

Figure 1. Primitive Scene of First MeOH Production - by Cynthia S. Boyd.

consequence, their policy makers determined that they must do two things, one of which was to derive liquid fuels such as methanol and related products and the second to derive ammonia. These must all be synthesized from internally available raw materials.

The motivation, of course, was to free Germany from the need to bring in nitrate for fertilizer and explosives from Chile and also to provide liquid fuels from internally available sources. It is apparent that these German developments were necessitated by the geographic situation of Germany and by the fact that economic and possibly aggressive tendencies were dictating such a move. Ammonia synthesis from elemental N_2 and H_2 was developed by Haber approximately in 1906 and the first patent on the process appeared in 1910. Following closely thereafter was the synthesis of methanol from CO and hydrogen which in turn was derived from coal available internally in Germany. We will subsequently talk about the first methanol units which were really purification units for gas being used down-stream for ammonia synthesis.

Early Historical Factors as they Relate to the Present

It is frequently desirable to look at the historical background of a process to see why the present operating conditions have been selected, but it is also important to look at these factors because there may be reasons for again adopting or adapting these early conditions. The earliest synthetic methanol was synthesized from carbon monoxide, carbon dioxide, and hydrogen which were derived from coal by the water-gas reaction. From the earliest days of methanol and ammonia synthesis, the interrelationship between adjacent ammonia and methanol plants was evident. In some of the very earliest methanol plants, the synthesis gas was most conveniently and economically produced as a carbon monoxide, carbon dioxide, hydrogen, and nitrogen mixture. This gas was produced in gas sets in which coal or coke was first heated to incandescence in a converter; then steam or steam plus air was passed over the incandescent coke to produce the aforementioned gases. This gas contained ash as well as sulfur compounds and other contaminants. The purification procedure consisted of removal of the ash and sulfur compounds by scrubbing.

In this combination process shown in Figure 2, the

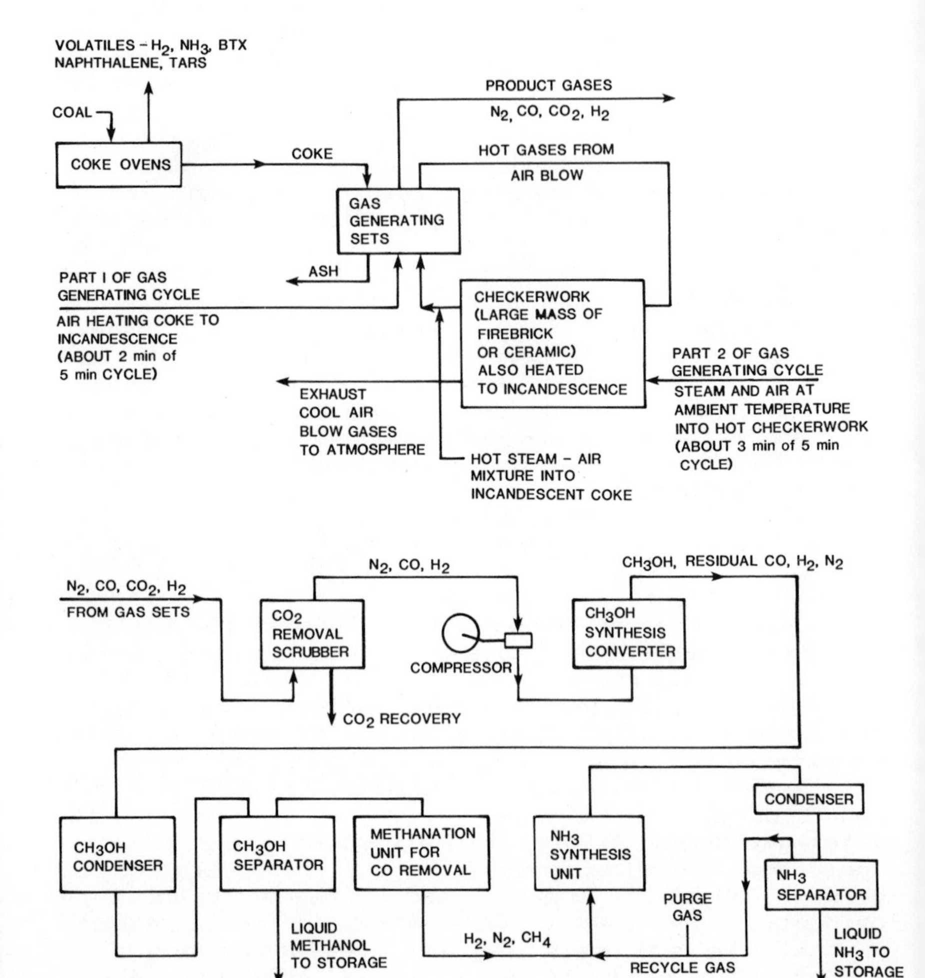

Figure 2. Flow sheet for a plant sequentially making methanol then ammonia from carbon monoxide, nitrogen and hydrogen derived from coal (ca. 1920's).

cleaned gas containing carbon monoxide, hydrogen, and nitrogen was first compressed to operating conditions which could have been as high as ca. 1000 atm. and was then passed over a methanol synthesis catalyst to remove the carbon monoxide and a part of hydrogen so that the gas effluent which then passed into the ammonia converter would have the proper 1:3 ratio of nitrogen:hydrogen. The methanol was condensed from the gas stream before it passed into the ammonia converter, and any residual carbon monoxide was removed from the nitrogen-hydrogen stream by methanation ($CO + 3H_2 \longrightarrow CH_4 + H_2O$). The pure ammonia synthesis gas was passed over the ammonia synthesis catalyst in a downstream converter. There were many problems with this type of operation, some relating to the seasonal nature of the sales demand for both methanol, used largely as antifreeze at this time, and ammonia, used as a fertilizer.

When methanol is synthesized from synthesis gas containing nitrogen (nitrogen is almost always present), there is always the simultaneous synthesis of organic amines. These are objectionable because they give an objectionable basic property to the methanol and cause an offensive odor.

The high temperatures and pressures used in early plants dictated specialized equipment which could be fabricated only by those who had experience in, and facilities for, munitions manufacture. This resulted in alloys and design often showing the common parentage with the munitions of the period between WW I and WW II. Since then, specific fabrication techniques have been developed which have made it possible to fabricate the larger size converters and interconnecting piping and the more easily assembled and disassembled facilities now required.

These earlier facilities were also plagued by corrosion due to the sulfur compounds, to carbon dioxide (as carbonic acid solution), or to hydrogen embrittlement, or to iron carbonyl formation from the high pressure-low temperature carbon monoxide containing gases. These problems were for the most part solved, but the iron carbonyl problem remains and seems to defy an economic solution.

The synthesis gas preparation phase of methanol synthesis is presently the most in need of major innovations and developments. As natural gas and liquid hydrocarbons become more costly and less abundant, the less easily processed sources of carbon must be used. First to be considered will be the sources which have been historically important, next the ones most widely utilized presently, and finally speculation as to the sources and processes of the future. The earlier sources included fermentation by-product gases (Commercial Solvents Corp.), coke oven by-product gas, and steel furnace off gases. The coke oven by-product gas is deficient in carbon monoxide and carbon containing molecules whereas the off gas from blast furnaces is high in carbon monoxide and deficient in hydrogen. These latter sources of hydrogen and carbon monoxide have been considered and to some extent used, but because of the many difficulties involved in their utilization, they both are of very minor importance presently. However, as the need intensifies for complete conservation of our energy resources, they may in the future by utilized rather than burned as is almost universally the present case.

The preparation of carbon monoxide and hydrogen from coke or coal by the blue-gas and blow-run gas technique was frequently used. This process was used by E. I duPont Nemours at the Belle, West Virginia plant when the plant was first placed in operation in the late 20's and continued as the source of synthesis gases for both methanol and ammonia until the late 40's. The procedure used and a diagram of the facilities are shown in Figure 3. In addition to the gas sets which were used for the gas generation, a complete set of coke ovens was also required for the conversion of the coal to coke. At one time the conversion of coal directly to synthesis gases was the basic procedure, but because of difficulties with compounds being volatilized from the coal which later condensed in the gas lines to plug them, the procedure was altered to using coke which avoided these problems. The duPont Company at that time advertised that the products of the plant were from coal, air, and water, which is something we find many people presently speculating about without realizing that this had as a matter of fact been an industrial process in the 20's. It could hardly be considered a viable process of the present, but there are certain features of the operation and process which present interesting backgrounds which we certainly must assess in the present light of needs to derive fuels and chemicals from domestically available fossil energy.

Steam-hydrocarbon reforming has become the most frequently used method for synthesis gas preparation, and usually natural gas, treated to remove all com-

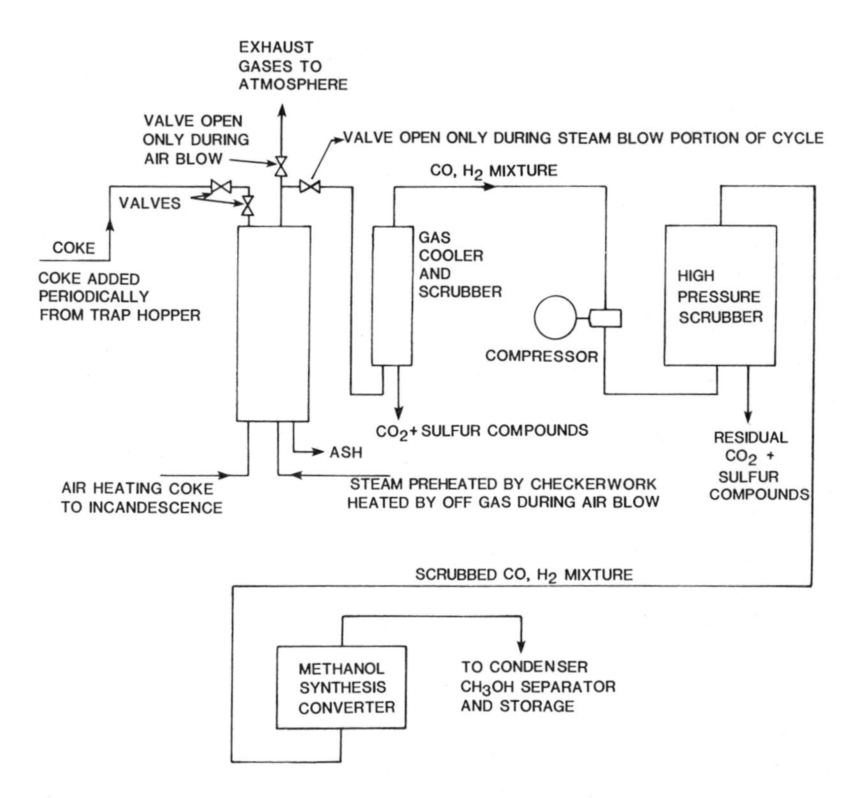

Figure 3. Flow sheet for a plant producing methanol from carbon monoxide and hydrogen mixtures derived from coal by the blue gas process.

ponents except methane and nitrogen, is the preferred hydrocarbon.

The steam-hydrocarbon reforming process is highly developed and will operate for months or even years without interruption, except for normal outages scheduled for boiler inspection, routine maintenance, and other attention which is placed on a definable schedule. The heat balance and utilization are well engineered ordinarily so that there is little waste, and what heat is unused on the furnace side of the reformer is subsequently recovered for use to generate steam.

The basic equation is:

$$CH_4 + H_2O = 3H_2 + CO$$

but for methanol synthesis the following relationship between the hydrogen and carbon monoxide is necessary:

$$2H_2 + CO = CH_3OH$$

Therefore, the following simultaneous operation is performed in the reformer by adding carbon dioxide into the feed gas:

$$CO_2 + CH_4 = 2CO + 2H_2 + H_2O$$

The latter reaction, of course, produces excess carbon monoxide relative to that needed for methanol, and as a consequence, in practice, the feed to a reformer comprises natural gas, steam, and carbon dioxide.

Synthesis Gas Purification

The steam-carbon dioxide-hydrocarbon conversion is conducted over a catalyst such as nickel (oxide) on alumina. This type of catalyst can be purchased in quite similar composition from a number of catalyst vendors. In the case in which the feed stock is processed over a catalyst as in steam-hydrocarbon reforming, it is essential that the gas be purified, at least to some extent, prior to its passage over the reforming catalyst, particularly if the catalyst is of the typical composition of supported and promoted nickel (oxide). In steam hydrocarbon reforming, the methane (natural gas) is usually detoxified using an adsorbent such as carbon on which is impregnated suitable chemical adsorbents such as elemental iron or copper. There are at least two of these metallized carbon desulfurizers in parallel with one on the line while the other is being regenerated. Regeneration is effected by passing steam or steam plus a small amount of oxygen (parts per million range) over the adsorbent in such a way that the sulfur is removed as sulfur dioxide or hydrogen sulfide. Heavy hydrocarbons simultaneously adsorbed by the carbon are removed as such to be flared, recovered, for sale, or combusted to derive useful heat. If there are solids such as sodium chloride and its associated minerals entrained with the natural gas, obviously these impurities must be removed; water scrubbing can be used.

Steam used in the operation must be free of solids since the entrainment of solids in the mist from boilers is not an infrequent cause of physically poisoning the reforming catalyst at least in the upper (upstream) part of the reformer tubes.

Hydrocarbons themselves can be troublesome. First, the steam-to-carbon ratio must be kept above an experience range, and second, unplanned for higher hydrocarbons must be avoided because higher hydrocarbons are more susceptible to dehydrogenation (cracking, causing carbon deposition) than are the lower ones such as methane, ethane, and even propane.

If, for some reason, the natural gas has a surge in higher hydrocarbon content, the effect can be disastrous in that the nickel catalyst will quickly become coated with a sooty carbon, which quickly deactivates it. A second result is that the carbon may be deposited interstitially in the catalyst body causing it to disintegrate to powder, which of course stops flow through the tube so that when naphtha is used instead of natural gas, the problems are similar but more severe, and proper modifications are made in the catalyst to make it resistant to severe poisoning and carbon deposition conditions. These catalysts usually contain lower amounts of nickel and may contain such promoters as uranium, potassium, or other alkali or alkaline earth oxides. Inasmuch as alkali metal oxides migrate from the catalyst at higher temperatures and pressures, it is necessary to operate the naphtha reformers at milder conditions than when natural gas is the feed. In such cases, it may be desirable to operate two reformers, one for the conversion of naphtha to lighter hydrocarbons, following this with a converter designed and operated more like the natural gas reformer previously described.

Throughout the past decade or two there have been several shifts to and from steam-hydrocarbon reforming vs. hydrocarbon partial combustion as the more economical procedure for synthesis gas generation. The equation for the partial oxidation reaction is as follows:

$$H_2O + C_nH_{2n+2} + O_2 \longrightarrow H_2 + CO + CO_2 + H_2O \text{ (Exothermal)}$$

The ratio of ingredients in the feed, that is, water, hydrocarbon, and oxygen, is adjusted so that the desired ratio of carbon monoxide plus carbon dioxide and hydrogen are obtained in the effluent gas. The temperature and pressure, of course, influence this ratio so that all of these factors must be considered when the feed gas ratios are established.

A number of potential sources of synthesis gases have been advanced in recent years, many due to the 1972 energy crisis in the U.S. These sources require much effort and development before they become applicable. However, the production of synthesis gas is a crucial consideration because of the large impact it has on the cost of the methanol product.

Catalyst Developments

In commercial operation catalyst life, as well as activity, must be considered. In the earlier installations of methanol synthesis plants, there was little gas purification of the scale and effectiveness that is presently used, and as a result the catalysts or operating conditions that were used were necessarily more rugged and severe than is presently the case. In other words, the fact that recent methanol installations operate under milder temperature and pressure conditions is attributable not only to improved catalysts but also to the fact that the synthesis gases presently employed have been detoxified to the extent that catalysts that have long been known are now practicable.

Mittasch and Schneider. The first patent for the synthesis of methanol was granted in Germany to Mittasch and Schneider in 1913 (1). The catalysts which they described were oxides of cerium, chromium, manganese, molybdenum, titanium, and zinc which had been "activated" by incorporating alkalies such as sodium and potassium carbonates. The products were methanol, higher alcohols and saturated and unsaturated hydrocarbons. Pressures and temperatures were 100-200 atmospheres and 300° to $400^{\circ}C$. These were the temperatures and pressures used almost exclusively in this and other early work.

Fischer and Tropsch. The next patent was issued to Fischer and Tropsch and was granted in 1922 (2), 9 years after the Mittasch-Schneider patent. The catalysts which were claimed by Fischer and Tropsch were elemental nickel, silver, copper and iron and most specifically iron plus cesium and rubidium-hydroxides. The products of the Fischer-Tropsch catalysts at that time were methanol, higher alcohols, other oxygenated products but no hydrocarbons either saturated or unsaturated. The name that was assigned to this product was "synthol." Fischer and Tropsch used a slightly higher temperature (420°C) and pressure (134 atm.) than Mittasch and Schneider employed.

Mittasch, Pier and Winkler. The next patent to issue was also in Germany and was granted to Mittasch, Pier and Winkler in 1923 (3). The catalysts that were taught for use at 380-420°C and 200 atm. pressure were as follows:

Catalyst A: copper oxide, zinc oxide, chromium oxide (Cr^{+3}) and optionally manganese oxide.

Catalyst B: zinc oxide, chromium oxide (Cr^{+6}).

Catalyst C: zinc oxide, chromium oxide (Cr^{+2}) and manganese oxide.

Catalyst D: zinc oxide and chromium trioxide as trivalent chromium.

It will be quickly recognized that these are the catalysts which very closely approach the modern methanol synthesis catalysts. The one containing manganese was similar to that used commercially 30-40 years ago for the synthesis of higher alcohols. Strangely the product described by the inventors is identified as pure methanol. We will consider subsequently the fact that the manganese does have a strong tendency for the production of higher alcohols which would cast some doubt on the identity of the products of all of these catalysts as being pure methanol.

Mittasch, Pier and Winkler. The next catalyst patent was also granted to Mittasch, Pier and Winkler and was issued in 1925 (4). Some of the catalysts which are taught are rather exotic in light of present knowledge and consist of A) Cr_2O_3 plus zinc oxide; B) zinc oxide plus uranium oxide; C) zinc oxide plus vanadium pentoxide; D) zinc oxide plus tungsten oxide; E) cerium oxide plus manganese oxide and lastly, F) copper oxide and zinc oxide. The product from reactions at 380-420°C and 200 atm. pressure was said to be methanol plus various other oxygenated products.

Lewis and Frohloch. Another modern catalyst composition was described by Lewis and Frohlich in 1928 (5). This catalyst had the composition of zinc oxide, copper oxide and aluminum oxide. The product that was obtained when synthesis gas was passed over this catalyst at 275-350°C and 100 atm. pressure was methanol.

Lazier. Lazier (6) invented and assigned to the DuPont Company the next catalyst to be described. This was covered by a patent issued in 1931 and the catalyst, itself, was zinc oxide, manganese oxide, and chromium oxide. Depending on the manganese ratio and operating conditions (350-475°C and 267-900 atm. pressure) this catalyst could produce 20 to 50% higher alcohols in addition to methanol.

Those familiar with modern methanol and methanol-higher alcohol synthesis processes will quickly recognize that these early investigators have pretty well disclosed the chemical composition of presently-used catalysts for these operations. The improvements that have been made since the early investigators identified the compositions have largely been in the way the components are put together and the physical characteristics of the catalysts. The early investigators quite often simply took oxides and then heated them to very high temperatures in the presence of an alkali. Melting or sintering was considered desirable. In fact, in the DuPont Company we had with us in the catalyst development area a person who had been with the Nitrogen Fixation Laboratory in the early 20's. He had his grounding in fused, promoted iron oxide ammonia synthesis catalyst. He was the type who felt that all catalysts, whether they were for methanol, higher alcohols or ammonia, all should be fused. The result was that "catalysts" were produced but all were quite poor physically and catalytically.

Despite the great strides in catalyst development, my personal feeling and recent observations have strengthened this feeling that there is still a great deal that could be done to further improve catalysts for methanol and higher alcohols services.

Since we have discussed catalysts, the next thing that we plan to discuss is the catalytic reactors and it would be out of character for me not, at this time, to emphasize the fact that when catalysts are being developed the catalytic scientists and the project engineer designing the converter must work very closely together. It has frequently been a costly situation when they do not work together until the plant design is essentially frozen, at which time it may be found that some of the characteristics of the catalyst, such as high exothermisity, were not adequately considered and the reactor is not designed to dissipate this extra heat load. An often neglected consideration is that the catalytic scientist and reactor designer must work closely during developmental stages, establishing abrasion resistance, density, particle size, gas distribution, porosity and heat dissipation.

Methanol Unit Designs and Types

The first commercial methanol units usually were designed in such a way that they could accommodate the gases which contained large amounts of nitrogen and relatively small amounts of CO or CO_2. The methanol process was really a purification process for the ammonia synthesis gas and the gas was first processed through the Claude type methanol unit which is shown in Fig. 4. You will see that there are first two converters in parallel, then a second series of converters in parallel, then a condenser system for the removal of the alcohol and a combining of the gas streams which then pass through the two additional converters making a total of 6. Finally the effluent passes to a second condenser system which removes the methanol. The alcohol-free gases pass to a methanator which purifies the ammonia synthesis gas by methanation of the carbon monoxide and carbon dioxide. After removal of the water vapor the mixed gas (N_2:$3H_2$) is passed through an ammonia synthesis system of essentially the same design as the previously described Claude unit. The ammonia is condensed and the unreacted gases are burned or are recycled to the synthesis gas process.

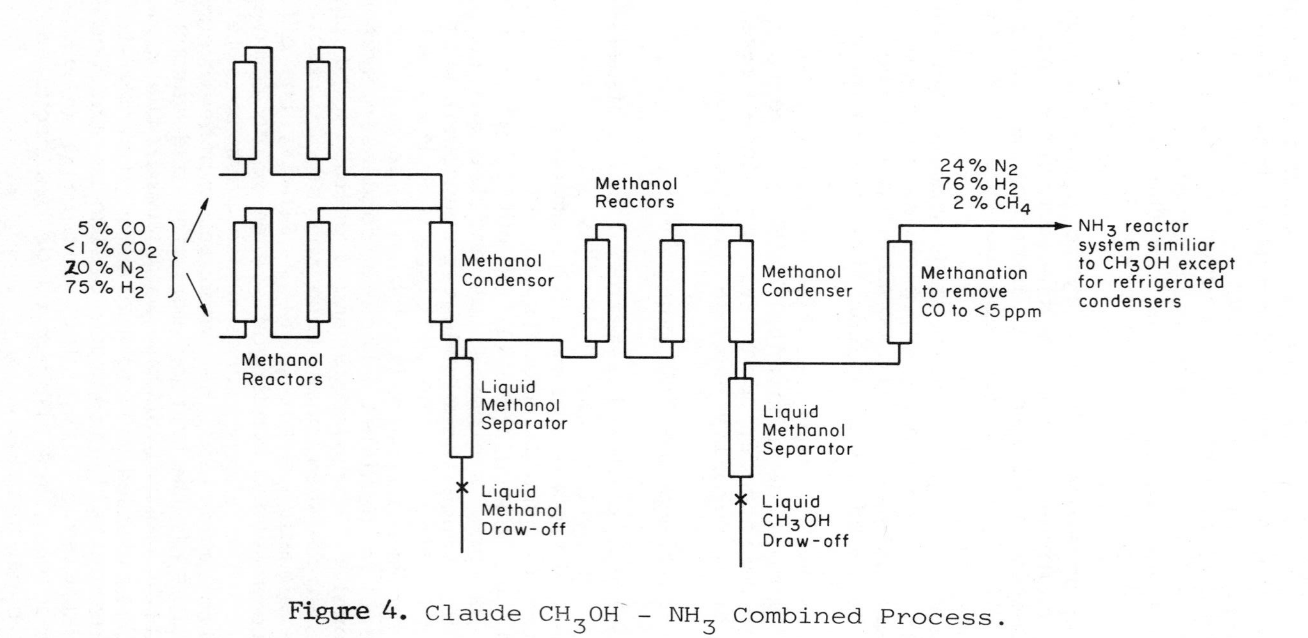

Figure 4. Claude CH_3OH - NH_3 Combined Process.

Casale Recirculation Unit

The second type synthesis unit is the modern recycling unit and is known as the Casale. The Casale recycle unit is shown in Fig. 5 and consists of a loop in which there is a gas mixture comprising approximately 10% CO, 3-5% carbon dioxide and the remainder hydrogen. The gas passes into the reactor (which will later be described) then emerges from the reactor into a heat recovery unit and thereafter a condenser; the cooled gases enter a methanol separator from which the methanol is withdrawn at the bottom and the unreacted gas discharges at the top. The unreacted gases pass to the recycle pump suction but on the way a tee with a valve permits the withdrawal of a certain amount of unreacted gases to prevent the build-up of excessive quantities of inert gases. The recycle pump discharges gas at a sufficiently high pressure that it will again enter the reactor. However, before entering the reactor, make-up gas (to compensate for alcohol formed and gas purged) is added as the stoichiometric composition, 33% CO, 67% hydrogen.

A problem relating to this system is the fact that large quantities of gas must be recirculated at high expenditures of energy. Consequently, it is essential to have as much reaction as possible occur per pass. A second, and related problem, is the obvious need to minimize pressure drop through the converter. This need is translated into optimizing catalyst bed depth and diameter and particle size and shape of the catalyst while obtaining maximum approach to equilibrium, Fig. 6.

Reactor Types

Claude. The reactors used with the Claude type unit are basically a cylindrical column of catalyst whereas reactor types used in the Casale unit are much more complex.

Multitubular. The older type is the multitubular reactor shown in Figure 7. Although this is mentioned as the old-type because it was used decades ago, it has recently come back into favor in the Lurgi units. Instead of the internal heat exchanger serving as a gas preheater, the reactor tubes are surrounded with a liquid which removes the heat and can be cycled through a boiler for heat recovery. There are many problems with this reactor in that the

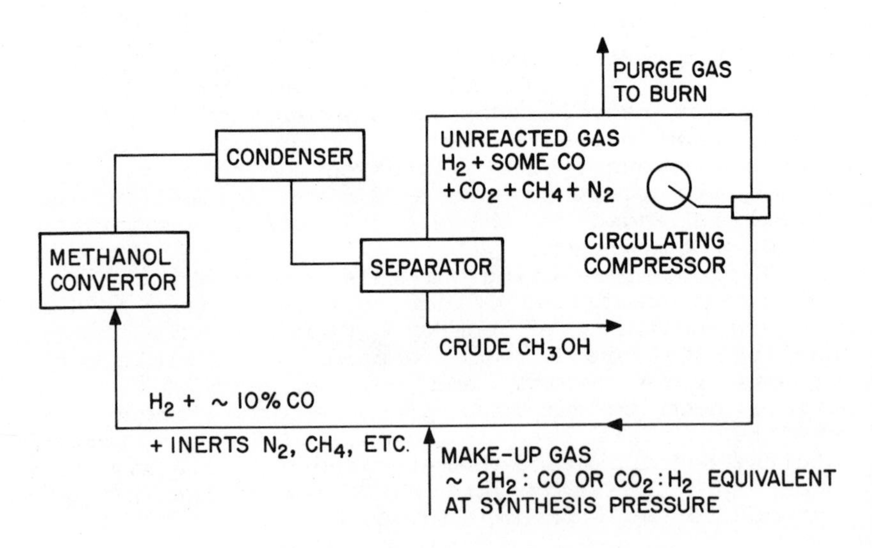

Figure 5. Casale MeOH Synthesis Loop.

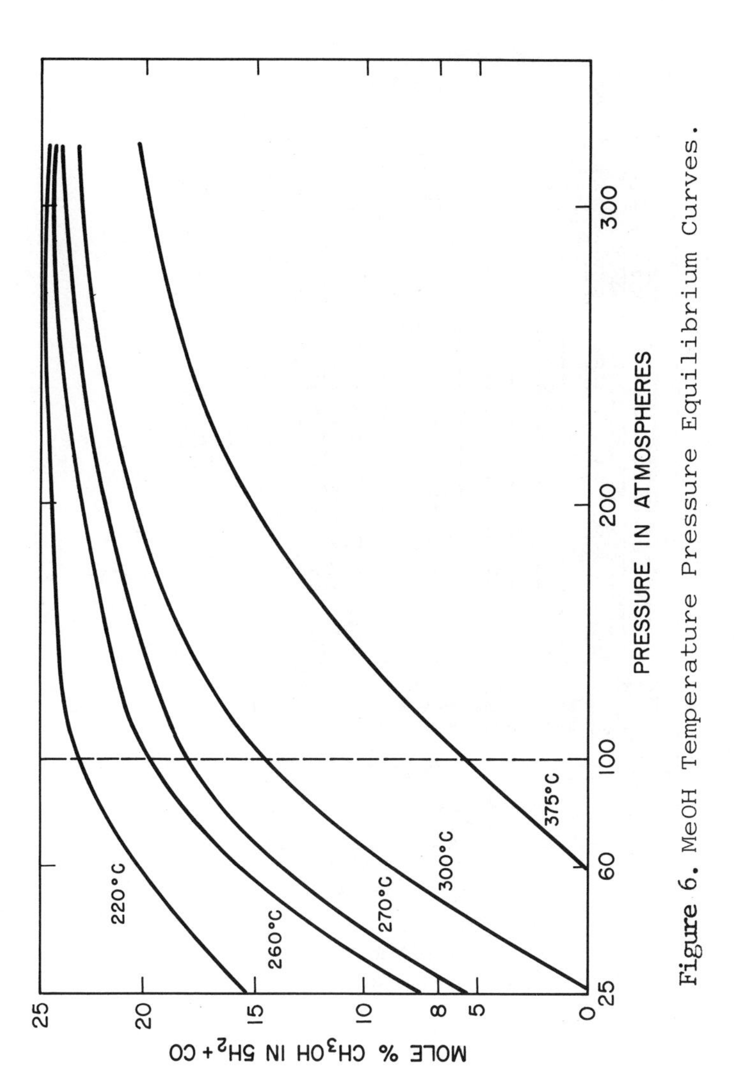

Figure 6. MeOH Temperature Pressure Equilibrium Curves.

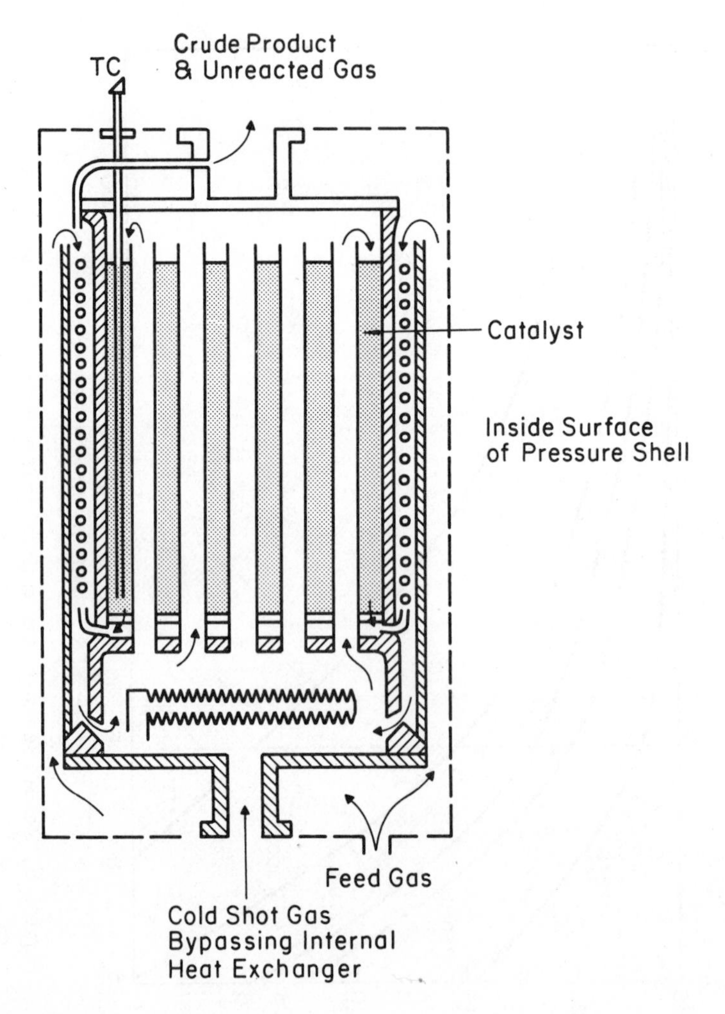

Figure 7. Typical Tubular Synthesis Reactor.

tubes tend to become warped and heat exchange between tubes becomes inferior. The major problem is the sealing of the tubes at the tube sheets. Improperly sealed tubes may result in gas passing through these poor seals from the catalyst side of the tube into the boiler with a possible over pressuring of the boilers.

The second reactor to be described is the tray type, Figure 8, which allows the catalyst to be put on individual trays at certain depths throughout the reactor. In the usual design of this reactor cold synthesis gases are added between trays to drive the reacting gases to a lower temperature favoring a better equilibrium.

The most recent catalyst converter (the Wentworth process, Figure 9), is yet to be installed in an operating plant but it is anticipated that within the next few years at least two plants employing this system will be in operation. This is also a tray type system but it is designed in such a way that catalysts most efficient for operation at given temperatures, are charged on each tray with the initial catalyst being low temperature, the next catalyst being medium temperature, the third catalyst being high temperature. Thereafter, the gas is removed from the converter and heat is recovered and the gas is returned to the converter at a temperature identical to that pertaining when the gas went into the first tray of catalysts. The gas proceeds from the first tray to the second tray which has the temperature and catalyst characteristics of the second tray in the first series. Thereafter, the gas is again cooled and heat recovered and the gas then passes back into the converter for a low temperature pass over the catalyst of the type in trays 1 and 4. This favors low temperature equilibrium and high conversion and thus minimizes recycling and is reported by the advocates as reducing energy consumption by approximately 30%.

Catalyst Life Then and Now

Figure 10 shows catalyst life of the early operations and the present operations which are generally set for approximately two years before recharge. The remarkable effect of catalyst and process developments can readily be seen.

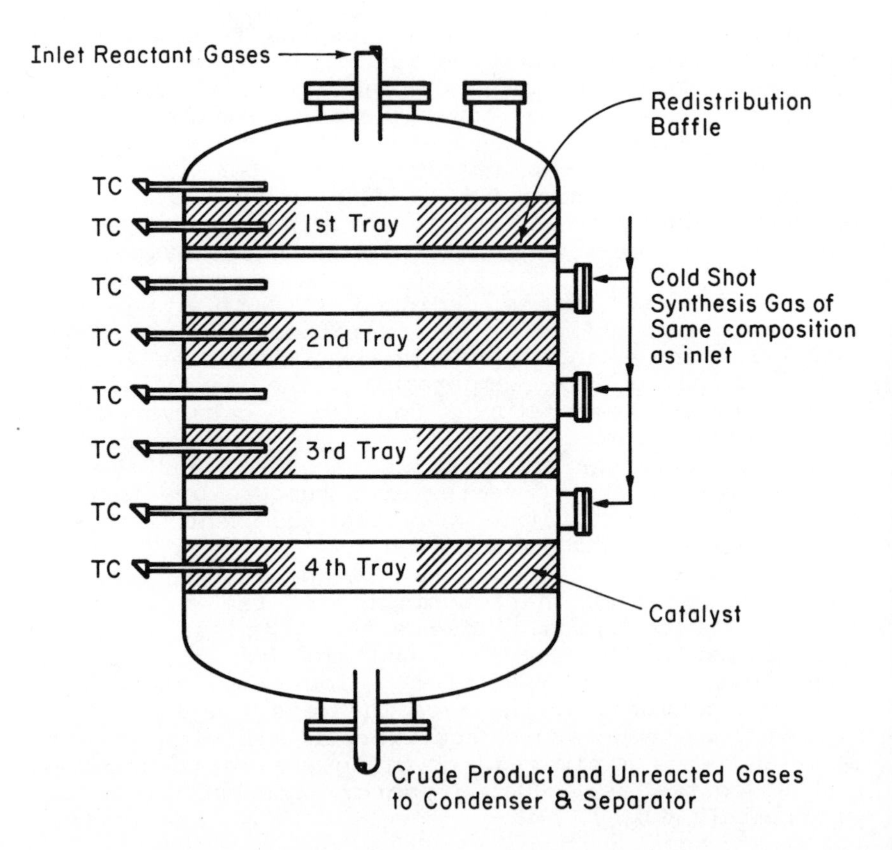

Figure 8. Tray Type Synthesis Converter.

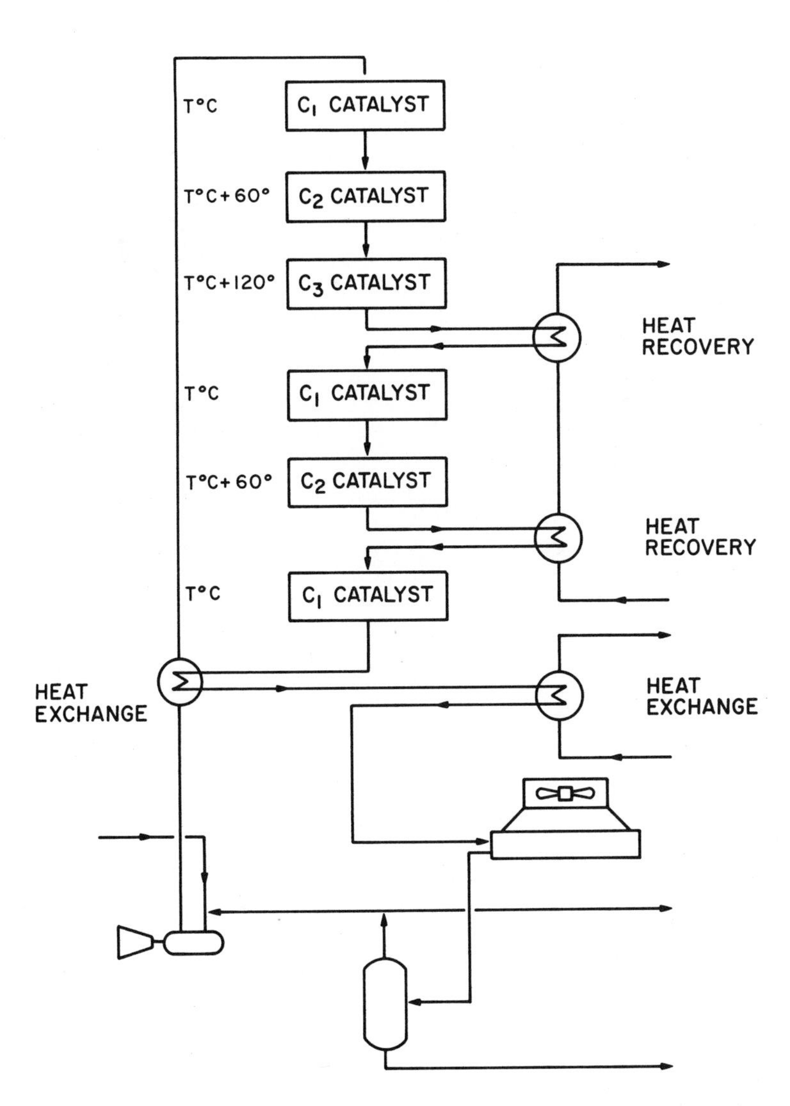

Figure 9. Preferred Example of New Reactor System.

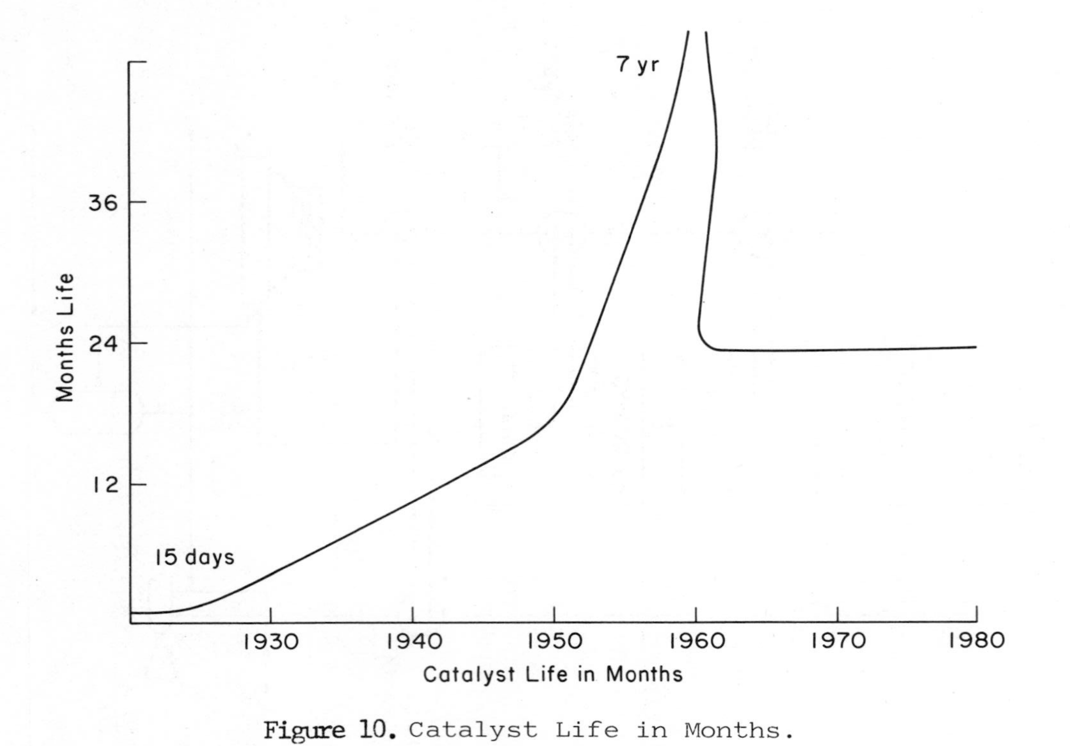

Figure 10. Catalyst Life in Months.

Gas Purity Then and Now

One of the initial problems with alcohol synthesis is the fact that gases were relatively impure. This required the use of high temperatures and relatively crude catalysts. These function at 450° whereas over the years the trend has been toward lower temperature and more active catalysts which are also more temperature sensitive. Present temperature trends are toward approximately $250^{o}C$, permissible when the synthesis gas purity is excellent as in the case presently indicated by Figure 11 plotting the sulfur level.

Operating Pressure Then and Now

Initially, the reactors in the days of the early methanol and ammonia syntheses operated at approximately 1000 atmospheres or 15,000 psi. This, over the years, has been reduced to 12,000, then to 5,000 and then to 750 but presently the trend is back up toward 2,500 psi (Figure 12). It is likely that, when single line 5,000 or 10,000 ton per day plants are constructed, the optimum pressure will be in the range of 2,500 to 4,000 psi. The economics of pressure and size are all closely related to many factors such as capital investment, reactor size, synthesis gas generating pressure, interconnecting piping size, method of generating synthesis gas, and problems of maintenance.

Methanol Prices Then, Now and in the Future

Figure 13 shows the price of methanol over the last 55 years. The present price quoted is about 72 cents but spot market price can be as low as fifty-eight to sixty cents.

Product Uses and Forecasts for the Future

The uses of methanol in relative descending order of volume are as follows:

Formaldehyde
Dimethyl terephthalate
Solvents
Methyl halides
Methyl amines
Methyl methacrylate
Inhibitor for formaldehyde
Acetic acid
MTBE, Fuels - increasing rapidly

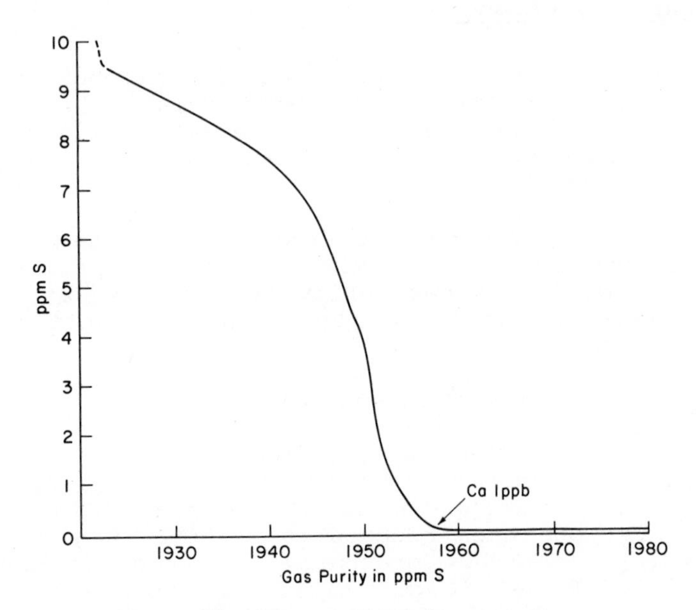

Figure 11. Gas Purity in ppm S.

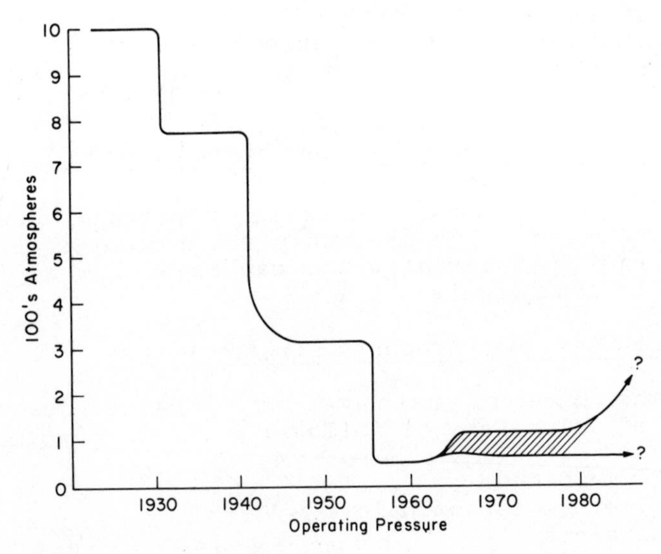

Figure 12. Operating Pressure Then and Now.

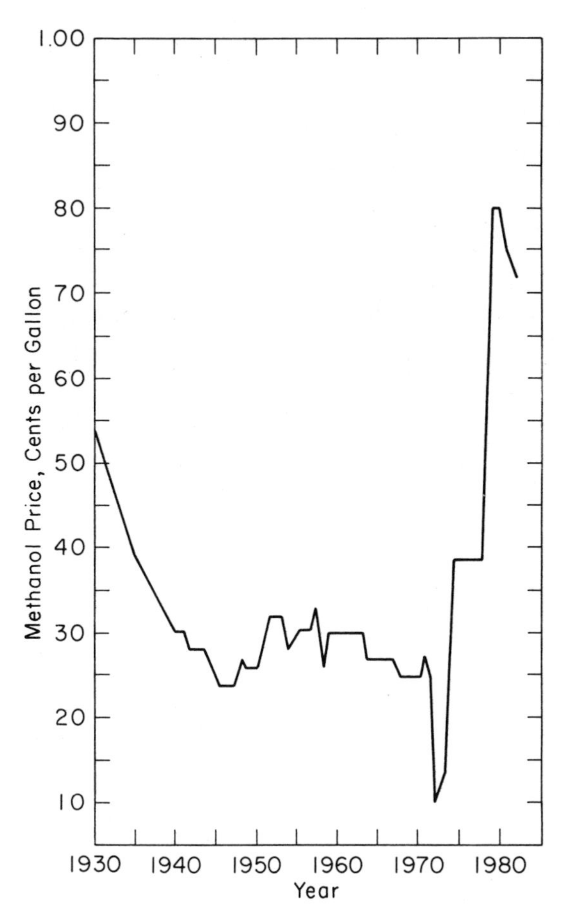

Figure 13. Methanol Price, Cents per Gallon.

Production Capacity

The annual production capacity for U.S., Canada and New Zealand is to become approximately 2,500 million gallons in 1983. This is an increase of about 140% in six years. This capacity is approximately 6.5 days equivalent of gasoline consumption in the U.S. annually. It is evident that if one of our liquid fuels is to be methanol in any major proportion, orders of magnitude increases in capacity will be called for. Current research and road tests of automobiles fueled with 100% methanol are very encouraging both economically and operationally. This is a separate and lengthy subject.

Literature Cited

1. A. Mittasch and C. Schneider, German Patent 293, 787, 1913 (U.S. Patent 1,128,804; Feb. 16, 1915).
2. H. Fischer and H. Tropsch, German Patent 411,216; 1922.
3. A. Mittasch, M. Pier and K. Winkler, German Patent 441, 433; 1923.
4. A. Mittasch, M. Pier and K. Winkler, European Patent 299, 714; March 26, 1925.
5. W. K. Lewis and P. K. Frohlich, Ind. and Eng. Chem. 1928, 20, 287.
6. W. Lazier, U.S. Patent 1,829,046; Oct. 27, 1932.
7. A. B. Stiles, A. I. Ch. E. J., 1977, 23, 362.

RECEIVED January 25, 1983

27

Applications of Magnetic Resonance in Catalytic Research

WALLACE S. BREY
University of Florida, Department of Chemistry, Gainesville, FL 32611

The early development of the use of magnetic resonance methods in catalyst research is surveyed. Various magnetic resonance parameters that may be utilized are introduced and the types of information that may be obtained by their measurement are described in relation to contributions by early workers in the field. Some comments on the limitations of NMR and EPR methods as well as indications of prospects for future progress are given, along with a brief mention of recent advances in the applications of specialized techniques of NMR. References are provided for a number of extensive and detailed review papers.

In this account we will attempt to trace some of the developments in the application of nuclear and electron magnetic resonance techniques to the study of catalysts and catalytic processes from the pioneering studies up to the middle 1960's. To place these developments in perspective, some account will be given of the invention of magnetic resonance methods for condensed matter and of the various types of apparatus that were available, especially for NMR where there has been substantial improvement in instrumentation with time. (We shall abbreviate nuclear magnetic resonance by "NMR" and electron paramagnetic resonance by "EPR". We prefer the latter designation to electron spin resonance, because electron orbital as well as spin moments may be involved.) Since many readers may not be familiar with the principles of magnetic resonance, qualitative explanations of basic aspects are included. Emphasis will be on work in the United States, but some contributions from elsewhere will be mentioned.

0097-6156/83/0222-0375$06.00/0

The Beginnings of Magnetic Resonance Experiments

In the 1920's and the 1930's, physicists performing experiments in the gas phase, usually at very low pressure, demonstrated unequivocally that electrons and some nuclei behave as minute magnetic dipoles. Pauli had proposed in 1924 that the hyperfine structure observed in atomic electronic spectra might result from a nuclear angular momentum and an associated magnetic moment, and in the next year Uhlenbeck and Goudsmit showed that fine structure in atomic spectra can be explained on the basis of interactions of the electronic orbital magnetic moment with an intrinsic electron magnetic moment, such as might be associated with the angular momentum of a "spinning" electrically charged particle, in accordance with the results of the well-known Stern-Gerlach experiment of 1924. This experiment involved the deflection of a beam of atoms containing unpaired electrons in an inhomogeneous magnetic field, and Rabi extended the method to the measurement of nuclear magnetic moments. Hyperfine effects from nuclear moments were also observed as perturbations on the rotational energy levels of small molecules (1).

In order to make measurements of electronic and spin magnetic moments in condensed matter--in solids or liquids--without the necessity for volatilization in a high-vacuum system, it was attractive to attempt resonance experiments, which would have the added advantage of a specific determination of the energy level pattern as deduced from the particular energies required to induce transitions. For electronic spins, this type of experiment is not quite so critical, because the magnitude of the magnetic moment is sufficiently large to give sensible contributions to the bulk magnetic susceptibility, but the magnitudes of nuclear moments--a thousand or more times smaller than that for the electron--cause the nuclear contribution to the magnetic susceptibility to be lost in the experimental uncertainty of the susceptibility measurements.

A particular incentive to measure accurately the magnetic moment of a nuclear species--more specifically, the nuclear gyromagnetic or magnetogyric ratio, equal to the ratio of the moment of the nucleus to its angular momentum--was the desire to test theories of nuclear structure, which were under development in the late 30's and early 40's. In Holland the physicist C. J. Gorter led a group which invested substantial effort in the search for nuclear resonance in bulk matter, only to fail because of the unfortunate choice of samples for which the relaxation time was very long (2). Electron resonance was to come first, with a report from Zavoisky in Russa in 1945 of success in this area (3).

Success in the observation of nuclear magnetic resonance in bulk matter was achieved almost simultaneously by two groups in the United States: Purcell, Torrey, and Pound at Harvard used a single-coil or bridge method with paraffin as the sample (4), and Bloch, Hansen, and Packard at Stanford employed an apparatus with separate transmitter and receiver coils with water as the sample (5). Bloch and Purcell shared the Nobel prize in recognition of these achievements.

The First Decade of Applications to Catalysis

The author first learned of the magnetic resonance experiment in 1951, when Pierce Selwood gave a talk on magnetic methods as applied to catalysts as part of a series of lectures on recent developments in catalysis sponsored by the Philadelphia Section of the American Chemical Society. Although Selwood did not mention nuclear resonance in the main part of the lecture, there was considerable excitement about this mysterious new type of experiment among some of the members of the audience, which resulted in a lengthy discussion period devoted almost entirely to an explanation of nuclear induction, as the experiment was then called, with the aid of a precessing pointer. A pioneer in the applications of all sorts of magnetic methods to catalytic systems. Selwood had published his first nuclear induction paper in 1949 (6). With R. B. Spooner, he utilized a double-coil apparatus with 20 Hz modulation, to measure by nuclear resonance absorption, according to the techniques of Bloembergen, Purcell, and Pound (7), the effects of supported paramagnetic catalysts on the relaxation times of the nuclei of liquid water. For MnO_2 on TiO_2, for example, it was found that a smaller fraction of the paramagnetic ions is available to the water when the concentration of active oxide on the support is greater, indicating larger microcrystals of the active oxide at higher concentrations, a result in accord with magnetic susceptibility data for this system. For iron oxide on γ-alumina, however, the relaxation results were quite different from the susceptibility results, and it was concluded that the iron atoms go into solid solution in the alumina upon heating, thus becoming inaccessible to the water as well as unavailable for catalytic processes. A comment in this paper is interesting: "The signals obtained are very small; their detection is a matter of difficulty owing to spurious effects." However, there is also a note of optimism in the proposal that the method would be suitable for "in situ" measurements of catalytic systems.

Selwood's next publication on NMR appears in the proceedings of a Faraday Society Disucssion held in Liverpool in April of 1950 (8). Incidentally, this volume contains an extended introduction by Sir Hugh Taylor in which he draws attention to two earlier Faraday Society General Discussions, one in 1922 in which the Lindemann and Langmuir theories were introduced, and one in 1932 in which the distinction between chemical adsorption and physical adsorption was clearly drawn. In his paper, Selwood presented data for the accessibility to water of the paramagnetic materials iron oxide, copper oxide, and chromia, each supported on alumina. The relaxation times of water in contact with the oxide were compared to those for water in solutions containing the same metal ions. As in the previous results, accessibility tended to decrease with increasing concentration of paramagnetic ions on the surface, but there is now also pointed out a direct parallel of the results for chromia with the activity of samples of this material for the conversion of n-heptane to toluene.

To understand Selwood's initial work a bit better, it is helpful for one to have a description of several aspects of nuclear "relaxation". Spin-lattice or T_1 relaxation processes are transitions from one spin state to another, tending to restore in time the spins in a system to thermal equilibrium according to a Boltzmann distribution, and induced by occurrences in the sample in the vicinity of the nucleus rather than by radio signals applied by the experimenter. For nuclei with spin quantum number 1/2, these transitions are produced by fluctuations in the magnetic field at the location of the nucleus, fluctuations most effective when they have a maximum frequency component at the nuclear resonance frequency in the fixed magnetic field of the spectrometer. They should, in other words, correspond fairly closely in frequency to the spectrometer operating frequency. Such fluctuations in field may occur as the result of rotation of the molecule containing the nucleus, when the effects of other magnetic nuclei in the molecule are felt, or by diffusion of the molecule in the presence of magnetic fields from other molecules. Relaxation is particularly rapid when there are paramagnetic species nearby, because of the large magnetic moment of the electron which generates magnetic fields effective at longer ranges; any appreciable concentration of species with unpaired electrons causes paramagnetic relaxation to predominate over other types of relaxation, except perhaps for nuclei with spin greater than 1/2. Nuclei with spins of 1 or more have electric quadrupole moments as well as magnetic dipole moments, and therefore undergo spin flips caused by varying electric fields associated with motions of electrons.

In 1956, Selwood published the results of studies by the spin-echo method, in which the response of the nuclear magnetization to a series of pulses is measured, for methanol, ethanol, water, and n-hexane adsorbed on γ-alumina, silica-alumina, MnO, and CuO, all presumably diamagnetic materials (9). He found for the proton relaxation time marked reductions over those in the liquid phase, but pointed out that many uncertainties remained, such as the effects of traces of paramagnetic impurities and of the shapes of pores in which the adsorbate molecules were located.

The year 1956 saw the First International Congress on Catalysis in Philadelphia, for which the proceedings were published in 1957 as Volume 9 in the Advances in Catalysis series. The author was fortunate in being able to attend this meeting; held in the Bellevue-Stratford hotel, it represented a gathering of leaders in catalytic research from throughout the world, including Russia. Only two papers referred to magnetic resonance, however. Selwood merely mentioned the promise of nuclear and electron paramagnetic resonance techniques as adjuncts to magnetic susceptibility methods, but John Turkevich described his work on the effects of the temperature of heating and of the adsorption of oxygen or nitric oxide on the intensity and bandwidth of resonances from sugar charcoal (10).

About this time, a number of workers were investigating the EPR spectra of porous carbons prepared in various ways, and a paper from Ingram's laboratory in England had been entitled, "Paramagnetic Resonance from Broken Carbon Bonds" (11). In 1958 Turkevich showed that the catalytic activity of charcoal is related to the characteristics of its EPR absorption (12). For the ortho-para hydrogen conversion at -196°, the rate constant was a maximum for samples treated at 600°, which have a maximum signal intensity and minimum linewidth. For hydrogen-deuterium exchange at 50°, the rate constant increases monotonically with preparation temperature up to 950°, paralleling an increase in linewidth which is attributed to an increase in electron mobility.

In the latter 50's, several groups applied NMR to study adsorption on metal oxides, although direct correlations with catalytic activity were not made. Aston's group at Pennsylvania State looked at water, CH_4, and CF_4 on TiO_2 (13-16), and workers at Bell Laboratories examined water on the same oxide (17). Analysis was based primarily on line widths: for long correlation times, or solid-like behavior, one expects very broad lines, while high mobility is associated with narrow lines. John Zimmerman and coworkers at Magnolia Petroleum Field Research Laboratories in Dallas began a series of more sophisticated relaxation measurements for water on silica gel which will be described further below, and Winkler at Leipzig was involved in investigations of water on alumina which were interpreted in terms of macropores and micropores in the adsorbent (18).

Developments Through the Early 1960's

The period of about five years beginning in 1958 represented a substantial increase in interest in magnetic resonance in catalysis as reflected by the number and diversity of publications. In 1958, there was a substantial leap forward in sensitivity and in operating convenience in commercial NMR instrumentation with the introduction by Varian Associates of a 14,000 gauss spectrometer operating at 60 MHz for protons. This instrument included an optically actuated field stabilizer and a phase sensitive detector. Earlier Varian instruments operated at 40 MHz for hydrogen and used diode detection so that nulling baseline leakage was not easy. The author was privileged to have one of the first of the new instruments in his laboratory in 1958; a "dual-purpose" or DP model, it could perform audio-modulated wide-line experiments as well as high-resolution experiments.

In 1960, the first review of magnetic resonance applied to catalysis appeared. Written by D. E. O'Reilly of Gulf Research and Development Laboratories in Pittsburgh, it was published as a section of Volume 12 of Advances in Catalysis (19). In addition to a comprehensive review of results of other workers in the field and a thorough discussion of principles and methods, O'Reilly included considerable detail about various results from the Gulf Laboratories, including some material not previously published. EPR studies of the oxidation states of chromium in chromia-alumina catalysts are extensively described and the results are related to activity for hydrogen peroxide decomposition and for dehydrocyclization of n-heptane reported by others. Some of O'Reilly's NMR results represent a change in emphasis in the study of catalytic phenomena-previous work concentrated on the resonances of adsorbed species--toward direct examination of the catalysts themselves. He examined carefully the lineshapes of the aluminum nuclear transitions in various forms of alumina and related the transition intensity to BET area for γ-alumina. The fluorine-19 resonance in fluorine doped alumina was investigated, as were the relaxation times of protons in partially dehydrated silica gel and silica-alumina; much of the last work had been reported earlier in a classic paper with Leftin and Hall in 1958 (20).

In Europe, radiofrequency spectroscopists--many of them now magnetic resonance researchers--met in Leipzig in September 1961 at the Tenth Colloque Ampère. The proceedings of this meeting included several papers dealing with catalyst systems (21). Work from Turkevich's laboratory on the catalytic properties of $LiAlH_4$, heated to produce active centers for H_2-D_2 exchange, was described. A Varian 40 MHz spectrometer was used in an attempt to observe Li and Al resonance; although success was limited, this was another early attempt to look directly at a catalyst. From Nicolau's laboratory in Berlin came a report of EPR signals of Pt catalysts supported on carbon, where the effect of heating produced parallel variation of signal characteristics and activity for hydrogenation. A group in Grenoble observed deuterium double-quantum transitions in zeolites in which the water of crystallization had been exchanged by D_2O. Although it had no particular reference to catalytic activity, this work is interesting because it is a very early application of a technique used extensively in much more recent studies of deuterium.

The first volume of the Journal of Catalysis appeared in 1962, and included two significant papers on the application of EPR to real catalyst systems. One by Poole, Kehl, and MacIver (22) extended EPR studies of chromia-alumina systems. The investigation of chromia-containing catalysts continued to be a very productive and well-ploughed field and has been described in several places (23).

In the other magnetic resonance paper in the Journal, Keith

Hall showed that several polynuclear hydrocarbons and phenylated amines react to form cation radicals on the surface of a silica-alumina catalyst (24). It was hoped to use hyperfine splitting to identify the radicals formed, but broadening of the lines by the surface obscured much of the hoped-for detail. Rough estimates of the spin concentration could be made and a significant effect of oxygen in formation of radicals and in determination of their concentration was observed, although the precise role of oxygen could not be assigned. Hall's work was part of an effort to answer two related questions which continued to perplex catalyst researchers--and were sometimes the cause of controversy among them--questions as to whether the acid sites on the catalyst surface are Lewis or Bronsted in nature, and whether the intermediates in catalytic reactions are carbonium ions or radicals. The work followed closely upon preliminary results of Rooney and Pink in England (25), who showed that perylene and anthracene when adsorbed on silica-alumina catalysts gave an EPR signal with hyperfine splitting similar to that found for the radicals in concentrated sulfuric acid, and who later concluded that cation-radical formation takes place at Lewis acid sites and showed that oxygen has a reversible broadening effect at room temperature (26). In addition, Fogo of Union Oil, in a short communication (27), had reported that the presence of oxygen on the surface facilitated formation of radicals from anthracene, and that hydrogenation of a silica-alumina catalyst reduced the formation of radicals from perylene.

Thus, as NMR spectroscopists were turning some of their attention to the material of the catalyst, the EPR folks were beginning to look at the species that might be found on the surface.

The effects of high-energy radiation, both in aiding the formation of radicals on the surface and in producing defect centers within the catalyst, were becoming of interest. In Russia, Kazanskii was working in this field (28, 29), and Paul Emmett collaborated with scientists at Oak Ridge to study the formation of hydrogen atoms in irradiated catalysts (30). It was argued that the mutual modification of constituent properties in silica-alumina catalysts was indicated by the fact that these materials gave more hydrogen atoms upon irradiation than did the separate components, and it was pointed out that the number produced should depend both upon acid strength of the hydrogen source and upon the stability of the trapping sites. It was suggested that, although NMR does not distinguish between the OH groups in the mixed catalysts and those in the pure components, part of the water content of the mixed materials must be in a special form capable of yielding hydrogen atoms upon irradiation.

We mentioned earlier the series of papers from Zimmerman's group in Dallas; a series which was continued with important contributions from Don Woessner and under the name of Socony Mobil Oil Company (31-37). The relaxation behavior of water on silica gel was measured by spin-echo methods, and it was established that the water exists in two different environments, with different restrictions on mobility. Whether or not the two phases can be distinguished depends upon the relation between the lifetime in each phase and the relaxation times in the two phases. A theory relating these magnitudes was developed for isotropic motion of the molecules. The treatment was then extended to anisotropic motion, assuming random reorientation of the proton-proton vector about a normal to the surface, and the predictions were found to be consistent with the experimental results. Studies of the silica-water system over an extended temperature range proved very profitable and permitted a much more detailed analysis of the motions in the two phases as well as an estimate of about 4.9 kcal/mol for the activation energy for interphase transfer. In 1966, Woessner applied these theories to an extensive study of benzene adsorbed on silica gel. Further developments in the general area have owed much to the theoretical analysis of the interrelations of motion, relaxation, and interphase transfer as worked out by Resing (38, 39).

An area of considerable potential in catalytic studies which was investigated in the late 50's and early 60's and still holds promise of further applicability involves the transfer of polarization from one magnetic species to another. Uebersfeld published a whole series of papers concerned with the effects observed when the EPR transitions in a paramagnetic material such as charcoal are pumped, and protons in the vicinity of the unpaired electrons are observed (40, 41). If the protons are linked to the electrons by hyperfine coupling, the proton resonance intensity is enhanced by the Overhauser effect; if the interaction is of the direct dipolar type, the "solid effect" leads to enhancement. Krebs also investigated these effects in detail and obtained information about the defect centers in charcoal through the enhancements by the solid effect of the resonances associated with chemically bound protons (42). The general approach of transferring spin information from one species to another has considerable promise for providing additional information about atomic and molecular arrangements in the vicinity of surfaces.

The Chemical Shift

By the early sixties, the various aspects of magnetic resonance that might be exploited in the study of adsorption and catalysis had been fairly well delineated, with one exception. Improvements in instrumentation, developments in theory, and better correlations with other experiments were to come, but, except for the applications of the NMR chemical shift, substantial beginnings had been made.

The desire of physicists to measure precisely nuclear magnetogyric ratios was, as we have seen, one of the impelling reasons for the invention of the NMR Experiment. Very shortly, however, it became evident that this effort was to be less fruitful than anticipated, when Knight observed that the resonance peak for cupric ion in solution was accompanied by another peak, which he came to realize was produced by metallic copper in the probe. The complication which entered the efforts at measurement of the nuclear properties is that the electrons surrounding the nucleus get in the way by altering the magnitude of the magnetic field which reaches the nucleus from that which is produced by the external magnet. Called by the physicists, perhaps a bit disdainfully, the "chemical shift", this feature of NMR was to become a keystone in the structural application of NMR for molecules in the liquid phase.

The magnitudes of chemical shift differences between nuclei in different environments are relatively small in comparison to typical linewidths of resonances in the solid phase, and so nuclei of different shifts in solids could not be distinguished by the techniques of the sixties. Adsorbed molecules fall in a borderline region between solid and liquid; depending upon their mobility and consequent line narrowing, it may or may not be possible to distinguish nuclei in different chemical locations in a molecule, although such distinctions would permit one to learn much about the relative restrictions on motion of different parts of a molecule as it is held on the surface. Wide-line NMR methods, using audio modulation and phase sensitive detection at the modulation frequency, gain sensitivity at the expense of resolution, as the resonance lines are broadened by the modulation. And pulse methods, used without Fourier transformation to unravel the frequency content of the response, give only one averaged relaxation time result for all nuclei of one species, such as all the hydrogen atoms, in a sample. Chemical shift distinctions would permit one to learn about the relative mobility of different parts of a molecule--and perhaps about the extent of the perturbation caused by the adsorption process--which might lead to a better model of the adsorbent-adsorbate interaction.

It is interesting to recall that, in the late 1950's, Paul Emmett, visiting the author's laboratories, told of his experience in being persuaded to send a sample of an adsorbent-adsorbate system to Varian for evaluation of NMR techniques. There came back an enthusiastic report of a substantial chemical shift difference between the adsorbate and the material in the liquid phase, to be followed shortly by another communication which stated that the difference disappeared when a magnetic susceptibility correction was applied to the results. This illustrates a problem which arises whenever one attempts to measure a chemical shift value in a heterogeneous system, the problem of a suitable reference and of eliminating magnetic susceptibility effects which are particularly troublesome in such systems of dubious geometry. Differences between chemical shifts in the same molecule, if they can be resolved, are soundly based, but absolute values may very well not be attainable. Although there are some reports to the contrary, instrumentation of this period probably did not permit resolution of shift differences for hydrogen for surface coverages of less than a monolayer, those of catalytic significance. However, in 1964, Kenneth Lawson, in these laboratories, was able to show by a different approach, utilizing temperature dependence studies, variations in restrictions on bonding of various parts of alcohol or amine molecules adsorbed on high-area thorium oxide (43). For example, as the system is cooled, the hydroxyl hydrogen contribution to the proton resonance of methanol is broadened at about -15°C to the point where it is not observed, although the methyl resonance continues as a relatively narrow line down to -115°.

Developments in NMR methods in recent years have made the prospects for extension of the use of chemical shift differences to characterize adsorption very exciting. For example, the use of Waugh-type pulse sequences on hydrogen has made it possible to average out most of the dipolar interactions between neighboring hydrogen atoms, although the experiment is rather difficult to carry out because of the high-power, short duration pulses which must be applied to the spin system (44).

For other nuclei than hydrogen, which have much greater chemical shift ranges, behavior of different parts of molecules may more readily be distinguished. In results from our laboratories, for which we have not been able to generate an unambiguous model, resonances for all four nonequivalent types of fluorine in perfluoropropylene adsorbed on aluminum oxide were resolved. The appearance of two different CF_3 resonances at slightly different chemical shifts suggests the presence of two distinct arrangements of the adsorbent-adsorbate complex.

Carbon-13 spectroscopy seems to have great potential for the study of heterogeneous systems (45, 46). Not only is the chemical shift range substantial, but the use of strong irradiation of hydrogens eliminates the dipolar broadening from that source and magic angle spinning averages out chemical shift anisotropy (47). Cross polarization from protons (48) can be utilized to enhance sensitivity, and the availability of high-field superconducting spectrometers may be helpful. Although unfavorable effects on relaxation times may under some circumstances be a disadvantage for high-field observation of spin-1/2 nuclei, high fields are beneficial for quadrupole nuclei such as aluminum. Magic-angle spinning is also particularly helpful in removing first-order quadrupolar interactions of these nuclei (49).

Although we have not yet quite attained Selwood's goal of following catalytic reactions by magnetic resonance while the reactions are occurring, there is still hope that the future will see substantial progress in experimental techniques such that his vision will become a reality. While we must continue to ask questions about the results of magnetic resonance--Are these results relevant to the catalytic process? Are the molecule or radicals giving rise to the spectra intermediates in the reaction or do they represent deadend byways?--answers will often be forthcoming from consideration of circumstantial evidence and from comparison with other spectroscopic and kinetic results, and much of interest and value surely remains to be learned.

Reviews of the Literature

For readers who may wish further details on the subject, we conclude by mentioning some reviews and collections of papers which may serve as a guide to the literature of the subject. In a recent volume on spectroscopic methods in catalysis, there is one chapter on NMR and one on EPR; both contain references to other earlier reviews (50). EPR in catalysis was reviewed by Kokes in 1968 (51) and by Tanford in 1972 (52). Nuclear relaxation studies of adsorbates were reviewed by Packer in 1967 (53), and in 1972, Pfeifer (54) provided a comprehensive account of the theory of relaxation and diffusion analysis for adsorbates together with a complete listing of publications up to that time, arranged alphabetically by authors. In 1972, an extensive article on NMR written by the French-Belgian group of workers in this field appeared in Catalysis Reviews (55). Finally, for listings of literature in the last few years, one may consult recent volumes of the Specialist Periodical Reports published by the Royal Society (56, 57, 58).

Literature Cited

1. Kopfermann, H. "Nuclear Moments," translated by E. E. Schneider, Academic Press, New York, 1958.
2. Gorter, C. J.; Broer, L. J. F. Physica 1942, 9, 591.
3. Zavoisky, E. J. Phys. (U.S.S.R.) 1945, 9, 211.
4. Purcell, E. M.; Torrey, H. C.; Pound, R. V. Phys. Rev. 1946, 69, 37.
5. Bloch, F.; Hansen, W. W.; Packard, M. E. Phys. Rev. 1946, 69, 127.
6. Spooner, R. B.; Selwood, P. W. J. Am. Chem. Soc. 1949, 71, 2184.
7. Bloembergen, N.; Purcell, E. M.; Pound, R. V. Phys. Rev. 1948, 73, 679.
8. Selwood, P. W.; Schroyer, F. K. Disc. Faraday Soc. 1950, 8, 337.
9. Hickmott, T. W.; Selwood, P. W. J. Phys. Chem. 1956, 60, 452.
10. "Advances in Catalysis", Vol. IX, Academic Press, New York, 1957.
11. Bennett, J. E.; Ingram, D. J. E.; Tapely, J. G. J. Chem. Phys. 1955, 23, 215.
12. Turkevich, J.; Laroche, J. Z. physik. Chem. 1958, 15, 399.

13. Fuschillo, N.; Aston, J. G. J. Chem. Phys. 1956, 24, 1277.
14. Fuschillo, N.; Renton, C. A. Nature 1957, 180, 1063.
15. Stottlemyer, Q. R.; Murray, G. R.; Aston, J. G. J. Am. Chem. Soc. 1960, 82, 1284.
16. Mays, J. M.; Brady, G. W. J. Chem. Phys. 1956, 25, 583.
17. Aston, J. G.; Bernard, J. W. J. Am. Chem. Soc. 1963, 85, 1573.
18. Winkler, H. Kolloid-Z., 1958, 161, 127.
19. O'Reilly, D. E. "Advances in Catalysis," Vol. XII, Academic Press, New York, 1960; 31-116.
20. O'Reilly, D. E.; Leftin, H. P.; Hall, W. K. J. Chem. Phys. 1958, 29, 970.
21. "Spectroscopy and Relaxation at Radiofrequencies," Proceedings of the Tenth Colloque Ampere, Leipzig; North-Holland, Amsterdam, 1962.
22. Poole, C. P.; Kehl, W. L.; MacIver, D. S. J. Catal. 1962, 1, 407.
23. Kokes, R. J. "Experimental Methods in Catalytic Research," Anderson, R. B., Ed., Academic Press, New York, 1968; 467-470.
24. Hall, W. K. J. Catal. 1962, 1, 53.
25. Rooney, J. J.; Pink, R. C. Proc. Chem. Soc. 1961, 70.
26. Rooney, J. J.; Pink, R. C. Trans. Faraday Soc. 1962, 58, 1632.
27. Fogo, J. K. J. Phys. Chem. 1961, 65, 1919.
28. Kazanskii, V. B.; Pariiskii, G. B.; Voevodskii, V. V. Disc. Farad. Soc. 1961, 31, 203.
29. Kazanskii, V. B.; Pecherskaya, Y. I. Kinetic Catalysis (U.S.S.R.), Eng. trans. 1961, 2, 417.
30. Emmett, P. H.; Livingston, R.; Zeldes, H., Kokes, R. J. J. Phys. Chem. 1962, 66, 921.
31. Zimmerman, J. R.; Holmes, B. G.; Lasater, J. A. J. Phys. Chem. 1956, 60, 1157.
32. Zimmerman, J. R.; Brittin, W. E.; J. Phys. Chem. 1957, 61, 1328.
33. Zimmerman, J. R.; Lasater, J. A. J. Phys. Chem. 1958, 62, 1157.
34. Woessner, D. E. J. Chem. Phys. 1961, 35, 41.
35. Woessner, D. E.; Zimmerman, J. R. J. Phys. Chem. 1963, 67, 1590.
36. Woessner, D. E. J. Chem. Phys. 1963, 39, 2782.
37. Woessner, D. E. J. Phys. Chem. 1966, 70, 1217.
38. Resing, H. A. J. Chem. Phys.1965, 43, 669.
39. Resing, H. A. Advan. Mol. Relaxation Processes, 1968, 1, 109.
40. Erb, E.; Motchane, J. L.; Uebersfeld, J. Compt. rend. 1958, 246, 2121.
41. Jacubowicz, M.; Uebersfeld, J. Compt. rend. 1959, 249, 2743.
42. Krebs, J. J.; Thompson, J. K. J. Chem. Phys. 1962, 36, 2509.
43. Brey, W. S.; Lawson, K. D. J. Phys. Chem. 1964, 68, 1474.

44. Kaplan, S.; Resing, H. A.; Waugh, J. S. J. Chem. Phys. 1973, 59, 5681.
45. Gay, I. D. J. Phys. Chem. 1974, 78, 38.
46. Michel, D.; Pfeifer, J.; Delmau, J. J. Magn. Reson. 1981, 45, 30.
47. Stejskal, E. O.; Schaefer, J.; Henis, J. M. S.; Tripodi, M. K. J. Chem. Phys. 1974, 61, 2351
48. Vaughan, R. W.; Schreiber, L. B.; Schwarz, J. A. "Magnetic Resonance in Colloid and Interface Science," Resing, H. A.; Wade, C. G., Eds., ACS Symposium Series 34, American Chemical Society, Washington, 1976; 275.
49. Fyfe, C. A.; Gobbi, G. C.; Hartman, J. S.; Lenkinski, R. E.; O'Brien, J. H.; Beange, E. R.; Smith, M. A. R. J. Magn. Reson. 1982, 47, 168.
50. Delgass, W. N.; Haller, G. L.; Kellerman, R.; Lunsford, J. H. "Spectroscopy in Heterogeneous Catalysis," Academic Press, New York, 1979.
51. Kokes, R. J. "Experimental Methods in Catalytic Research," Anderson, R. B., Ed., Academic Press, New York, 1968; 436-475.
52. Lunsford, J. H. "Advances in Catalysis," Vol. 22, Academic Press, New York, 1972; 265-344.
53. Packer, K. J. "Progress in Nuclear Magnetic Resonance Spectroscopy," Emsley, J. W.; Feeney, J.; Sutcliffe, L. H., Eds., Vol. 3, Pergamon Press, Oxford, 1967; 87-128
54. Pfeifer, H. "NMR Basic Principles and Progress," Diehl, P.; Fluck, E.; Kosfeld, R., Eds., Vol. 7, Springer, New York, 1972; 53-153.
55. Derouane, E. G.; Fraissard, J.; Fripiat, J. J.; Stone, W. E. E. Catalysis Reviews, 1972, 7, 121-212.
56. "Specialist Periodical Reports. Spectroscopic Properties of Inorganic and Organometallic Compounds," Vol. 14, Royal Society of Chemistry, London, 1981; 102-105.
57. "Specialist Periodical Reports. Nuclear Magnetic Resonance," Vol. 9, Royal Society of Chemistry, London, 1980; 143-146, 268-269.
58. "Specialist Periodical Reports. Electron Spin Resonance," Vol. 6, Royal Society of Chemistry, London, 1981; 119-121, 232.

RECEIVED October 29, 1982

28

Chain Growth and Iron Nitrides in the Fischer–Tropsch Synthesis

ROBERT B. ANDERSON

McMaster University, Department of Chemical Engineering, Institute for Materials Research, Hamilton, Ontario, Canada

Developments in the Fischer-Tropsch synthesis at the Bureau of Mines from 1945 to 1960 include a simple mechanism for chain growth and the use of iron nitrides as catalysts. The chain-growth scheme can predict the carbon-number and isomer distributions for products from most catalysts with reasonable accuracy using only 2 adjustable parameters. Iron nitrides are active, durable catalysts that produce high yields of alcohols and no wax. During the synthesis, the nitrides are converted to carbonitrides.

At the end of World War II, widespread fear of petroleum shortages led to the authorization of the U.S. Bureau of Mines to initiate a large research and development program on fossil fuel-to-oil processes. A laboratory was established at Bruceton, Pa., under the direction of the late Henry H. Storch (1). Dr. Storch was an authority on catalysis and coal and an outstanding research administrator. This chapter is dedicated to his memory.

In research on the Fischer-Tropsch synthesis, FTS, at the Bureau, mechanisms of the growth of the carbon chain and the use of iron nitrides as catalysts were developments that were not anticipated by previous German work. Herington (2) in 1946 was the first to consider chain growth in FTS. He defined a probability β that the chain will desorb rather than grow at the surface, where

$$\beta = \phi_n / \sum_{n+1}^{\infty} \phi_i = (1-\alpha)/\alpha \qquad (1)$$

and α is the probability of chain growth and ϕ_i is the moles of product of carbon number i. Equation 1 provides an unambiguous

0097–6156/83/0222–0389$06.00/0

description of the growth process and is still used occasionally (3). Friedel and Anderson (4) showed that if β was constant over a range of carbon numbers, in this range

$$\phi_{n+1}/\phi_n = \alpha, \quad (2)$$

and

$$\phi_n = \phi_i \ \alpha^{n-i} \quad (3)$$

Equation 3 was also derived by a kinetic scheme for adding carbon atoms one at a time to the growing chain when the ratio of the parameters for propagation to termination is constant. Equation 3 may be summed in various ways; for example, Manes (5) showed how the yields of gasoline, diesel oil and wax change with α. The diesel-oil fraction, C_{12}-C_{18}, of the condensed hydrocarbons was 20 to 30%, relatively independent of the value of α.

Anderson, et al., (6,7) extended the chain growth process to account for the production of straight carbon chains and chains with methyl branches; ethyl-substituted species had not yet been found in synthesis products. Branching was postulated to be a part of the chain growth as depicted by the network in Table 1, in which carbons are added one at a time to the end or penultimate carbons at one end of the chain as indicated by the asterisks.

Table 1. Chain Growth Network

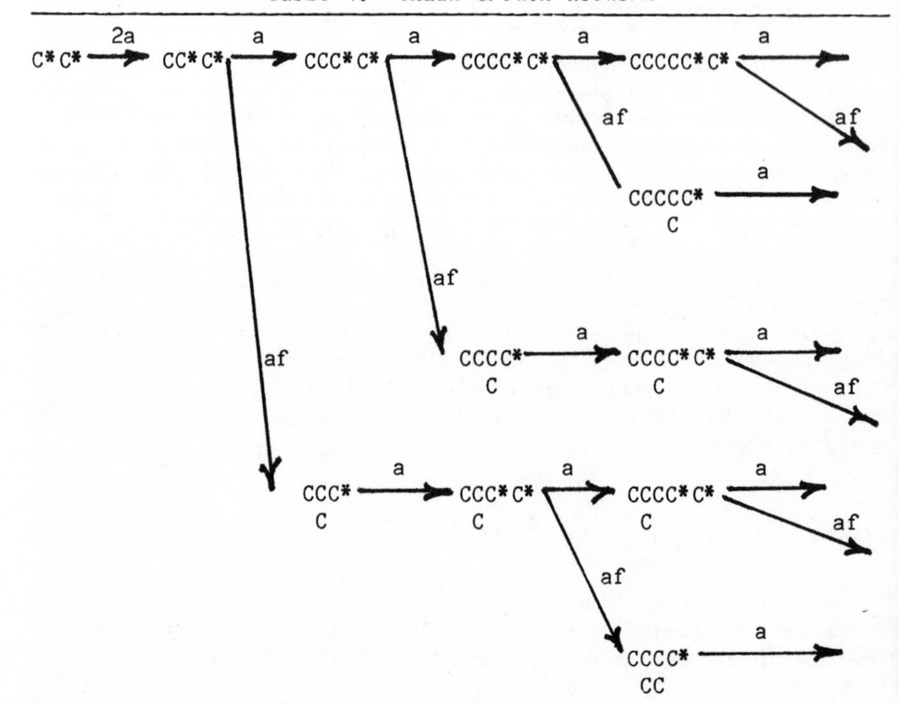

Addition to the penultimate carbon does not occur if it is already attached to 3 carbons. If the kinetics constants are arranged as shown in the network, where for straight chains $\phi_{n+1} = \phi_n a$ and for branched chains $\phi'_{n+1} = \phi_n af$, the ratio of the moles of a monomethyl branched chain to straight chains in a given carbon number is f or 2f, the ratio of dimethyls to straight molecules f^2 or $2f^2$.

The carbon number distribution is given by

$$\phi_n = \phi_2 \; a^{n-2}(1 + (n-3)\; f + \frac{(n-4)(n-5)}{2}\; f^2) \qquad (4)$$

for the range of n from 4 to 8. Terms for trimethyl and more highly substituted species must be included for larger carbon numbers.

One of distributions from the thesis of Achtsnit (8) for products from a $Co-ThO_2$-kieselguhr catalyst with $2H_2+1CO$ feed at 187°C and 1 atm was used to illustrate the usefulness of equation 4, as shown in Table 2. This product had been hydrogenated to saturate the olefins to simplify the analysis, and only data for normal and monomethyl branched molecules were reported. The amount of dimethyl species should be about 10% of the monomethyls. The agreement between calculated and experimental values are sastifactory for C_4 and above, often within the overall uncertainties of the analysis. If the growth parameter for C_2 is taken as 2a as shown in Table 1, the value in parentheses, 6.95, is obtained. The values for C_1 are widely different, and it is possible that methane is produced in other ways and should not be included in the calculation.

Equation 2 was rediscovered in 1976 and perhaps impertinently given the name "Schulz-Flory equation" honoring the famous polymer chemists (9 to 14). A more appropriate name might have been the "Bureau of Mines equation".

Recently, some have suggested that only straight chains are produced in the synthesis reactions and branched molecules arise from the incorporation in the synthesis of olefins, particularly propylene (12,15). Ethylene and propylene do incorporate on Co catalysts, but not on Fe. Lee and Anderson (16) used equation 4 to calculate the moles of propylene that must be incorporated to produce the branched molecules per mole of propylene produced. Here equation 4 was used only to represent the distribution data. The values of propylene incorporated per propylene produced are large numbers approaching or exceeding 1.0; for example, for data in Table 2, 0.782, if the summation is to n = 14 and 0.914 if the summation is to infinity. Thus, the incorporation of propylene cannot be the only source of branched molecules for Co, and with Fe, this mechanism is definitely not the source of methyl branches.

Table 2. Prediction of products from a cobalt catalyst at 187°C, $2H_2$ + 1CO feed, and 1 atm, (1). The parameters are a = 0.742 and f = 0.082.

Carbon number and structure[a]	Exptl. mole %	Calc.	Carbon number and structure	Exptl., mole %	Calc
1	44.9	18.74	10N	1.31	1.28
			2M	0.238	0.210
2	5.41	13.90(6.95)	3M	.226	.210
			4+5M	.269	.315
3	8.22	10.31			
			11N	1.01	0.948
4N	7.98	7.65	2M	0.159	.155
2M	0.500	0.627	3M	.157	.155
			4M	.146	.155
5N	5.89	5.67	5M	.133	.155
2M	1.14	0.932			
			12N	0.788	0.703
6N	4.37	4.21	2M	.110	.115
2M	0.921	0.690	3M	.117	.115
3M	.518	.345	4M	.108	.115
			5+6M	.149	.173
7N	3.07	3.12			
2M	0.595	0.513	13N	0.609	0.522
3M	.691	.513	2M	.080	.086
			3M	.087	.086
8N	2.35	2.32	4M	.089	.086
2M	0.325	0.380	5+6M	.145	.171
3M	.506	.380			
4M	.246	.190	14N	0.484	0.387
			2M	.062	.063
9N[a]	1.72[b]	1.72[b]	3M	.070	.063
2+4M[a]	0.428	.564	4M	.061	.063
3M[a]	.275	.282	5M	.066	.063
			6+7M	.073	.095

[a] 9N denotes n-nonane; 2+4M, a composite fraction of 1-methyloctane and 4 methyloctane; and 3M, 3 methyloctane.

[b] These values set equal in the calculation

In 1948, Jack (17) reported the following reaction sequences on iron powders at 300° - 400°C as shown in Table 3.

Table 3 Reaction Sequences for Nitrides, Carbonitrides and Carbides of Iron.

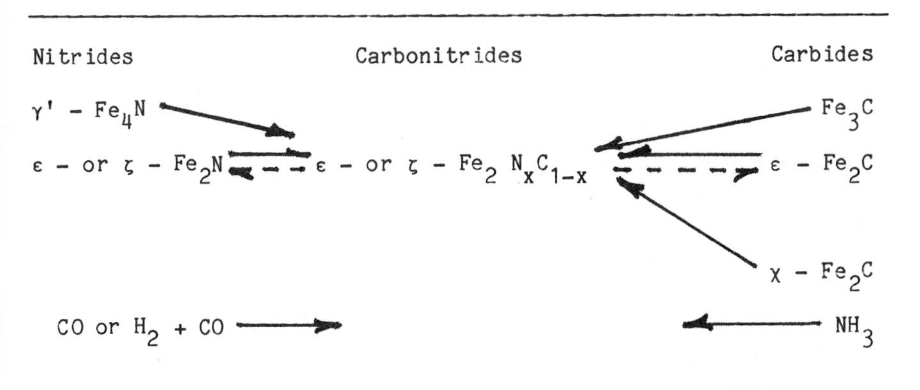

The dashed arrows correspond to these reactions that do not occur readily. These sequences were subsequently shown to proceed more rapidly or at lower temperatures on porous iron catalysts of moderate surface area (18) than on the iron powders.

The iron carbonitrides seemed to be the best candidates for the initial tests in the Fischer-Tropsch synthesis (FTS); however, apparatus for preparing them was not available and the nitrides were used in the first tests. The ε-Fe_2N was rapidly hydrogenated in pure H_2 as shown in Figure 1, but this reaction as well as a variety of hydrogenolysis reactions, are inhibited by CO in the FTS. The nitrogen was removed only slowly during the FTS and the nitrogen lost was largely replaced by carbon to form an ε-iron carbonitride as shown in Figure 2 (19 to 22). Iron catalysts are oxidized during the FTS, presumably by water produced in the reaction. Figure 3 shows that the nitrided catalysts also oxidized more slowly than reduced catalysts (22). In the experiments described in the present paper two fused magnetite catalysts were used: D3001 containing K_2O and MgO and D3008 containing K_2O and Al_2O_3. Both catalysts were nearly completely reduced in H_2 at 450°C and some reduced samples were nitrided in NH_3 at 350°C.

In synthesis tests with D3001 and D3008 with $1H_2$ + 1CO gas at 7.8 atm and an hourly space velocity of 100, the activity of the reduced catalysts decreased rapidly (23), for D3001 a 2-fold decrease in 8 weeks and for D3008 a 3-fold decrease in 4 weeks. Initially the nitrides were about twice as active as the reduced

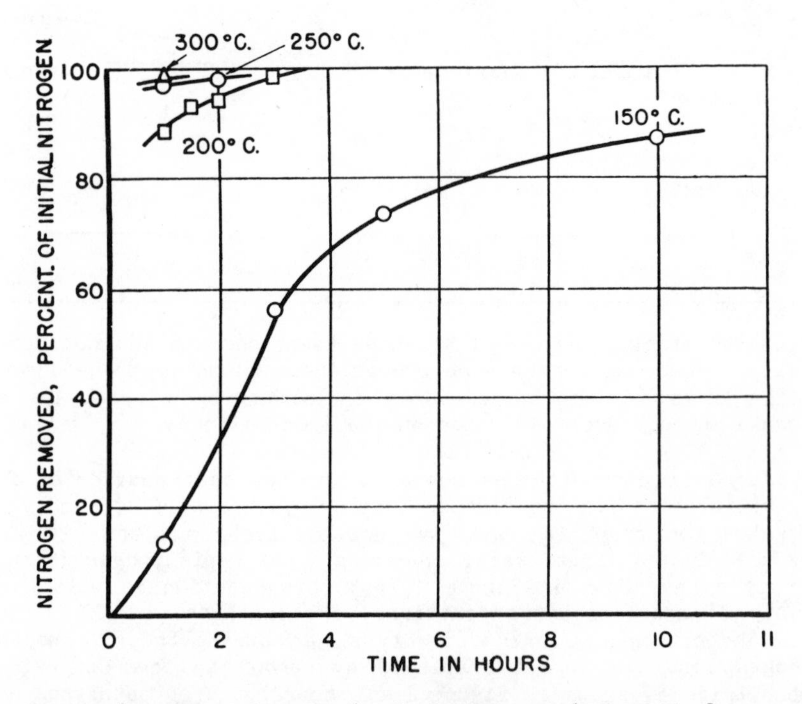

Figure 1. Rate of removal of nitrogen with pure H_2 from a sample of D3001 converted to ε-Fe_2N. (Reproduced from Ref. 18. Copyright 1953, American Chemical Society.)

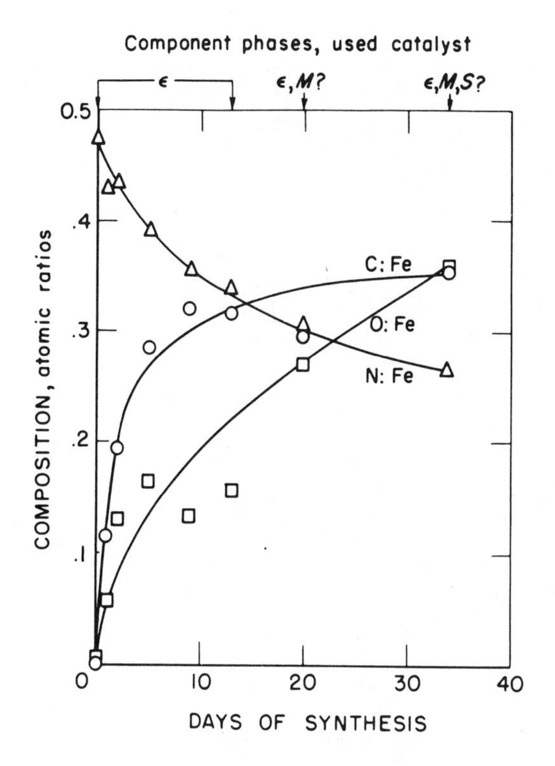

Figure 2. Composition changes of nitrided catalyst D3001 during synthesis with $1H_2$ + 1CO gas at 21.4 atm. Phases found in the used catalyst by x-ray diffraction are given at the top where ε denotes ε-nitride or carbonitride, M magnetite, and S $FeCO_3$ or $MgCO_3$ (22).

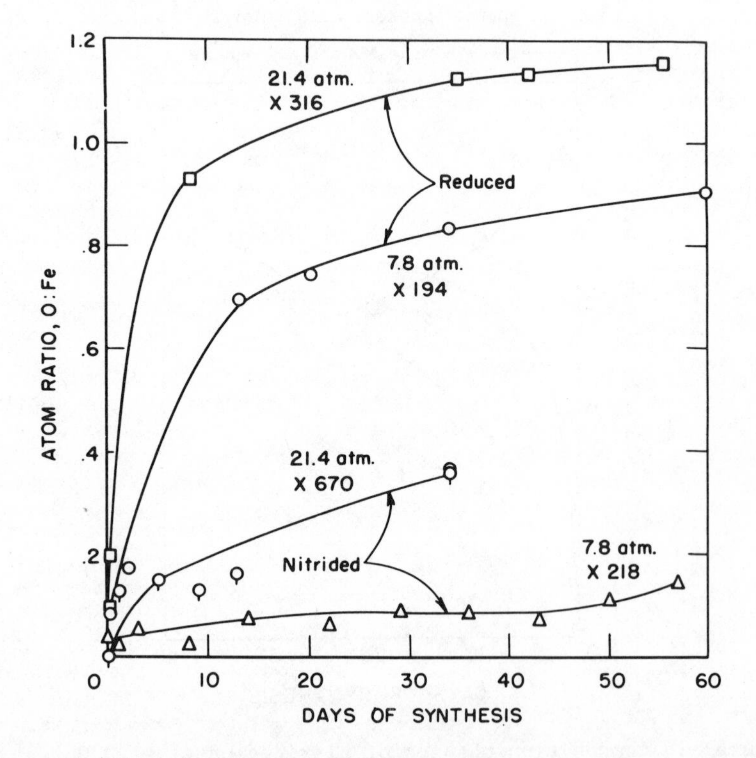

Figure 3. Oxygen content of reduced and reduced-and-nitrided catalyst D3001 during synthesis with $1H_2 + 1CO$ gas at 7.9 and 21.4 atm (22).

catalysts. The activity of D3001 was constant for 10 weeks, and D3008 lost about half its activity in 7 weeks. At 21.4 atm the initial activity of the nitrides was more than twice those at 7.8 atm. After an initial decrease the activity of both nitrided catalysts at 21.4 atm remained constant until the tests were terminated voluntarily after 7 or more weeks (23).

Selectivities are shown in Figure 4, where the yields are given in weight % on a CO_2- and H_2O-free basis. The blocks in the histogram starting from the top denote gaseous hydrocarbons and on the bottom, fractions from a 1-plate distillation representing gasoline (<185°C), diesel oil, soft wax and hard wax (>464°C). In the gaseous hydrocarbon blocks = denotes the % olefins (23). The gasoline and diesel oil fractions were examined by infrared, and estimates were made of olefin content as bromine number, Br, weight % hydroxyl group by OH and weight % carbonyl by CO. For these two catalysts nitriding sharply decreased the yield of wax without increasing the gaseous hydrocarbons. The <185°C fraction was doubled by nitriding and the OH content of this fraction is increased 10-fold indicating a large increase in the production of alcohols.

The selectivity and enhanced activity of nitrided iron catalysts was due to the presence of nitrogen in the catalyst. Treating an operating nitrided catalyst with H_2 at 385°C removed the nitrogen, and the activity decreased and the selectivity changed to that of a reduced catalyst (19). Catalysts nitrided to γ'-Fe_4N were converted to ε-carbonitrides during the synthesis and its catalysts properties were the same as samples pretreated to ε-Fe_2N. Catalysts containing ε-iron carbonitrides were prepared by first pretreating the sample with NH_3 and then with CO and also by the sequence CO and NH_3 (24). The available data suggest that catalysts pretreated to ε-iron carbonitrides by either of the sequences have no advantages over nitrides used in the synthesis.

Three reviews have been made (22,25,26); the latest (26) summarized Bureau of Mines tests of nitrided iron in fluid-bed, slurry, and oil circulation reactors of small pilot-plant size. In these units the nitrides operated nicely, having the longest life of any catalyst used in these reactors. The yields of C_1 and C_2 hydrocarbons and oxygenates are often larger than some would like.

Nitrided iron catalysts are like other iron catalysts in sensitivity to poisoning by sulfur compounds and in the relative usage of H_2 and CO. For all iron catalysts water is the main primary product and CO_2 is produced by the water-gas shift. Nitrided iron produces low of wax, and they can be used in fluid-bed/or entrained reactors at temperatures below 260°C, where reactions leading to deterioration of the catalyst are slow. At temperatures above 280°C the nitrogen is removed rapidly, and the catalytic properties change to those of the reduced catalyst.

Only one paper on iron nitrides in the FTS has been published since 1970 (27); thus this field seems wide open for new researchers with new ideas.

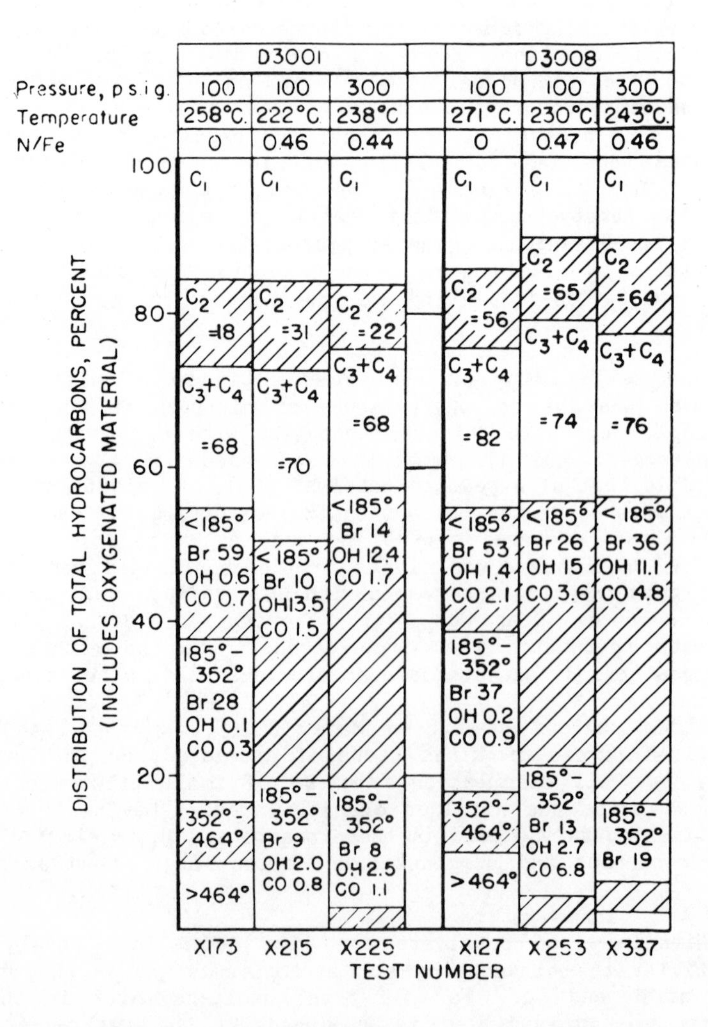

Figure 4. Selectivities of reduced and reduced-and-nitrided catalysts D3001 and D3008 in synthesis with $1H_2 + 1CO$ gas. (Reproduced from Ref. 23. Copyright 1952, American Chemical Society.)

Literature Cited

1. R.B. Anderson, Fuel 1962, 41, 295 (Obituary, Dr. Henry H. Storch).
2. Herington, E.F.G., Chemistry & Industry 1946, 347.
3. Pichler, H., Schulz, H., and Elstner, M., Brennstoff - Chemie, 1967, 48, 3.
4. Friedel, R.A., and Anderson, R.B., J. Amer. Chem. Soc., 1950, 72, 1212, 2307.
5. Manes, M., J. Amer. Chem. Soc., 1952, 74, 3148.
6. Anderson, R.B., Friedel, R.A., and Storch, H.H., J. Chem. Phys., 1951, 19, 313.
7. Storch, H.H., Golumbic, N., and Anderson, R.B., The Fischer-Tropsch and Related Syntheses, John Wiley & Sons, 1951, 600 pp.
8. Achtsnit, H.D., Dissertation, University of Karlsruhe, 1973.
9. Anderson, R.B., J. Catal., 1978, 55, 114.
10. Anderson, R.B., J. Catal., 1979, 60, 484.
11. Henrici-Olive, G., and Olive, S., Angew. Chem., Intl. Ed. Engl., 1976, 15, 136.
12. Henrici-Olive, G., and Olive, S., J. Catal., 1979, 60, 481.
13. Madon, R.J., J. Catal., 1979, 57, 183.
14. Madon, R.J., J. Catal., 1979, 60, 485.
15. Schulz, H., Rao, B.R., and Elstner, M., Erdoel Kohle, Erdgas, Petrochem., 1970, 23, 651.
16. Lee, C.B., and Anderson, R.B., unpublished results.
17. Jack, K.H., Proc. Roy. Soc. (London) 1948, A195, 34, 41, 56 .
18. Hall, W.K., Dieter, W.E., Hofer, L.J.E., and Anderson, R.B., J. Amer. Chem. Soc., 1953, 75, 1442.

19. Anderson, R.B., Shultz, J.F., Seligman, B., Hall, W.K., and Storch, H.H., J. Amer. Chem. Soc., 1950, 72, 3502.

20. Anderson, R.B., and Shultz, J.F., U.S. Patent 2 629 728 (February 24, 1953).

21. Shultz, J.F., Seligman, B., Lecky, J. and Anderson, R.B., J. Amer. Chem. Soc., 1952, 74, 637.

22. Shultz, J.F., Hofer, L.J.E., Stein, K.C., and Anderson, R.B., Bureau of Mines Bulletin 612, 1963, 70 pp.

23. Shultz, J.F., Seligman, B., Shaw, L., and Anderson, R.B., Ind. Eng. Chem., 1952, 44, 397.

24. Shultz, J.F., Abelson, M., Shaw, L., and Anderson, R.B., Ind. Eng. Chem. 1957, 49, 2055.

25. Anderson, R.B., Adv. Catal., 1953, 5, 355.

26. Anderson, R.B., Catal. Rev.-Sci. Eng., 1980, 21, 53.

27. Borghard, W.G., and Bennett, C.O., Ind. Eng. Chem., Prod. Res. Dev., 1979, 18, 18.

APPENDIX

CATALYSIS AT THE BUREAU OF MINES IN PITTSBURGH AND BRUCETON

1930 TO 1960

by Robert B. Anderson, McMaster University

In the 1920's catalysis was studied at the Central Experiment Station of the Bureau of Mines in Pittsburgh (1,2). Gas mask charcoals had been investigated there in World War I. In 1931 Dr. A.C. Fieldner, Head of the Coal Branch, persuaded Dr. Henry H. Storch to transfer from a Bureau station in New Brunswick, N.J., to Pittsburgh to become Chief of the Physical Chemistry Section. The rapport between Fieldner and Storch over the next two decades led to important contributions to chemistry and coal utilization. For example, Dr. Louis Kassel performed excellent researches in thermodynamics, and kinetics. Investigations of coal hydrogenation, CH, in autoclaves were begun early in this period and by 1940 had been expanded to pilot plant scale. A little-known but very successful research was on the design of high pressure vessels, closures and valves for use in coal hydrogenation. Research on the Fischer-Tropsch synthesis, FTS, was started again in 1943.

In the early 1940's there was considerable concern regarding petroleum shortages in the United States, and the research on coal-to-oil processes was greatly accelerated when the Congress passed the Synthetic Fuels Act in 1944, establishing within the Bureau of Mines, the Office of Synthetic Liquid Fuels. Storch was appointed Chief of its Research and Development Branch, and a new laboratory was built at Bruceton, PA., just south of Pittsburgh. Dr. Martin A. Elliott was Assistant Chief; he continued research in coal gasefication and was invaluable to the management of the Branch. In less than two years the personnel was increased from 50 to 300. Here, Dr. Storch showed his abilities as a research administrator by selecting a team of exceptionally competent young scientists and engineers to head the research groups. The Branch moved to the new Bruceton laboratories in 1947-48, and the next decade was a dynamic period in which research flourished.

Storch became Chief of the Synthetic Liquid Fuels Branch in Washington in 1951, but delayed moving until 1953. Elliott became Chief of coal-to-oil work at Bruceton in 1951, but departed for Academia in 1952, and R.B. Anderson was his replacement. In 1954 Storch left the Bureau to become Director of the Basic Research Department of American Cyanamid, Stamford, Conn.

In coal hydrogenation, Dr. L.L. Hirst was in charge of early work until he transferred to the Bureau of Mines at Morgantown, W.V., in 1948 to direct the coal gasification work. He was followed by E.L. Clark and in 1953 by Raymond Hiteshue. Polyfunctional molybdenum catalysts supported on HF-treated clays were developed for vapor phase hydrogenation of CH oils. In about 1947, Dr. Sol W. Weller transferred to this section and with Dr. M.E. Pelipetz studied catalysts for the primary hydrogenation of coal.

Research in the Organic Chemistry Section involved the structure of coal, the chemistry and separation of the complex mixtures of organic molecules from CH, metal carbonyls, and homogeneous catalysis. The chemistry and structure of carbonyls and related compounds were determined, and the kinetics and mechanisms of the hydroformylation reactions was elucidated. The section was directed by Dr. Milton Orchin until 1953 and subsequently by Dr. Irving Wender. Dr. Heinz Sternberg made valuable contributions to this research.

FTS research resumed in 1943 with catalyst preparations by Norma Stern Golumbic and larger scale work by Charles O. Hawk and Homer E. Benson. Dr. Larry J.E. Hofer began studying carbides of iron-group metals in the Physical Chemistry Section. After arriving in 1944 and 1945, Drs. Robert B. Anderson and Sol W. Weller initially spent about a year with the Physical Chemistry Section, and then Weller transferred to CH catalyst research and Anderson to the FT Catalyst Testing Section. Dr. R.A. (Gus) Friedel became Chief of the Spectroscopy Section; here he contributed to the FTS by determining isomer and carbon number distributions for FT products and by initiating studies of schemes for the growth of the carbon chain. Hofer's research led to an understanding of the Fe_2C carbides and their magnetic properties. Two highlights of the work of the Catalyst Testing Section are described in the preceeding account of the growth of the carbon chain and the use of iron nitrides. In this research J. Floid Shultz and Dr. W. Keith Hall made important contributions.

Investigations of the FTS in small pilot plants at Bruceton was initially under the direction of Dr. J.H. Crowell, followed by Homer E. Benson in 1947 and Joseph H. Field in 1958. Benson and Field were major contributors to the development of the oil circulation and the hot-gas-recycle FTS processes, and the hot carbonate CO_2-scrubbing process. Martin D. Schlesinger studied the FTS in slurry and fluidized-bed reactors.

The election of November 1952 and the advent of a Republican Administration portended changes in the research at Bruceton. At other locations programs were drastically altered, and work at the demonstration plant at Louisiana, Missouri, was stopped. Other maladies that eventually led to the demise of the Bureau of Mines might have been discerned by the careful observer. Research was becoming increasingly mixed with politics. Nevertheless, research at a substantial level has continued to the present time at the Bruceton laboratory despite several major changes in the organization and the research plans and objectives.

Literature Cited

1. Yant, W.P., and Hawk, C.O., J. Amer. Chem. Soc., 1927, 19, 1454.

2. Smith, D.F., Hawk, C.O., and Golden, P.L., J. Amer. Chem. Soc., 1930, 52, 3221.

RECEIVED November 29, 1982

Discovery and Development of Olefin Disproportionation (Metathesis)

ROBERT L. BANKS
Phillips Petroleum Company, Bartlesville, OK 74004

Heterogeneous catalyst studies and factors that contributed to the discovery of the olefin disproportionation (metathesis) reaction are described. Also provided is a personal commentary on the developments of heterogeneous catalyst technology associated with this intriguing reaction and its commercialization.

One of the most fascinating reactions of hydrocarbons to emerge in recent years is the catalyzed disproportionation or metathesis of olefins. First disclosed in 1964 (1), this versatile reaction has opened up a new and exciting field of hydrocarbon chemistry. It has been studied in research institutions throughout the world, resulting in more than 2000 publications, and has been the topic of four international symposiums (2). Commercially, it is used for the interconversion of light olefinic hydrocarbons, the backbone of today's petrochemical industry, and for the synthesis of olefins for the specialty chemicals market (3, 4).

Observed in 1959 during an investigation in our laboratory of a new heterogeneous catalyst composition (1, 5, 6), this novel reaction is now known to be catalyzed by a number of both heterogeneous and homogeneous systems; the latter were disclosed in 1967 by Calderon and coworkers at Goodyear (7) and in 1968 by Zuech at Phillips (8). We referred to the new reaction, which can be visualized (Figure 1) by the breaking and reformation of two olefinic bonds (the type and number of bonds remaining unchanged), as "olefin disproportionation." However, the scope of the reaction rapidly broadened to the extent that the term "disproportionation" did not strictly apply to all cases, prompting the uses of a variety of names (e.g., "olefin reaction"). The term "olefin metathesis," introduced by the Goodyear workers (7), covers the broad scope of the reaction and is now commonly used.

Presented in this paper is a review of the heterogeneous catalyst studies in our laboratory that led to the discovery of the olefin disproportionation reaction, accompanied by personal comments on the development of heterogeneous catalyst technology associated with this intriguing reaction.

0097–6156/83/0222–0403$06.00/0

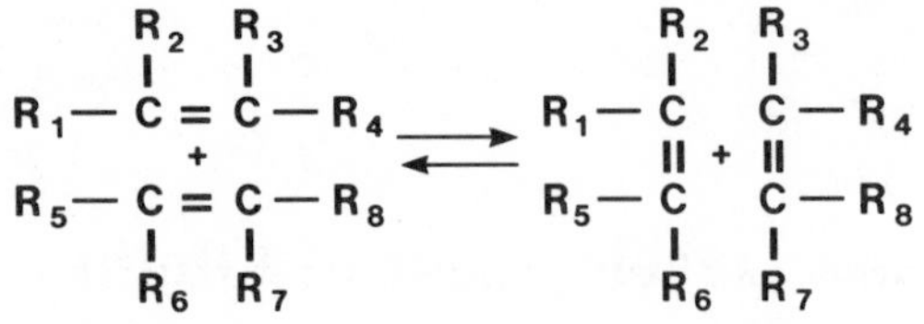

$$
\begin{array}{ccccc}
 & R_2 & & R_3 & \\
 & | & & | & \\
R_1- & C & = & C & -R_4 \\
 & & + & & \\
R_5- & C & = & C & -R_8 \\
 & | & & | & \\
 & R_6 & & R_7 &
\end{array}
\rightleftharpoons
\begin{array}{ccccc}
 & R_2 & & R_3 & \\
 & | & & | & \\
R_1- & C & & C & -R_4 \\
 & \| & + & \| & \\
R_5- & C & & C & -R_8 \\
 & | & & | & \\
 & R_6 & & R_7 &
\end{array}
$$

Figure 1. New olefin reaction.

Following World War II, process research at Phillips had become essentially research in catalysis and I became involved in heterogeneous catalyst studies directed toward improving petroleum refining technology and finding more efficient ways to utilize petroleum resources. The goal of our specific research project in 1959 was to develop an effective heterogeneous catalyst that would replace mineral acids used to catalyze the olefin-paraffin alkylation reaction. That work, in turn, utilized techniques evolving from our earlier development of a supported chromium oxide polymerization catalyst (Phillips Marlex Polyolefin Process), especially the procedures to eliminate traces of catalyst poisons from hydrocarbon feed and the experimental system.

During the screening of potential heterogeneous catalyst compositions for alkylation activity, Group VI transition metal hexacarbonyls became available in experimental quantities and we speculated that if these zero-valent metal compounds could be supported on a high surface area substrate, they might exhibit unique catalytic properties. The key factor was to support the carbonyl compounds without destroying their integrity. A procedure, now commonly used in homogeneous-heterogeneous catalyst preparations, was developed that consisted of impregnating under vacuum a precalcined (500 to 600°C) support with a non-aqueous (e.g., cyclohexane) solution of the metal carbonyl and removing, by nitrogen flushing followed by a vacuum treatment at a temperature slightly below the decomposition temperature of the metal carbonyl, the hydrocarbon solvent from the catalyst (1).

Molybdenum hexacarbonyl supported on high surface area gamma alumina was the first catalyst prepared by the new technique. In the initial test of that catalyst for alkylation activity, we recovered a small amount of liquid product equivalent to less than one per cent of the butene-isobutane feed mixture. However, BGLC (Before Gas Liquid Chromatography) analysis of this seemingly insignificant amount of product revealed that it consisted almost entirely of 2-pentene rather than products of alkylation or dimerization reactions (i.e., C_8 hydrocarbons). This result was both unexpected and puzzling: N-Butenes/Isobutane → 2-Pentene? Thus, the experiment was repeated. Detailed analyses of the total effluent from the reactor showed that butenes disappeared and that propylene, in addition to pentenes, was a reaction product. Of significance to us was that propylene and 2-pentene were formed in nearly equimolar amounts:

$$\text{N-Butenes} \xrightarrow[\text{Catalyst}]{Mo(CO)_6 \cdot Al_2O_3} \underset{51\%}{\text{Propylene}} + \underset{40\%}{\text{Pentenes}} + \underset{9\%}{C_6+}$$

We concluded that the olefin component of the feed mixture had been disproportionated to homologs of shorter and longer carbon chains (1).

Additional studies with various olefinic hydrocarbons verified that a new catalytic reaction had been discovered in which the olefinic bonds are catalytically cleaved and recombined in a highly specific and efficient manner to form new olefinic products.

Early Catalyst Developments

The discovery of a new reaction catalyzed by a new catalyst composition prompted the immediate testing of other supported transition metal carbonyls. Tungsten carbonyl supported on alumina catalyzed olefin disproportionation but was less active than the molybdenum homolog. Chromium carbonyl behaved as a polymerization catalyst. No advantage was observed via substituting part of the carbonyls with Π-coordinating ligands. However, an interesting phenomenon emerged from those studies. When ethylene was contacted with $Mo(CO)_6 \cdot Al_2O_3$, cyclopropane was a reaction product, Table 1 (1):

Table 1. Ethylene reactions over group VI metal hexacarbonyls (5)

Product, wt %[a]	Alumina impregnated with			Type of Reaction
	$W(CO)_6$	$Mo(CO)_6$	$Cr(CO)_6$	
1-Butene	71	26	3	Dimerization
2-Butene	8	26		Isomerization[b]
Propylene	21	28		Disproportionation[b]
Cyclopropane		8		Ring closure
Methylcyclopropane		12		Ring closure
Solid Polyethylene			97	Polymerization

a. Tests at 120°C and 500 psig.
b. Secondary reactions.

This observation was consistent both with the mechanistic scheme we initially considered (four-carbon metallocycle) and the currently favored metallocarbene mechanism (2). To shed some light on the nature of the active species, carbon monoxide pressure was monitored; the partial loss of CO's occurred under both preparation (solvent removal) and reaction conditions.

At that stage of the catalyst studies (Spring, 1960), Grant C. Bailey and I considered the possibility that metal oxides, corresponding to the carbonyls, might also exhibit similar catalytic properties. We found that not only did supported molybdenum and tungsten oxides promote the disproportionation of olefins, but they were more active than the corresponding carbonyls. One of the most active of the metal oxide catalysts was a commercial cobalt molybdate catalyst (3% CoO/8% MoO_3/ 89% Al_2O_3), making it apparent that the disproportionation reaction had been occurring unnoticed in routine hydrocarbon processes.

Emphasis in the laboratory rapidly shifted to the development of the more active oxide catalysts. Catalyst compositions and catalyst activations and pretreatments were investigated. Various olefinic hydrocarbons were tested and the ranges of reaction conditions (e.g., Figure 2) were established for patent coverage and for preliminary economic evaluation.

Evaluation

The initial investigation of cobalt molybdate catalysts provided data for conceptual designs and economic appraisals of several of many suggested applications of olefin metathesis. Economics for converting propylene to ethylene and butenes were found to be favorable and it appeared that the proposed process could be applied commercially in several areas: e.g., balancing olefin production from naphtha cracking units, producing high-purity ethylene for chemical plants in remote locations, and combining with alkylation units to produce higher octane gasoline. Thus, the disproportionation of propylene was selected for the initial development of olefin metathesis technology. Because three olefins were involved, the name "Triolefin Process" was chosen.

Development

Early in 1962, following Phillips Management's decision to develop the Triolefin Process, laboratory studies were resumed and expanded. In addition to conducting a detailed investigation of cobalt molybdate catalyst systems, an extensive search for other catalyst compositions active for olefin metathesis was made. Concurrent with these investigations were studies designed to expand the scope and explore other applications of olefin metathesis reactions. Pilot plant development of Triolefin Process technology was initiated about six months after laboratory studies had been resumed.

The search for new catalyst compositions, conducted by L. F. Heckelsberg, produced a large list of heterogeneous compositions effective for metathesis reactions (9). A major finding of Dr. Heckelsberg was that molybdenum and tungsten oxides supported on high surface area silica were very active for olefin metathesis at temperatures several hundred degrees higher than the optimum temperature for the alumina-based catalysts (10). Interesting properties were exhibited by supported rhenium oxide catalysts:

while rhenium oxide supported on alumina is very active at room temperatures, on silica it displayed two temperatures at which activity maxima occur. Although most catalysts that were active for metathesis contained a transition metal, it was found that magnesium oxide, when treated with carbon monoxide, catalyzed the disproportionation of propylene without the benefit of a promoter (6). These new catalysts broadened the effective temperature range for metathesis reactions, allowing the selection, for specific applications, of a reaction temperature which favors metathesis over competing side reactions (e.g., high temperature needed for isobutylene metathesis to be the predominant reaction) (6).

Following Heckelsberg's discovery of high-temperature metathesis catalysts (Spring, 1963), emphasis in pilot plant and laboratory development of the Triolefin Process was switched from the cobalt molybdate catalyst to silica-based catalysts. This change facilitated catalyst regeneration (partly eliminating costly heating-cooling cycles) and significantly reducing reactor size (equilibrium conversions were achieved at space rates an order of magnitude higher). The silica-based catalysts were also more resistant to poisons; in most cases the activities of these catalysts were fully restored when the catalyst poison was removed from the feed (Figure 3) (10). One disadvantage of the silica-based catalysts was that they had not been commercially produced and procedures and specifications for commercial production had to be developed. Since silica is extremely difficult to form into tablets or extrudates, a key hurdle was the development of a technique for forming particles of satisfactory crushing strength without impairing activity or selectivity.

Early laboratory studies showed that the primary disproportionation products of propylene were ethylene and *cis*- and *trans*-2-butene; 1-butene and C_5+ product extrapolated to zero at zero contact time (Figure 4) (1). The reversibility of metathesis reactions was also demonstrated. An interesting selectivity phenomenon was observed for the disproportionation of butenes: the conversion of a mixture of 1-butene and 2-butene was higher than the conversion of either olefin by itself. Furthermore, while the disproportionation of 1-butene favored the production of ethylene and hexene, a butene mixture rich in 2-butene tended to produce propylene and pentenes, indicating that the predominant reaction occurred between the 1-butene and 2-butene.

The results of the above and similar studies provided the basis for a key contribution that aided the development of olefin metathesis technology at Phillips: D. L. Crain, in 1963, recognized that product distributions obtained by disproportionation of various olefins and mixtures of olefins could be explained by a concerted "four-center" mechanistic scheme (Figure 5A). This mechanism was proposed in a later publication of Bradshaw and coworkers of British Petroleum (11). Although the simple "four-center", or "quasi-cyclobutene," scheme is no longer the

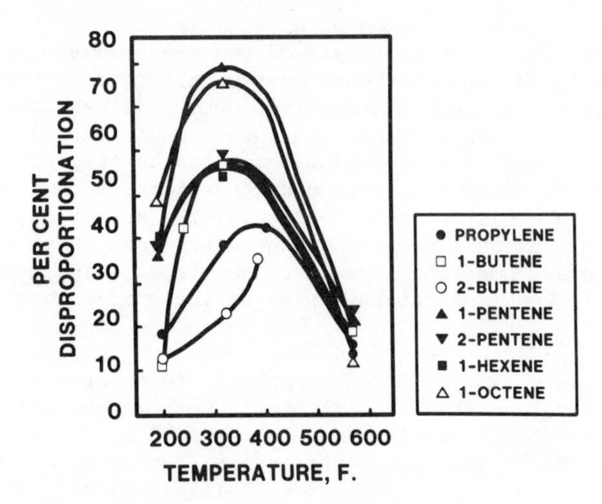

Figure 2. Effect of temperature on disproportionation conversion over cobalt molybdate-alumina. (Reproduced from Ref. 1. Copyright 1964, American Chemical Society.)

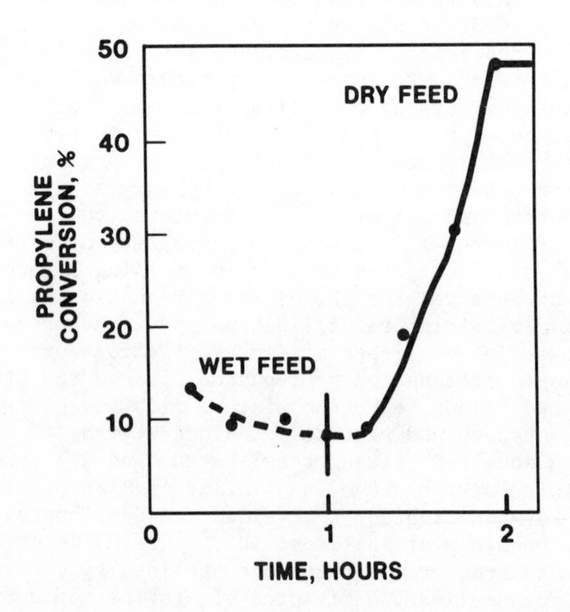

Figure 3. Temporary effect of catalyst poisons on silica-based catalysts. (Reproduced from Ref. 10. Copyright 1968, American Chemical Society.)

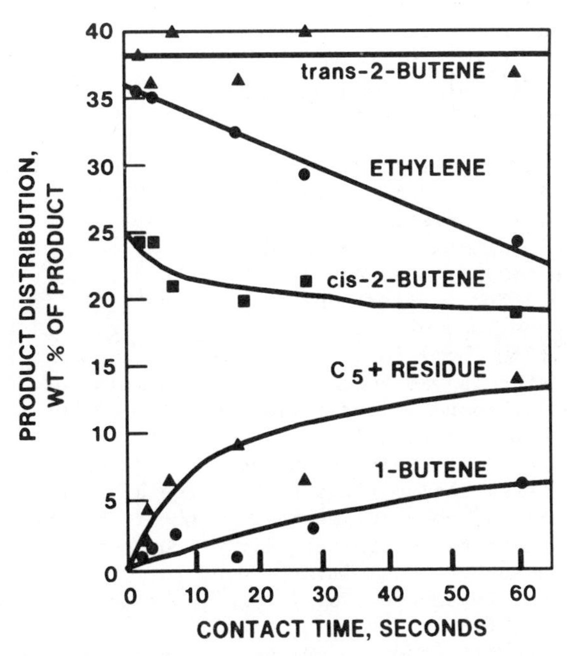

Figure 4. Propylene product distribution at zero contact time. (Reproduced from Ref. 1. Copyright 1964, American Chemical Society.)

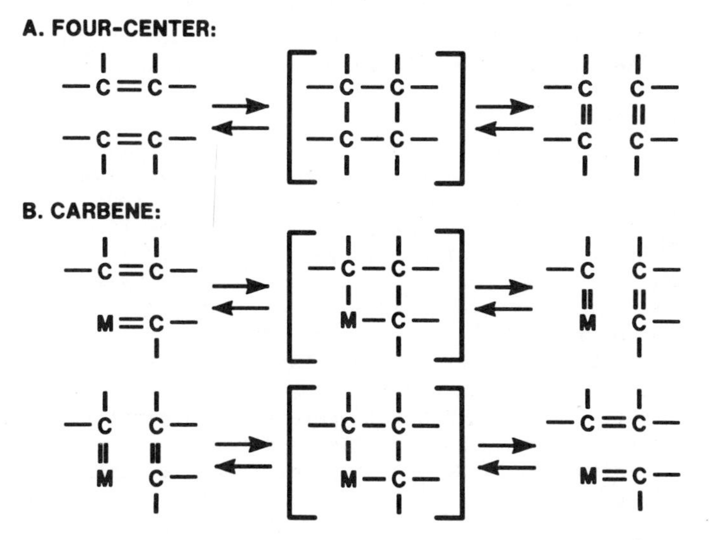

Figure 5. Mechanistic schemes for olefin metathesis.

accepted mechanism for olefin metathesis reactions, its early use in guiding and planning catalyst modification studies and in extending olefin metathesis to other olefinic hydrocarbons was extremely beneficial.

The elucidation of the role of double-bond isomerization activity in metathesis process is an example of the helpfulness of the four-center mechanism. As the scheme predicted, in certain applications the elimination of double-bond isomerization activity (acidic isomerization sites were destroyed by various mild caustic treatments) prevented secondary metathesis reaction resulting in very high selectivity to specific products (5). In contrast, in other applications (e.g., linear olefin and neohexene processes) to obtain a high level of productive metathesis, the mechanistic scheme indicated a need for enhanced isomerization activity; this was accomplished by addition of a very selective double-bond isomerization catalyst to the scheme (5). G. C. Ray and D. L. Crain used the scheme to postulate that metathesis reactions could be extended to the cleavage of cyclic olefins with ethylene to yield alpha-omega diolefins (Figure 6). The four-center scheme implies that in theory the number of metathesis reactions is limited only by the number of compounds containing carbon-carbon multitype bonds.

Pilot plant development of Triolefin Process technology, requiring about four years, including establishing preferred feed composition, purification techniques, catalyst composition, catalyst activation and regeneration procedures, process conditions, and cycle length. Catalyst life was also determined in repeated cycles and certain design premises (e.g., temperature differential in the reactor) were demonstrated to be valid. Pilot plant runs demonstrated that equilibrium conversions and selectivities in the high 90's were feasible (5).

Commercial Applications

Commercialization of olefin metathesis was accomplished in 1966 (12). Shawinigan Chemical Ltd., at their Varennes complex near Montreal, Quebec, brought the Phillips Triolefin Process on stream. With an excess of propylene at that location, Shawinigan had a need for polymerization-grade ethylene and high-purity butenes. This presented an ideal opportunity for application of the Triolefin Process. The commercial unit was operating at full capacity two weeks after start-up and its performance exceeded that predicted by laboratory and pilot plant studies. Part of the successful operation can be attributed to the very effective elimination of catalyst poisons from the commercial unit. In 1972, operation of the Shawinigan plant was terminated due to a change in economic climate.

Another industrial use of the versatile olefin metathesis reaction was commercialized by Shell in 1977. As part of their high-olefin process, SHOP, they are producing detergent range (i.e., C_{11}-C_{14}) olefins via cross-disproportionation of heavy and light olefin fractions over a metathesis catalyst operating in the 80-140°C temperature range (13).

The latest (1980) commercial application of olefin metathesis is Phillips Neohexene Process (4). Neohexene, an intermediate in the synthesis of a perfume musk, is produced by cross-metathesis of diisobutylene with ethylene (i.e., ethylene cleavage) over a bifunctional (double-bond isomerization/metathesis) catalyst system (Figure 7).

Potential Uses

Technology for a number of applications of olefin metathesis has been developed (3, 4). At Phillips, potential processes for producing isoamylenes for polyisoprene synthesis and long-chain linear olefins from propylene have been through pilot plant development. In the area of specialty petrochemicals, potential industrial applications include the preparation of numerous olefins and diolefins. High selectivities can be achieved by selection of catalyst and process conditions. The development of new classes of catalysts allows the metathesis of certain functional olefins (4, 14). The metathesis of alkynes is also feasible (15).

Fundamental Contributions

Worldwide studies over the past two decades of olefin metathesis catalysts and reactions have played a significant role in the advancement of the science of both heterogeneous and homogeneous catalysis. The influence of the heterogeneous catalyst substrates, metal components, ligands (including chelating polyolefins added to the feed, Figure 8), and catalyst modifications and pretreatments have provided insights into heterogeneous catalysis. Rate-temperature profiles show that maximum metathesis activities occur at approximately 100°C lower than the maximum H-D exchange rates over the corresponding substrate, indicating a relationship between proton mobility and metathesis activity (6).

Generally, catalytic models are based upon studies involving homogeneous systems, with little, if any, basis for the extrapolation to heterogeneous catalysts. However, in the case of olefin metathesis, the homogeneous catalysts are strikingly similar to their heterogeneous counterparts; a vast majority of these catalysts contain molybdenum, tungsten, or rhenium as halides, oxides, or carbonyls. Further similarities in reaction conditions and susceptibility to photo-assistance make this reaction ideal for establishing relationships between heterogeneous and homogeneous catalysts. This was demonstrated recently by the results of very detailed kinetic and mechanistic studies, conducted by several groups of investigators, which

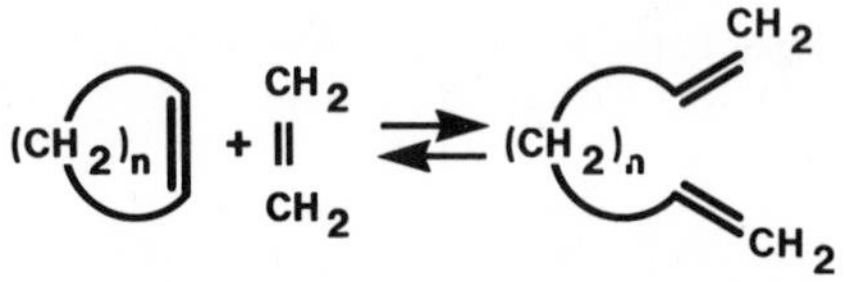

$$(CH_2)_n\text{[cyclic olefin]} + \begin{matrix} CH_2 \\ \| \\ CH_2 \end{matrix} \rightleftharpoons (CH_2)_n\text{[}CH=CH_2\text{, }CH=CH_2\text{]}$$

Figure 6. Synthesis of alpha-omega diolefins.

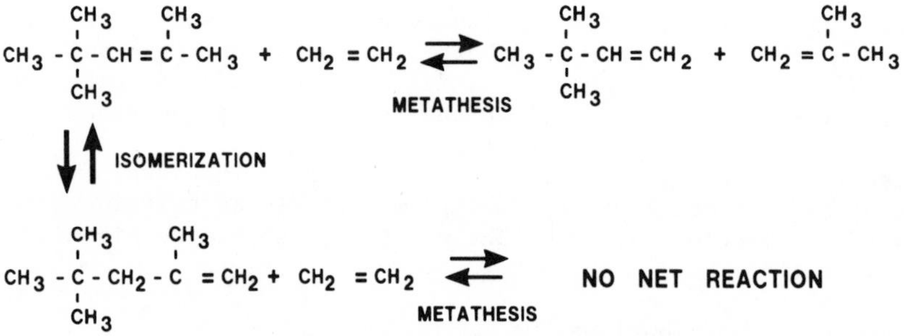

$$CH_3-\underset{CH_3}{\overset{CH_3}{C}}-CH=\overset{CH_3}{C}-CH_3 + CH_2=CH_2 \underset{\text{METATHESIS}}{\rightleftharpoons} CH_3-\underset{CH_3}{\overset{CH_3}{C}}-CH=CH_2 + CH_2=\overset{CH_3}{C}-CH_3$$

$\updownarrow$ ISOMERIZATION

$$CH_3-\underset{CH_3}{\overset{CH_3}{C}}-CH_2-\overset{CH_3}{C}=CH_2 + CH_2=CH_2 \underset{\text{METATHESIS}}{\rightleftharpoons} \text{NO NET REACTION}$$

Figure 7. Neohexene process chemistry.

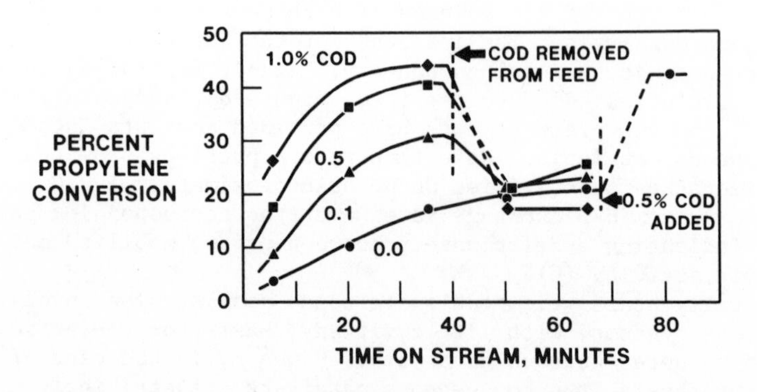

Figure 8. Effect of 1,5-cyclooctadiene on activity of tungsten oxide-silica catalyst. (Reproduced from Ref. 5. Copyright 1979, American Chemical Society.)

indicate a single-step metallocarbene mechanism (Figure 5B) for both the homogeneous and heterogeneous catalyzed metathesis reactions (2).

The potential relationship of various types of olefin reactions in an area of fundamental interest. While molybdenum- and tungsten-containing catalysts promote metathesis of olefins, chromium-containing catalysts tend to polymerize them. The similarity of these three Group VI transition metals leads to the speculation that various olefin reactions might proceed through mechanistic schemes that only have subtle differences (16). A very general philosophy concerning mechanisms in the intimately related fields of metathesis and polymerization is now emerging.

Reflection

It is interesting to reflect on those factors that might have contributed to a particular scientific discovery. In the case of olefin disproportionation, Hanford's law (rephrased) certainly applies: "If you want to find something new, you have to try something new" (17). We were investigating a newly available compound for catalytic behavior, and devised a new technique for preparing new catalyst compositions. Also, we were utilizing relatively newly developed purification methods to significantly reduce the level of catalyst poison in the feed and experimental system. Improved analytical techniques were available. A possible factor was the relatively mild activity of the supported molybdenum hexacarbonyl catalyst; heterogeneous catalysts now known to be very active for promoting metathesis reactions were used in earlier catalyst research, both at Phillips and elsewhere, but, because of competing reactions, the metathesis reaction had not been recognized. However, perhaps the most significant factor was "serendipity."

Acknowledgments

In addition to those acknowledged in the text and references, many other people at Phillips have made important contributions to the development of heterogeneous metathesis catalysis. Contributions by other research groups, particularly those in the academic sector, to this new field are credited in various review articles, e.g., reference (2).

Literature Cited

1. Banks, R. L.; Bailey, G. C. Ind. Eng. Chem., Prod. Res. Develop, 1964, 3, 170-173.
2. Banks, R. L. "Catalysis" (Specialists Periodical Reports) Kemball, C., Ed.; The Chemical Society, London, 1981 Vol. 4, 100-129.
3. Banks, R. L. J. Mol. Catal. 1980, 8, 269-276.

4. Banks, R. L.; Banasiak, D. S.; Hudson, P. S.; Norell, J. R. J. Mol. Catal. 1982, 15, 21-33.
5. Banks, R. L. Chemtech, 1979, 9, 494-500.
6. Banks, R. L. Amer. Chem. Soc. Pet. Chem. Prepr. 1979, 24(2), 399-406.
7. Calderon, N.; Chen, H. Y.; Scott, K. W. Tetrahedron Letters 1967, 34, 3327-3329.
8. Zuech, E. A. Chem. Commun. 1968, 1182.
9. Heckelsberg, L. F.; Banks, R. L.; Bailey, G. C. Ind. Eng. Chem. Prod. Res. Develop. 1969, 8, 259-261.
10. Heckelsberg, L. F.; Banks, R. L.; Bailey, G. C. Ind. Eng. Chem. Prod. Res. Develop. 1968, 7, 29-31.
11. Bradshaw, C. P. C.; Howman, E. J.; Turner, L. J. Catalysis 1967, 7, 269-276.
12. Chemical Week, July 16, 1966, 77 and July 23, 1966, 70.
13. Freitas, E. R.; Gum, C. R. CEP, January 1979, 73-76.
14. Pennella, F.; Banks, R. L.; Bailey, G. C. Chem. Commun. 1968, 23, 1548-1549.
15. Boelhouwer, C.; Verkuijlen, E. Amer. Chem. Soc. Pet. Chem. Prepr. 1979, 24(2), 392-393.
16. Banks, R. L.; Bailey, G. C. J. Catalysis 1969, 14(3), 276-278.
17. Chemtech, 1980, 10(3), 129.

RECEIVED November 29, 1982

30

The Development of Automotive Exhaust Catalysts

GEORGE R. LESTER

Universal Oil Products, Inc., Corporate Research Center, Des Plaines, IL 60016

The story of the development of automotive exhaust conversion catalysts is at once a simple and complex one. The major challenge is to keep it simple while remaining true to the complex interplay of technological achievements with political ideals and geographic, demographic, economic, and bureaucratic realities which culminated in the successful mass application of catalysts to the control of automotive exhaust pollution.

Our study begins in California in the late forties, when the combination of the rapidly increasing population of humans and automobiles with geography and personal expectations focused political and scientific attention onto the photochemical reactions in the atmosphere between the hydrocarbons and nitrogen oxides which were being emitted in automobile exhaust, and on the effects of the reaction products on public health.

Public demand for relief from the increasingly frequent smog incidents caused the California Legislature in 1947 to allow the formation of county or regional air pollution control districts; as these districts were formed they began to gather data on the air quality and sources of pollution, and to encourage scientific investigation into the origins of the periodic episodes of widespread eye irritation and respiratory discomfort. In 1952, Professor A. J. Haagen-Smit of the California Institute of Technology published his studies (1) which showed that some hydrocarbons and nitrogen oxides endemic to automobile exhaust reacted in sunlight to produce oxidants, including ozone, which were already known to, among other things, cause cracking of rubber and irritation of the eyes. [Cal-Tech is located in Pasadena, which is situated in the northwestern corner of the Los Angeles "basin", and which generally feels the full impact of auto exhaust-related smog.] This study, along with a concurrent investigation by the Los Angeles Air Pollution Control District (2) which showed that aerosols and mists could be produced photochemically by the polymerization of the photo-oxidation products of some of the exhaust hydrocarbons, laid the scientific basis for a serious

0097-6156/83/0222-0415$06.00/0

examination of the composition and health impact of automotive exhaust, and must be considered to be the impetus for the eventual development of the automotive catalytic converter.

In response to these (and other) studies, the Automobile Manufacturers Association in 1953 organized a Vehicle Combustion Products Committee; the Coordinating Research Council established a Composition of Exhaust Gases Group; and the individual manufacturers of automobiles, gasoline and gasoline additives initiated their own investigations. These efforts (3) included studies of the effects of individual driving habits and overall traffic patterns on vehicle operation and emission parameters, and of the exhaust composition as a function of engine design and operating parameters. They included the development of instrumentation for these measurements, investigations of the fundamental processes which determine the concentration of hydrocarbons and nitrogen oxides in exhaust, and examinations of methods to reduce pollutant levels by engine design, by fuel modification, and by exhaust after-treatment, including "non-flame reactors", thermal afterburners, and catalytic converter systems.

The insights gained as a result of the non-catalyst studies as well as those performed in various government agencies and academic institutions during this period built the scientific and engineering foundations for the non-catalytic methods that were eventually to be developed to reduce noxious emissions to meet the early (through 1974), relatively moderate, emission standards, and were also to prove invaluable in the eventual development of catalytic systems.

Catalytic oxidation to remove hydrocarbons was only one of a number of methods - including refrigeration, filtration, odor masking agents, and centrifuging - considered by the AMA Exhaust System Task Group (4), which was formed in 1955. In spite of much concern about the wide range of exhaust composition and conditions to which an exhaust after-treatment system would be exposed - including temperatures of 150-1500°F, flow rates of 10-250 standard cubic feet per minute, concentrations of H_2 and CO of from zero up to 7% and to 13%, respectively, and the potentially adverse effects due to the presence of lead anti-knock compounds, sulfur dioxide, and phosphorus and alkaline earths from lubricants - some preliminary experiments were performed using experimental and commercial (for non-automotive applications) catalysts from outside sources, with at worst disasterous results, and with at best "some promise". However, even the most durable catalytic converter tested did not "silence engine noise satisfactorily", and hence would probably require a separate noise muffler. Other problems anticipated by these early workers included the need to automatically regulate secondary air to help maintain the proper catalyst temperature, a probable requirement for insulation to not only improve catalyst performance but also for passenger comfort and to protect the underfloor brake fluid lines, difficulty in finding suitable space for the catalyst on the vehicle, and of course the cost of original installation and of the anticipated periodic replacement (4).

In a concurrent effort, Ford workers screened a number of catalysts and focused on vanadium oxide catalysts for hydrocarbon conversion. A major advantage (at the time) claimed for this catalyst was its inefficiency in oxidizing carbon monoxide, which eliminated the need for a secondary air pump to provide oxygen; this feature also would permit the use of lower-cost steels since the heat release associated with CO oxidation would be avoided (5). Although this catalyst was further developed as part of a system (6, 7, 8), it was to become decidely less attractive when California adopted exhaust standards in 1959 which included carbon monoxide (9).

In the same period, researchers at General Motors evaluated a catalyst (provided by Oxy-Catalyst, Inc.) and found "useful" activity for hydrocarbons and carbon monoxide for 12,000 miles; however, it required 2 large catalyst beds containing almost 50 pounds (22 kg) of catalyst (10).

By 1962, it appears that the vehicle makers had "exhaustively" evaluated the catalytic converter, and had come to some understanding of its potential and limitations. In the absence of compelling federal legislation establishing emission standards stringent enough to demand such costly and (so far) unproven technology, the exhaust pollution control efforts of the industry were directed toward less revolutionary and more predictable approaches such as engine modification and shorter maintenance intervals.

Consequently, during the period of 1962-1967, there was very little visible automobile industry interest in catalysis, but groups at the research and engineering centers of several auto manufacturers were established during this era to generate a better in-house understanding of catalysis as it might be applied to control automobile pollution (11), and perhaps to be better able to respond to potential legislative demands for the use of catalytic converters as "add-on" or retrofitted devices.

The potential implications of the Haagen-Smit study had not been lost on catalyst manufacturers, and the patent literature as early as 1953 (12) was beginning to reflect their interest in this (staggering) potential market. However, the divergent interests of the automobile and catalyst industries during the hearings which were a prerequisite to the establishment of government regulations on vehicle emissions probably limited the degree of interindustry cooperation compared to that which was to be seen years later when stringent nationwide emission standards were actually adopted. Even so, a variety of "off-the-shelf" catalysts were provided to vehicle manufacturers as candidates for their evaluation and for use in emission control system development efforts during the mid-fifties. During this period of relatively independent activity, a patent was issued (13) which seems to describe quite well many of the important features of the monolithic catalysts which eventually were to capture a major share of the automotive catalyst market, although the small-channel "honeycomb" ceramic supports for these catalysts had yet to be developed.

Back where it all began, the increasing vehicle population and the increasing frequency of "red-eye" days led the California Legislature in 1959 to require the State Department of Public Health to develop and publish standards for the quality of air and for vehicle exhaust emissions before February 1, 1960 (14). These standards, which were adopted on December 4, 1959, were based on the judgement that in order to achieve desirable air quality standards, 80% reduction of hydrocarbon (HC) and 60% reduction of carbon monoxide (CO) emissions were needed. In terms of the test cycle which was also established, the standards were 275 ppm (volume) of hydrocarbons as hexane (or 0.165% as C_1) and 1.5% (volume) of CO.

There was some surprise that a CO standard was established, since it is not involved in the sensory perception of smog; however, the studies of urban pollutant concentrations showed that the levels of CO frequently exceeded those known to cause significant impairment of important bodily functions in sensitive segments of the population, specifically those with impaired respiratory or circulatory capacity. Further, although the auto was not the sole contributor to smog, source inventory data indicated that the only significant source of atmospheric CO was motor vehicle exhaust (14).

After these standards were established, along with an appropriate test procedure, the California Legislature created a Motor Vehicle Pollution Control Board (MVPCB) and gave it the responsibility to issue certification to control devices meeting certain criteria of performance, durability, cost and ease of inspection. It also provided that, one year after certification of two or more devices, all new vehicles to be registered in the state would have to be equipped with such devices.

As this legislation was being considered and passed, potential catalyst suppliers initiated serious programs to develop such devices, usually in conjunction with muffler suppliers. To be certified, the devices, when installed, had to have met Fleet and Life testing to demonstrate satisfactory performance on at least 25 vehicles for 12,000 miles using regular (fully leaded) gasoline. Tests required besides the emission standards included tests for effects of the device on vehicle driveability and safety; sensitivity to backfire, water immersion, and mountain driving; impacts on noise, odor, and production of nitric oxides; and to insure that it was "failsafe" in the event of malfunctions. In spite of the awesome nature of this challenge, three catalytic devices (and one thermal afterburner) were certified by the Board on June 17, 1964, and were approved for installation on new 1966 and subsequent model year cars. The catalytic devices, developed by UOP-Arvin, Grace-Norris and Cyanamid-Walker, consisted of combined catalytic converter-noise mufflers with air pumps to supply the required oxygen, and were projected to add an annual overall cost (averaged over five years) of $26-$40 (1964 dollars) (15).

The development of these catalysts occurred in an atmosphere of tight secrecy, and little has been published on those efforts. All three were of the particulate type; one contained noble metal(s) while the other two were "noble-metal promoted base-metal catalysts" (16). The UOP noble metal catalyst was provided several years later to GM and to Ford for evaluation with unleaded fuel and for other studies, and it can be deduced that it was supported on low-density (about 0.32 g/mL apparent bulk density) 1/8 inch spheres (17) and may have contained about 0.47 troy ounces of platinum per cubic foot (*ca.* 0.16 weight percent), with the platinum concentrated in a subsurface shell some distance below the exterior surfaces of the spheres for improved poison resistance (18). The Cyanamid catalyst was later studied by Ford (16, 19); it apparently was an extrudate (1/8" diameter x 1/8" long) of about 0.67 g/mL ABD, with about 125 ppm (weight) of palladium and 5 weight % each of CuO and V_2O_5 on a 5% SiO_2-95% Al_2O_3 support of about 200+ M^2/g surface area (19). The author is unaware of published reports which might shed further light on the nature of the Grace catalyst.

Despite the efforts reflected in the California certification, these devices were not to be used in September of 1965, as had been legislated. The automobile manufacturers were unwilling to use such "add-on" devices on their vehicles which were to be registered in California, and accelerated their efforts to meet the 1966 California standards by engine modifications including improved carburetors, distributor adjustment, and intake manifold design changes. The California Legislature cooperated by modifying the certification procedure in 1965 to permit rapid certification of these modified vehicles, and as a result, the market for the certified catalytic devices vanished. Although the catalyst companies which had spent millions of dollars in this apparently futile cause were severely disappointed, one can reflect in hindsight that the June, 1964 certification of the catalyst devices did trigger the establishment of the 1966 California Standards. This was a very significant step toward the eventual establishment of California and Federal standards so strict that they could only be met cost-effectively with catalysts. However, at the time, the catalyst industry had good reason to believe that the California legislature had simply caved in under the pressure of the automotive industry, and to wonder if the catalyst developers had not paid a high price for a lesson in economics and political science. For their part, the automakers had avoided the unacceptable position of selling and warranting vehicles equipped with expensive devices made by third parties, with which they had little positive first-hand experience, and about which it was easy from their own limited experience to imagine significant potential safety and operational problems.

Although the 1965 California Legislative action brought many catalyst development efforts to a halt, the studies of air quality in California and in urban areas throughout the country led to the recognition that air quality standards needed to become more stringent, and that federal as well as state actions was needed to establish and enforce those standards. This led to passage of the 1967 Federal Clean Air Act, which established emission standards of all 1968 model year (MY) vehicles at the same levels as the 1966 California standards, and established the regulatory and bureaucratic mechanisms to mandate schedules for determining acceptable future air quality standards and the appropriate exhaust emission standards to permit them to be realized.

This law (the "Muskie Bill") caused a spirited resurgence of interest in automotive exhaust catalysts throughout the catalyst, refining, automotive, and gasoline additive industries. Catalyst suppliers who had tried to sell their engine dynamometers at salvage value in 1965 brought them back up to speed, and substantial catalyst research and evaluation groups were created by the major auto manufacturers. The same engine modification technology used to meet the 1966 California standards was used to satisfy the 1968 Federal requirements, and would indeed eventually be pushed further, along with air pumps, spark retardation, thermal afterburners, and exhaust gas recirculation to meet the progressively tighter standards through the 1974 MY. However, it was realized by many that these approaches had significant negative impacts on vehicle performance and driveability and on fuel economy which could become prohibitive as emission standards became more stringent.

The major problem anticipated with the use of catalysts, based on the earlier experience of catalyst makers and the auto industry, was the rapid deactivation of the catalyst in the presence of the non-hydrocarbon components of gasoline and lubricants. Particular concerns were the lead compounds and the halide-containing lead scavengers added to gasoline to prevent engine knock at the higher engine compression ratios favored for improved thermodynamic efficiency of the engine. Although the catalytic devices developed to meet the early California "demand" had functioned with the use of "fully-leaded" gasoline, "acceptable" performance had anticipated annual or biennial catalyst replacement, while at the same time meeting standards much less stringent than those anticipated for the 1970's.

It was also realized that catalysts might be expected to deactivate, even with unleaded fuel, as a result of exposure to high temperature over a long period of time; thus there was clearly a need to determine the relative rates of degradation due to "thermal" and to "poison" deactivation. Studies were initiated at Ford (19, 20) and GM (17) in 1967 to answer this and associated questions relative to the potential of automotive catalysts.

In the Ford test, vehicles were operated with catalytic converters located either near the engine or at the rear of the car to afford different operating temperatures, and using fuel of varying lead additive contents, ranging from "zero" ($\sim$0.05 g Pb/gal) to 3 g Pb/gallon ("fully-leaded"). The catalyst was the Cyanamid $Pd\text{-}CuO\text{-}V_2O_5\text{-}SiO_2\text{-}Al_2O_3$ extrudate catalyst referred to earlier; 5% vermiculite (in the unexpanded form) was mixed in the bed in an effort to minimize the catalyst attrition which might occur due to the difference in coefficients of thermal expansion of the catalyst and the metal container. Mileage was accumulated by employees in home-to-work transportation.

This study exposed a number of attrition and other catalyst deterioration problems which were probably related to the sensitivity of this particular catalyst to higher temperatures. It was observed that the presence of the lead additive caused a four-fold reduction in half-life for hydrocarbon conversion, although the CO deactivation rate was independent of the lead level. It is likely that the presence of low melting and highly reactive (with the support and lead compounds) V_2O_5 in the catalyst contributed to the substantial problems of attrition and catalyst loss, and these tests might have been a significant factor in the eventual Ford commitment to noble-metal monolithic catalysts. It was fortunate for the catalyst interests that shortly after the Ford report on the Cyanamid catalyst, engineers at GM (17) presented their studies of a supported noble metal catalyst. For the GM study, two 1968 model vehicles were equipped with an air pump and with converters containing the UOP noble metal-low density spherical catalyst which had been certified in California. These vehicles were operated for 50,000 miles on lead-free fuel, although one car was inadvertently fueled with leaded gasoline at 36,000 miles. It was found that there was no loss of catalyst on either vehicle, and the hot cycle efficiencies for hydrocarbon and CO after 50,000 miles were 67% and 90%, respectively, for one car; both were about 90% for the car which had seen lead at 36K miles.

After the primary tests, one car was operated an additional 18,000 miles on a high speed schedule (Belgium block road conditions), which included mostly wide-open-throttle accelerations. The catalyst deactivation during this operation appeared to be no more severe than that observed during the "normal" mileage accumulation. Additional studies on the other 50,000 mile vehicle included intentionally "shorting" of one or two spark plugs for up to fifteen minutes to generate catalyst bed temperatures up to 1810°F (at full throttle), compared to the normal range of 900-1100°F; it was found that the catalyst performance was only slightly affected by this treatment.

It is difficult to overstate the significance of the GM lead-free tests in the evolution of catalytic control of vehicle emissions. They demonstrated that, when operated on unleaded fuel, catalytic systems could be made durable, dependable, resistant to engine system mis-operation and even mis-fueling, and that they need not substantively impact engine operation or performance, including fuel economy. Although it was cautioned that the use of noble metals in automotive catalysts was "not practical for large volume usage because of the high cost and limited supply" (17), Ford workers in a prepared discussion (21) of the GM paper cited figures from a platinum supplier (22) which disputed that concern.

Just seven months after the presentation of the GM tests, in January of 1970, Ed Cole, President of General Motors, in a speech at the annual meeting of the Society of Automotive Engineers, called for a "comprehensive systems approach to automotive pollution control" including consideration of the removal of lead additives from fuels, to make "advanced emissions control systems feasible", and in February, GM announced that all of its cars beginning with the 1971 models would be designed to operate on fuel of 91 Research Octane Number, leaded or unleaded. This was achieved by GM and other domestic manufacturers by reducing the compression ratio to about 8.5, and the petroleum industry responded by marketing unleaded or "low-lead" fuels for use with these vehicles. Responding to the impending legislative efforts, which were to culminate in the 1970 Clean Air Act, Mr. Cole later told the American Petroleum Institute of GM plans to install control systems including a catalytic converter on all new vehicles by 1975, which would require unleaded gasoline (23).

At that time, the U.S. Congress was preparing to pass Public Law 91-604, the Clean Air Act Amendment of 1970, which called for a 90% reduction of HC and CO by January 1, 1975, and a similar reduction of nitrogen oxides (NO_x) by the 1976 model year. These standards were eventually formalized at the levels of 0.41 g HC/mile and 3.4 g CO/mile for 1975 and 0.4 g NO_x/mile for 1976. [NO_x is the common term used to represent the total exhaust mixture of nitric oxide (NO), nitrogen dioxide (NO_x), and dinitrogen tetroxide (N_2O_4).] The Act authorized the EPA to request a study by the National Academy of Sciences on the technological feasibility of achieving these standards, and allowed the Administrator of the EPA to grant a year delay in enforcement of the standards, depending on the results of such studies. Similar laws were passed in California and in Japan, although the Japanese standards were based on a test procedure somewhat different than the test developed for use in the U.S.

The Clean Air Act of 1970 greatly accelerated the catalyst development efforts which had been revitalized in 1967. Urgent decisions were required on the configuration of the support, the choice of base metal versus noble metals as the active components, and the optimum compositions of the selected active ingredients; these decisions had to be made in terms of the new 1975-76 emissions standards and in anticipation of the use of unleaded fuels. A further demand on catalyst and system developers had arisen from the regulatory/legislative attention which had been focused on automobile-derived NO_x during the mid-sixties, and significant effort was being devoted during this period to catalytic and non-catalytic methods of controlling these nitrogen oxides.

A recent review article (16) discusses the pertinent characteristics of the monolithic (or honeycomb) ceramic support and the high area alumina pellet supports. Although only pelleted catalysts had been successfully certified in California in 1964, ceramic honeycomb supports (24) with very attractive properties had been developed and were an intriguing alternative to pellets. Both types of supports had advocates based on preliminary considerations of relative durability, maintenance and/or refill capability, resistance to catalyst poisons, lightoff, resistance to attrition, space requirements, converter shape, high temperature resistance, heat and mass transport, or the back pressures associated with given levels of performançe. Very reasonable prejudices based on earlier experience with one type or the other were undoubtedly also a factor in the preliminary selection of systems for further study. Finally, the choice of active catalyst type (base or noble metals) affected the selection of support indirectly, but possibly decisively, in some cases. As a result, most automobile manufacturers and potential catalyst providers mounted extensive development programs to produce the "best" decisions relative to support type; not surprisingly, different final selections were eventually made. When the dust settled, GM and several foreign suppliers chose particulate supports while Ford, Chrysler and many others decided to use monolithic supports for the 1975 model year vehicles.

The choice between base and noble metals as the active components was an equally important decision. The preliminary decisions were generally based on corporate judgements on the cost and availability of noble metals weighed against the lower specific activity of base metals for both saturated hydrocarbon and CO oxidation (which implied larger converter volume and space requirements) (25) and the greater sensitivity to thermal degradation of supported base metals relative to the noble metals (16). The importance of the cost-availability factors led the major automobile manufacturers and catalyst developers to devote extensive effort to the development of acceptable base metal catalysts which could overcome these problems; at least to a large extent, these efforts were quite successful. However, a fact of which catalyst developers had not been aware earlier eventually mandated the choice of noble metals for the 1975 vehicles; this was the discovery in 1971 that the more effective base metal catalysts were poisoned by the sulfur present even in unleaded gasoline (26). The fuel sulfur was emitted from the engine as SO_2 and reacted (in the oxygen-rich exhaust) with some of the active base metal components to form relatively inactive sulfates at the temperatures (<600°C) experienced in the converters. Although the sulfate-loaded catalysts were still somewhat active for exhaust hydrocarbon oxidation, they were inactive for CO oxidation (26, 27). Sulfur poisoning was found to be less severe above 600°C because of the thermal instability of the base metal sulfates, but the low space velocity operation (and hence large converters) required for base metals prohibited the location of the converter close enough to the exhaust manifold to ensure high temperature operation.

This limitation of base metals caused even those organizations which had been most committed to this approach to turn their attention to noble metals by mid-1972.

The selection of the optimum noble metal composition and loading level was diligently investigated during this period; these efforts proceeded concurrently with the individual efforts of the catalyst and substrate suppliers to optimize the catalyst and support formulations. In some cases, noble metal supply arrangements made by the auto companies may have defined the noble metal mixture and loading per vehicle, and the research objectives appeared to be limited to the preparation of the best possible catalyst with that noble metal content. In other cases, this variable was left as a degree of freedom, and the noble metal used per vehicle was allowed to vary among the engine families.

The interest in noble metals narrowed early to platinum, palladium and rhodium, with the two former elements receiving the major emphasis. It became clear that the greater "nobility" of platinum caused it to be more resistant to compound formation and/or chemisorptive poisoning than palladium, while palladium, by virtue perhaps of its enhanced reactivity with the support, was less prone to temperature-induced sintering and was thus more stable to anticipated or unanticipated high temperature exposures (30). The catalysts used in 1975 vehicles generally ended up with less than 0.1 troy ounces (3.1 g) of noble metals per vehicle; the most common noble metal mixtures were 71-29 and 67-33 (weight ratios) platinum-palladium, although a significant numbers of vehicles were equipped with catalysts containing only platinum, and a few contained platinum-rhodium (*ca.* 90-10) mixtures.

The data required to make these important decisions was obtained from combinations of laboratory, dynamometer-controlled engine, and full vehicle durability tests. The laboratory and engine tests were designed to permit judgements to be made relative to likely performance in the Environmental Protection Agency test procedure over 50,000 miles of vehicle operation, and the development of these accelerated catalyst tests was an essential part of the overall catalyst development programs (30, 31). However, tests of the entire system of which the catalyst is only one of many important components were eventually required to certify the vehicles using that system to the satisfaction of the government (30, 31).

Much of the data generated in the 1970-73 period was the subject of conflicting interpretation in subpoenaed testimony relative to the technical viability of catalytic converters in hearings before the EPA held to consider requests by automakers for delays in enforcements of the 1975-76 emission standards. As a result of these hearings (in April 1972 and March 1973), and of the EPA-commissioned report in February 1973 by the Committee on Motor Vehicle Emissions of the National Academy of Sciences, the EPA Administrator decided to ease for one year the 1975 HC standard from 0.41 to 1.5 g/mile, and of the CO standard from 3.4 to 15 g/mile (32); subsequent congressional actions (the Clean Air Act Amendments of 1974 and 1977) were to delay the imposition of the original 1975 HC standard until the 1980 model year, and the 3.4 g CO/mile standard until 1981 (except in California) (16). [The California legislature reacted by imposing "interim" standards generally intermediate between the original and "interim Federal" standards (16).] The "original 1976" NO_x standard of 0.4 g/mile suffered similarly, but more severely, in these actions; the Federal standards did not tighten to 1.0 g/mile until 1981, and it is unlikely the 0.4 g/mile level will ever be more than a "research goal" except perhaps in California.

The relaxation of the 1975 Federal standards had various impacts on the automobile manufacturers who had been proceeding with plans for the use of catalysts in 1975 in the event that it was necessary to meet the tight standards. Some large manufacturers proceeded to use catalysts extensively, while optimizing the engine calibrations to allow substantial improvements in fuel economy (12-28%) over the 1974 models (31). One chose to reduce the initial cost impact by using catalysts to treat only half of the exhaust of V-8 engines, which limited the use of engine calibrations to maximize fuel economy. This decision resulted in fuel economies at or below the 1974 level (33), which may have had a significant sales impact in view of the rapid increases in gasoline prices during the energy crisis following the embargo on mid-East oil in 1973. The situation was corrected by a number of mid-year model changes, including the use of catalysts on both banks of the V-8 engines.

Following the establishment of the 1975 "Interim Standards" in March of 1973, decisions were made to use catalysts in many of the vehicles which would begin EPA certification testing in October of 1973 for sale in September, 1974. Final decisions on catalyst suppliers were made and preparations to manufacture the large quantities of catalysts which would be required were well underway in late 1973. With little warning, the plans to use catalysts were placed in jeopardy by the EPA itself, when it was announced that a review was being conducted to determine whether the catalytic converters might not do more harm to human health than good (34).

It later developed that in 1971-72, EPA-sponsored studies at Dow of vehicle particulate emissions (35, 36) had shown increases in the mass of particulates from catalyst-equipped engines, and analysis of a Dow filter (by Ford, which had provided the catalyst) in late 1972 showed the presence of sulfuric acid. Ford reported this result in a February, 1973 letter to the EPA, which brought it to the attention of other vehicle manufacturers and initiated studies in-house and at Exxon to measure this effect more quantitatively as a function of catalyst type and vehicle operation modes. Estimates of potential roadside concentrations of sulfuric acid, based on an EPA computer program for the dispersion of gases at various locations away from the source, suggested that they might be, in "worst meteorological cases", several hundred micrograms per cubic meter ($\mu g/m^3$), which would have a significant adverse health impact (34).

Although this report and the resultant hearings on this issue held by the Senate Public Works Committee on November 5 and 6, 1973 did not prevent the use of catalysts on 1975 vehicles, it certainly precipitated a considerable budget increase for the EPA unit at Research Triangle Park (RTP) in North Carolina (34), which had "gone public" with this concern, and resulted in the expenditure of many million of dollars by automobile and catalyst manufacturers and unidentified other taxpayers. Early studies were presented at the February 1974 SAE meeting and at a symposium at RTP in April of 1974. Proponents of catalysts pointed out that alternative calculations based on the use of surrogates (CO, Pb), for which emission levels and roadside concentrations were both known, resulted in roadside sulfate predictions between twenty to one hundred times lower than those predicted by the EPA "HIWAY" computer model (39).

"Risk-benefit analyses" and "public health impact estimates" were the basis of "draft reports" released by the EPA in January 1975 (40, 41) (after they had been unofficially "leaked"). These "issue papers", which either included no references (41) or made references primarily to internal EPA memos, studiously overlooked completely the EPA's own clinical studies on the health effects on animals of exhaust from non-catalyst and from catalyst-equipped engines, which had been reported to the scientific public (42) and at the EPA Symposium on "Impact of Mobile Emission Controls" in April of 1974 (43). Those studies indicated no adverse effects on animals exposed to 1:10 diluted catalytic converter-treated exhaust during a period (1 week) which produced mortality rates of greater than 70% in infant rats exposed to similarly diluted non-catalyst-treated exhaust. It was established that the high mortality was not simply due to the levels of CO in the untreated exhaust.

In any event, EPA Administrator Russell Train had been considering requests by automakers for another delay of the original 1975 emission standards which were now, according to the 1974 Amendment to the Clean Air Act (32), due to apply for the 1977 MY. On March 5, 1975 he announced his decision (44) to retain the current Interim Standards on HC and CO for 1977 and recommended that Congress continue those standards through 1979 and delay the original 1975 standards until 1982; further, he also proposed that a federal standard for sulfuric acid emissions be established for 1979 models. Less than a month later, a new draft study (45) was released (46) by the EPA which showed that actual sulfuric acid emissions from 1975 production cars were about half the levels projected in the January "Issue Papers". It also challenged the "HIWAY" dispersion model calculations on expected roadside sulfate concentrations by analysis of the results from nine mutually independent sets of measurements including CO and Pb surrogate models and the carboxyhemoglobin (blood CO-levels) model which suggested that the "HIWAY" model "over-predicted" sulfate levels by factors of two to ten.

The Train decision and the conflicting EPA analyses of the impact of catalysts on sulfate exposure and public health were the basis of stern questioning in the hearings on emission standards which were being held by the Senate Public Works Subcommittee on Environmental Pollution and the House Interstate and Foreign Commerce Subcommittee on Public Health and Environment. At these hearings, conflicting data were also presented by public interest groups, catalyst suppliers, automobile makers, and by oil refiners; comments from the latter group were primarily directed to proposals to solve the sulfate problem by significantly reducing the sulfur content of gasoline. Costs were estimated by Exxon at 3 billion dollars (46).

Out of the resulting confusion, General Motors, which was reported to have spent $300 million on catalytic converters and was considered to be "the auto industry's major advocate" of catalysts (46) announced on May 9, 1975 its plans to "conduct an extensive research experiment to obtain data on actual atmospheric effects resulting from exhaust emissions of a large fleet of automobiles with catalytic converters" (47). Participation of the EPA was invited, and other automakers were invited to also provide catalyst equipped vehicles for a 400 car fleet to be driven over a course at GM's Milford, Michigan Proving Ground. Vehicle emissions and the resultant concentrations of exhaust components in the vicinity were to be measured to permit the development and verification of models for the dispersion of exhaust pollutants from actual heavy freeway traffic under a range of meteorological conditions, including particularly the "worst meterology" conditions under which the EPA "HIWAY" Model was most suspect (47).

This test was conducted on 17 days in October of 1975, and the papers based on the measurements by GM, the EPA, and other participants were presented at a symposium (48) hosted by the EPA at Research Triangle Park on April 12, 1976. These studies showed that the "HIWAY" model overpredicted down-wind roadside pollutant concentrations by large factors (15-50) under "worst-meteorology" conditions, and actually could underpredict "upwind" levels, because the model did not recognize the local mixing near the exhaust pipe outlet due to the turbulence created by the movement of the vehicles and the rise of the heated exhaust plume due to convection. The inaccuracies resulting from ignoring these effects were shown to be most pronounced under the "zero-wind, unstable" (or "worst-meteorology") conditions.

In recognition of the significance of this experiment, an opening statement was made at the symposium by T. A. Murphy, Deputy Assistant Administrator of the EPA, in which it was indicated that the EPA had dropped its proposal for a 1979 sulfate emission standard. However, although the statement repeated that the Administrator's decision to delay the '77 HC and CO standards was "solely because of his concern as to adverse health impact of sulfuric acid emissions", there was no suggestion of reconsideration of that decision or of his recommendation for further delays through 1979. When the smoke cleared, the 1977 amendment to the Clean Air Act, while accepting the delay of HC and CO standards for the period of 1978-79 as proposed by Train (44), created tighter standards than he had proposed for 1980 and re-established the original 1975 HC and CO standards for 1981, along with a 1.0 g NO_x/mile standard.

This history should not conclude without a few comments on some of the impacts of the development of automotive emission control catalysts. Catalysts have certainly been established as predictable and dependable automotive components, and are likely to be a method of choice for control of automotive pollutants as long as any emission standards exist for passenger vehicles, and particularly as long as fuel economy is an important consideration. The significant improvements in fuel economy observed in 1975 model vehicles were only possible because the engines could be calibrated to better fuel economy while catalysts were utilized to control the pollutants. It is doubtful that the fleet average fuel economy schedule that was legislated in response to the 1973 energy "crisis" could have been established, even with the pre-1975 emission standards, without the use of catalysts. The cost effectiveness of catalysts is proven by their comprehensive use since 1975 by an industry were "a dime a car is worth a million dollars" is a common expression. The emission standards made possible with catalysts have been estimated to prevent as many as 4,000 deaths and four million illness-restricted days per year in the U.S. urban population (49); similar impacts in Japan can be presumed. The requirement for unleaded fuel in catalyst-equipped cars permitted a more orderly "phase-down" of the overall lead content of refineries when the EPA independently became convinced that lead represented a direct health hazard. The greatest single use of platinum in the U.S. is for emission control catalysts, accounting for over 57% of the annual consumption, or about 750,000 troy ounces per year (in 1979) (50); other catalyst usage accounted for only about 230,000 troy ounces. Finally, the high visibility of the scientific and political arguments which accompanied the introduction of catalysts for automotive emission control has created a potpourri of true and untrue impressions about catalysts in the general public, with the result that there are now at least 100 million more experts in catalysis than there were ten years ago (51).

In this chapter, the emphasis has been placed on the chain of events leading to the use of catalysts on nearly all U.S. and Japanese automobiles in the 1975 model year. The sulfate issue has been covered through the events of 1976 because it threatened to result in the removal of catalysts from the emission control system, and was just reaching full bloom in September of 1974 when the 1975 catalyst-equipped cars were hitting the showrooms. The concurrent development of catalysts for the control of NO_x by direct catalytic reduction or by simultaneous catalytic conversion of HC, CO and NO_x, which were spurred by the inclusion of NO_x standards in the

1967 Clean Air Act is another story for another occasion. Several recent well-written and well-referenced reviews of the technical aspects of automotive catalysis are available (16, 48, 52), and are recommended to those who are interested in more detailed technical considerations of this chapter in the history of catalysis.

"Inhale! State's Air is Cleanest in Years" - Headline, Chicago Tribune, (C. Bukro), January 4, 1983. This chapter is dedicated to the thousands who made this headline possible.

Literature Cited

1. Haagen-Smit, A. J. Ind. Eng. Chem. 1952, 44, 1342-6.
2. Mader, P. P.; MacPhee, P. D.; Lofbert, R. T.; Larson, G. P. Ind. Eng. Chem. 1952, 44, 1352-5.
3. Twenty Society of Automotive Engineers (SAE) papers discussing these early studies were compiled in Volume 6 of the Technical Progress Series (TP-6) of the SAE, by the SAE Vehicle Emission Committee, 1964 [L.C. Card No. 64-16836].
4. Nebel, G. J., paper presented at Society of Automotive Engineers Meeting, Seattle, Paper No. 174, August, 1957.
5. Cannon, W. A.; Hill, E. F., Welling, C. E., presented at Society of Automotive Engineers Meeting, Seattle, Paper No. 174, August, 1957.
6. Cannon, W. A.; Welling, C. E., paper presented at Society of Automotive Engineers Meeting, Detroit, Paper No. 29T, January, 1959.
7. VanDerveer, R. T.; Chandler, J. M., paper presented at Society of Automotive Engineers Meeting, Detroit, Paper No. 29S, January, 1959.
8. Schaldenbrand, H.; Struck, J. H., paper presented at Society of Automotive Engineers Meeting, Detroit, Paper No. 486E, March, 1962.
9. Standards for Ambient Air and Motor Vehicle Exhaust Hydrocarbons, adopted December 4, 1959, California Department of Public Health.
10. Nebel, G. J.; Bishop, R. W., paper presented at Society of Automotive Engineers Meeting, Detroit, Paper No. 29R, January, 1959.
11. Klimisch, R. L. "Oxidation of CO and Hydrocarbons over Supported Transition Metal Oxide Catalysts", Proceedings of the 1st National Symposium on Heterogeneous Catalysis for Control of Air Pollution, Philadelphia, Franklin Institute, November, 1963.
12. U. S. Patent No. 2,664,340, Houdry, E. J., December 29, 1953.
13. U. S. Patent No. 2,742,437, Houdry, E. J. (to Oxy-Catalyst, Inc.), April 7, 1956.

14. Hass, G. C., paper presented at Society of Automotive Engineers Meeting, West Coast Meeting, Paper No. 210A, August, 1960.
15. Sweeney, M. P. J. Air Poll. Control Assoc. 1965, 15, (1), 13-18.
16. Kummer, J. T. Prog. Energy Combust. Sci. 1980, 6, 177-99.
17. Schwochert, H. W., paper presented at Society of Automotive Engineers Meeting, Chicago, Paper No. 690503, May, 1969.
18. U.S. Patent No. 3,259,454, Michalko, E. (to UOP Inc.), July 5, 1966.
19. Weaver, E. E., paper presented at Society of Automotive Engineers Meeting, Paper No. 690016, January, 1969.
20. Su, E. C.; Weaver, E. E., paper presented at Society of Automotive Engineers Meeting, Paper No. 730594, May, 1973.
21. Jones, J. H.; Weaver, E. E., prepared discussion of a paper presented at the Society of Automotive Engineers Meeting, Chicago, Paper No. 690503, May, 1969.
22. Brookes, H. R., address at American Metal Market Seminar on Platinum and Palladium, New York, October, 1968.
23. Cole, Edward N., address to Annual Meeting of the American Petroleum Institute, New York, November 17, 1970.
24a. Johnson, L. L.; Johnson, W. C.; O'Brien, D. L. Chem. Engr. Prog. Symp. Ser. 1961, 35, 55.
24b. U.S. Patent No. 3,437,605, Keith, C. D. 1969; No. 3,441,382, Keith, C. D. 1969 (both to Engelhard).
24c. U.S. Patent No. 3,518,206, Sowards, D. M.; Stiles, A. B., 1970 (to DuPont).
24d. Bagley, R. D.; Doman, R. D.; Duke, D. A.; McNally, R. N., paper presented at Society of Automotive Engineers Meeting, Detroit, Paper No. 730274, January, 1973.
25a. Barnes, G. J. "A Comparison of Platinum and Base Metal Oxidation", Paper No. 7 in "Catalysts for the Control of Automotive Pollutants", Advances in Chemistry Series 143, ACS 1975, pp. 72-84.
25b. Schlatter, J. S.; Klimisch, R. L.; Taylor, K. C. Science 1973, 179, 798-9.
26. Hunter, J. E., paper presented at Society of Automotive Engineers Meeting, Paper No. 720122, January, 1972.
27. Fishel, N. A.; Lee, R. K.; Wilhelm, F. C. Environ. Sci. Technol. 1974, 8, 280.
28. Yu Yau, Y. F. J. Catalysis 1975, 39, 104.
29. Klimisch, R. L.; Summers, J. C.; Schlatter, J. C. Paper No. 9 in "Catalysts for the Control of Automotive Pollutants", Advances in Chemistry Series 143, ACS 1975, pp. 103-115.
30. Lester, G. R.; Brennan, J. F.; Hoekstra, J. Paper No. 3, ibid.

31. Taylor, K. C. "Automobile Catalytic Converters", GMR-4190, PCP-192, General Motors Research, Warren, MI, October 8, 1982.
32. Decision of the Administrator, EPA, March, 1973.
33. Austin, T. C.; Hellman, K. H., paper presented at Society of Automotive Engineers Meeting, Paper No. 740970, October, 1974.
34. Shapely, D. Science 1973, 182, 368-371.
35. Moran, J. B.; Manary, O. J.; Fay, R. H.; Baldwin, M. J. EPA-OAP Publication APTD-0949 (July, 1971) [NTIS PB-207312].
36. Gentel, J. E.; Manary, O. J.; Valenta, J. C. EPA-OAWP Publication APTD-1567 (March, 1973).
37. Beltzer, M.; Campion, R. J.; Peterson, W. L., paper presented at Society of Automotive Engineers Meeting, Paper No. 740286, March, 1974.
38. Lester, G. R., prepared discussion of a paper presented at the Society of Automotive Engineers Meeting, Paper No. 740286, March, 1974.
39. Zimmerman, J. R.; Thompson, R. S. "A Simple Highway Air Pollution Model (HIWAY)", EPA Publication, EPA Meteorology Laboratory, Research Triangle Park, NC (1973).
40. Finklea, J. F.; Nelson, W. C.; Moran, J. B.; Akland, G. G.; Larsen, R. I.; Hammer, D. I.; Knelson, J. H. "Estimates of the Public Health Benefits and Risks Attributable to Equipping Light Duty Motor Vehicles with Oxidation Catalysts", National Environmental Research Center, EPA, RTP, NC, January 20, 1975.
41. Anonymous "Estimated Public Health Impact as a Result of Equipping Light Duty Motor Vehicles with Oxidation Catalysts", Office of Research and Development, Office of Air and Waste Management, EPA, January 20, 1975.
42. Hysell, D. K.; Moore, W. Jr.; Malanchuk, M.; Garner, L.; Hinners, R. G.; Stara, J. F. "Comparison of Biological Effects in Laboratory Animals of Exposure to Automotive Emissions Emitted With and Without Use of Catalytic Converter", 67th Annual Meeting of Air Pollution Control Association, Denver, June, 1974.
43. Hysell, D. K.; Moore, W. Jr.; Hinners, R.; Malanchuk, M.; Miller, R.; Stara, J. F. "The Inhalation Toxicology of Automotive Emissions as Affected by an Oxidation Exhaust Catalyst", Symposium on Health Consequences of Environmental Controls: Impact of Mobile Emission Control, NIEHS, NIH, PHS, HEW; and NERC, ORD, EPA, Durham, NC, April 17-19, 1974.
44. Anonymous "Exhaust Rules for '77 Autos are Delayed a Year Due to Sulfuric Acid Problem", Wall Street Journal, March 6, 1975, p. 12; EPA Press Release, March 5, 1975.
45. Anonymous "Evaluation of Sulfur Acid Exposures from LDV (Light Duty Vehicles) Emissions", EPA, April 3, 1975.
46. Magida, A. J. National Journal Reports, April 12, 1975, pp. 552-558.

47. Anonymous, Press Release, General Motors Corporation, May 9, 1975; Cadle, S. H.; Chock, D. P.; Levitt, S. B.; Monson, P. R. "A Proposal to Study Automotive Sulfate Dispersion from a Simulated Freeway", GM Research, Warren, MI, April 29, 1975.
48. Symposium, "The General Motors/Environmental Protection Agency Sulfate", Dispersion Experiment, Research Triangle Part, NC, April 12, 1975 (Chairman, Stevens, R. K. EPA). See also Cadle, S. H.; Chock, D. P.; Heuss, J. M.; Monson, P. R. GMR-2107, General Motors Research, Warren, MI.
49. Report of Coordinating Committee on Air Quality Studies, National Academy of Sciences, September 6, 1974.
50. Chaston, J. C. Platinum Metals Review 1982, 26, 3-9.
51. Lester, G. R. "Back-of-the-Envelope-Calculation, J. Reas. Est. 1983, 1, 1.
52. Hegedus, L. L.; Gumbleton, J. J. CHEMTECH 1980, 10, 630-642.

RECEIVED April 5, 1983

Attempts to Measure the Number of Active Sites

RUSSELL W. MAATMAN
Dordt College, Department of Chemistry, Sioux Center, IA 51250

There have been attempts to count active centers ever since their existence was postulated in 1925. Normally site densities, where the site density is the number of active centers per unit area, are thought to be near the maximum value, 10^{15} cm^{-2}, but in some cases values which are several orders of magnitude smaller have been suggested. A direct method of determining site density is one which depends on results of kinetic studies. Several direct methods, including one using transition state theory, are described; results are presented. Many indirect methods, along with results, are also discussed.

Solid catalysts were used commercially and in laboratories many years before 1925. But it was in that year that H.S. Taylor suggested that there are distinct, identifiable active centers on the catalyst surface (1). By 1929 Taylor claimed to have kinetic and adsorption evidence demonstrating the existence of these active centers (2). Evidence for the existence of such centers accumulated, so that in a 1969 review Boudart (3) could claim that active centers have been seen by an electron microscope.

Ever since the existence of active centers has been recognized it has been held that there can be several kinds of active centers on one surface. Taylor stated this specifically in 1926 (4). To take one example: It is not at all unexpected that the claim has been made for several kinds of catalytically active alcohol adsorption sites on alumina-containing catalysts (5, 6).

Yet a detailed study of active centers could not be begun before the catalyst surface area could be measured. If it is important to distinguish the active centers from the rest of the surface, it is also important to determine how much surface there is. Various methods of determining surface area have been used; but without doubt the advent of the easy-to-use BET method in 1938 (7) was a milestone in the history of the study of active centers.

0097–6156/83/0222–0435$06.00/0

The question concerning the number of active centers a given catalyst possesses is obvious. Often it has uncritically been assumed that all surface atoms of a given kind are catalytically active. Thus, if the cross-sectional area of a surface atom is $10 Å^2$ and the surface consists of only such atoms--as in a metal--and all are active, then there are about 10^{15} sites cm^{-2}. We shall henceforth refer to the number of active centers per square centimeter as the site density.

But many workers have pointed out (see, for example, Reference 8) that the active fraction of the surface can be very small; that is, the site density can be very low. By "very low" some workers mean 0.1%; others, as low as $1 \times 10^{-4}\%$. In any case, it is important to know the site density; the site is, after all, a reactant, and reactions cannot be well understood if the concentration of one reactant is not known. We shall see that several methods of determining site density have been used. But the matter is quite complicated. Two examples indicate how complicated the question can be: We might find that the site density is a function of experimental conditions, as it no doubt is when hydrogen spills over from metal to support (9, 10, 11). Again, an already complicated matter can be made more difficult by site synergism, an effect which has been exhibited (12).

There is no universally correct answer to the question, "What are the active sites?" In many cases it may be correct to answer that the sites are all the surface platinum atoms: in another, all four-coordinated surface aluminum atoms; and so forth. For a long time other answers have also been given. As early as 1934 Frost and Shapiro (13) concluded in a review that sites are often points of crystal lattice deformation. In this connection, Erbacher showed in 1950 (14) that catalytic sites could be annealed. In 1953 Thon and Taylor (15) pointed out that the active site could be a dissociatively adsorbed particle, for example, an adsorbed hydrogen atom. In the 1950's Eischens and Pliskin carried out their classic work of obtaining the infrared spectrum of adsorbed molecules and published a survey of the subject in 1958 (16). When they observed the adsorbed molecule, they could conceivably be observing the reacting molecule and, as a result, we could be much closer to knowing the nature of the catalytic site.

There are other types of active sites. In 1955 Kobozev (17) pointed out that sometimes--especially with certain metals--the catalytic properties could be due to the properties of atoms, but that in other cases--as with some metal oxides--the catalytic property might be associated with the entire crystal. We showed later (18) that in at least one case, silica-alumina catalyst, it is very easy to err concerning the nature of the site. Thus, evidence for Bronsted acidity could be interpreted as evidence for the presence of aluminum which can be ion-exchanged when the catalyst is placed in salt solution. Looking at all these examples, we conclude that that there are many different sources of activity on the solid catalyst surface.

What is the relation between the number of sites and the reaction rate? For over half a century workers have assumed that there is an intimate relation, but that qualifications must be made. For example, in 1931 Schwab and Rudolph (19) concluded for a nickel catalyst in liquid phase hydrogenation that only edges and corners are active. They based their conclusion on the observation that the rate was not proportional to the catalyst area, an area they obtained by a method no longer used.

Workers still have this attitude toward the need for knowing the number of active sites. Often catalytic scientists make conclusions based on small differences in site density (20). A half century after Taylor's 1925 postulate Monte Carlo simulations of catalytic surfaces are made (21); these simulations often depend upon the value of the site density. Again, many workers base conclusions on turnover frequencies, also dependent on site densities.

We shall attempt to show how workers have used experimental data, especially kinetic data, to estimate site densities. But there are problems. Typical among the problems which arise are these: Some rate data cannot be used for a reaction model, if the warnings of the workers themselves are to be taken seriously (22). Also, the Arrhenius plot even for simple reactions might be not only non-linear; it might contain a maximum (23). Again, there can easily be a systematic error in determining the initial reaction rate, used so often in reaction elucidation and site density determination (24).

There are both direct and indirect ways of determining site density. All the direct ways are kinetic. Some of the kinetic methods use trasition state theory (TST) while others do not. We shall look first at the direct, non-TST methods.

Direct, Non-TST Methods

There are limitations in using a kinetic method in elucidating any part of the catalytic reaction. Thus, it is almost always assumed that a rate-limiting step exists, even though such an assumption might lead to difficulty. There are also limitations on the physically-possible value of the site density. The upper limit is 10^{15} cm^{-2}, as explained earlier. The lower limit is determined by the fact that the turnover frequency cannot be greater than the number of molecules which collide with the site. For gas reactants this criterion leads to a lower limit of 10^{5}-10^{7} cm^{-2} in most systems (25).

Some methods of determining site density depend upon kinetics results but are still non-TST. An analysis of Boudart's facile reaction concept, described in 1966, provides a means of determining site densities in many systems (26). If a reaction is not demanding, that is, if its rate is proportional to the number of (for example) surface platinum atoms, then it is highly likely that all such atoms are active, and the site density is the surface density of those atoms. Another method utilizes the sticking coefficient of gas molecules on the catalyst surface. In 1963 Ashmead, Eley and Rudham (27) studied ortho-para hydrogen conversions over Nd_2O_3; in their analysis of the high-temperature reaction they used collision numbers and sticking coefficients and determined the fraction of the surface which is active, thus determining the equivalent of the site density.

Ethylene oxide polymerizes over metal oxides, hydroxides, and carbonates. In 1964 Krylov and coworkers (28) reported weighing the amount of adsorbed polymer; since they knew its average molecular weight, they could calculate the number of reactant molecules adsorbed on active sites, thus providing the site density.

Krylov and Livshits discussed in 1965 (29) the possibility of using optical rotatory properties of the catalyst surface, in their case the surface of magnesium tartrate, as it catalyzed the polymerization of propylene oxide, to determine the properties--and thus possibly the number--of active sites.

For some reactions, such as ortho-para hydrogen conversion, it is sometimes possible to show that the active site is a paramagnetic surface species. When such sites are counted, the site density is obtained (30, 31). In 1968 Shigehara and Ozaki (32) studied H_2-D_2 and o-H_2-p-H_2 reactions over copper; they found $\sim 10^6$ times more sites at high temperatures than at low, suggesting that the site density may increase from 10^8-10^{10} cm^{-2} to $\sim 10^{15}$ cm^{-2} as the temperature is increased.

Carter and coworkers in 1965 (33) examined alumina using infrared spectroscopy as the surface catalyzed ethylene hydrogenation reaction. By knowing the number of molecules in the system, the rate, and the total number adsorbed on active and inactive sites, they were able to conclude that one possible explanation of their results was that the site density was very low.

The hydroxyapatite-catalyzed dehydration of 2-butanol, examined by Bett and Hall in 1968 (34), was shown to be amenable to a special method of obtaining the site density. They determined the number of product molecules which desorbed from the surface at infinite flow rate using a microcatalytic reactor. From this number they deduced the number of active sites.

In the quenching of ethylene polymerization over supported Cr_2O_3 by labelled methanol, $^{14}CH_3OH$, Yermakov and Zakharov in 1969 (35) were able to determine how many alcohol molecules were on active sites by ^{14}C counting, thus providing the site density.

If one is virtually certain that a certain surface species is a reaction intermediate, measurement of the concentration of that species can yield the site density. For example, in 1969-1971 Gonzalez and coworkers (36, 37, 38) related formic acid decomposition over transition metal formates to surface formate concentration.

In 1980 Franco and Phillips (39) were able to show that in

the hydrogenation of benzene, using the results for a large number of runs carried out under significantly different conditions, that they could calculate the site density by a very complicated kinetic method without, however, using transition state theory.

Sometimes relative values of the site density have been obtained, even though only a minimum number of assumptions concerning the model were made. Thus, regardless of the kinetic theory used, the site density seems to be associated with the pre-exponential factor of the Arrhenius equation. In 1970 Baddour and coworkers (40) noted that this factor decreased seven orders of magnitude during the run as CO oxidized over Pd. Possibly the catalyst was changing rapidly, so that the site density decreased significantly.

The TST Method, a Direct Method

Eyring's earlier work on the theory of absolute reaction rates was published in the classic *The Theory of Rate Processes*, by Glasstone, Laidler, and Eyring, in 1941 (41). Chapter VII, "Heterogeneous Processes," has been a standard reference for catalytic chemists for over forty years.

Briefly, in the Eyring theory for any reaction, homogeneous or heterogeneous, there is an evaluation of the pre-exponential factor A in the Arrhenius equation

$$\text{Rate} = A \exp(-E/RT)$$

A contains the frequency factor kT/h and a factor which is a function of reactant concentrations. For a catalyzed reaction the catalyst is one of the reactants.

Glasstone, Laidler, and Eyring assumed that they could know all the quantities on the right of this equation whenever the activation energy could be determined, and that therefore they could predict the rate. They were partially successful. Obviously, one can err if he does not have the correct function of the concentrations, as would be the case if the wrong mechanism were used. This would lead to an incorrect assumption for the value of the site density.

One can, assuming the mechanism is known, turn the equation around and then calculate the site density. We and others have done just that (42, 43). Calculations have been made for both saturated and unsaturated surfaces, for both unimolecular and bimolecular reactions, for reactions in which the adsorbed molecule is immobile and also where it is mobile, and for reactions carried out at both low-conversion and high-conversion conditions.

But in the 40-year period in which TST has been used, questions concerning its validity have arisen. Some criticize the derivation of the theory (44). It cannot be used when the reaction is very fast (45). Occasionally the transmission coefficient is several orders of magnitude smaller than unity, the value usually assumed (46). Some workers maintain that there are too many questions concerning TST for it to be used with confidence to analyze the admittedly complicated heterogeneous catalytic systems (47). Others have postulated unusual surface energy relations (48) which, if they do exist, would make it very difficult to use TST. As we shall see later, postulating surface mobility for the adsorbed species does remove some difficulty.

Consider some specific TST examples. First, confidence in the TST method is strengthened if site densities calculated are the same as those determined by other methods. Glasstone, Laidler, and Eyring in their 1941 book (41) report 15 TST calculations for several different kinds of reactions--adsorption, desorption, and various other unimolecular and bimolecular reactions. They assumed a site density of 10^{15} cm^{-2} in each case. In 12 of the 15 cases this assumption was justified. This is one kind of confirmation of TST validity.

Also, one might expect that many metal catalysts would have a very high site density, perhaps near 10^{15} cm^{-2}. In 1966 Horiuti and coworkers (49), using a method we showed (42) is almost equivalent to the TST method, examined the reported data for 66 metal-catalyzed reactions. They found their assumptions to be valid. (Miyahara, one of Horiuti's coworkers, and Kazusaka later (50) modified the method to take variable orders into account.) Again, a very high site density is required for the postulated mechanism of isopropanol-acetic acid esterification over silica gel, as reported by Fricke and Altpeter (51). We showed (52), that such a high value for the site density is obtained using TST.

But what about site densities much lower than 10^{15} cm^{-2}? If such a value is obtained using TST, is it reliable? It can be reliable if there has been confirmation of the use of TST in catalysis. Most of the direct confirmations of TST have been for fairly high site densities. Two examples are given:

In the dehydration of 2-butanol carried out by Bett and Hall (34), referred to earlier, the site density was found to be 10^{13}-10^{14} cm^{-2} both by a non-TST method and by TST. On TiO_2, F-center electrons are the sites for the dehydration of phenylpicrylhydrazine; Misra (53) in 1969 found 10^{11} F-centers cm^{-2}, almost the same value calculated using TST.

Because there are physical limitations on allowable values of the site density, different mechanisms for a given reaction may be postulated and the site density calculated for a mechanism can be used to decide if that mechanism is a feasible one. In this way, many conceivable mechanisms have been eliminated. For example, in 1970 Cha and Parravano (54) used such calculations to exclude certain mechanisms in the gold-catalyzed transfer of oxygen between CO_2 and CO. We have reviewed the question of using the calculated site density as a criterion for the validity of a mechanism (25).

TST calculations have led to some unexpected or abnormal values of site density. Glasstone, Laidler, and Eyring reported some such results in their 1941 book (41). In 1964 Hayward and coworkers reported (55) for the desorption of hydrogen from nickel, even though hydrogen covered the surface, that TST calculation indicated that only one site in 10^8 was active; apparently hydrogen migrated to special desorption sites.

Using TST we calculated in 1967 (23) that the site density of silica-alumina for the cracking of cumene might be as low as 10^7 cm^{-2}. For t-butylbenzene cracking, Bourne and coworkers in 1971 (56) showed using TST that their silica-alumina catalyst had a site density of about 10^5 cm^{-2}. For cumene cracking over some metal faujasites Richardson calculated in 1967 (57) site densities as low as 10^6 cm^{-2}, assuming a surface area of a few hundred square meters per gram.

Such low values for the site density have been found in many cases in which the slow step occurs on the surface; we have reviewed this question (58). We have also reviewed the situation for other kinds of reactions (25, 42) and have shown for reactions in which the slow step is adsorption that site density values obtained by TST are often impossibly large. Both these problems--too low a site density for surface reactions and too high a site density for adsorption reactions--are removed if one assumes that the surface species retains some of its gas phase rotational or translational degrees of freedom.

For example, in 1962 Brouwer (59) assumed a high site density in the isomerization of 1-butene over alumina, but found that the surface species lost about 40 e.u. upon reaction. It is possible that the site density was somewhat smaller than he assumed. Yet some of the large entropy loss could have been real, in which case the species possessed some freedom of motion before reaction. Similarly, Lawson in 1968 (60) found an entropy loss of 25 e.u. in the decomposition of formic acid over silver. Knozinger and coworkers in 1972 (61), also assuming high site densities, found the entropy loss of the surface species to be large, in many cases more than 20 e.u., in catalyzed alcohol decomposition reactions.

That the surface species does possess some rotational or translational motion has been shown in many instances. Sata and Miyahara in 1974 in 1976 (62, 63) found it necessary to assume that the adsorbed C_2H_5-radical could rotate. In 1978 Aldag, Lin, and Clark (43) devised a partition function for a mobile adsorbed species and used it in the analysis of a catalyzed reaction. Others have also attacked this problem. According to Eley and Russell (64), the activated complex possessed some surface rotational and translational freedom in $^{28}N_2$-$^{30}N_2$ equilibration over rhenium.

Thus, some workers have assumed that the surface species is immobile and they have used TST to calculate site densities; others have assumed that the site density is about as large as is physically possible and have used TST (or some variation) to calculate the entropy loss of the surface species as it reacts.

We have recently (65) suggested that there is a third possibility. In this approach, we remove the need for making either of these assumptions by using an experimentally-determined quantity not usually used, the fraction of sites occupied under steady-state conditions.

Naturally, site density calculations are difficult or impossible when the reaction is very complex. Even then an attempted calculation may indicate that the reaction is complex; such is the case with the dehydrogenation of cyclohexane over platinum (66) and palladium (67) supported on alumina. Thus, even after 40 years, the TST method of calculating site densities helps us gain insight into solid-catalyzed reactions.

Indirect Methods

An indirect method is one in which it is assumed that the group of active centers is that group of surface sites which has a certain characteristic, say, the ability to adsorb a certain kind of molecule. Thus, counting those sites is the same as determining the site density. But as long ago as 1929 Taylor (2) stated that sites on the same surface could qualitatively or quantitatively differ with respect to catalytic activity. Sometimes one is justified in assuming that a certain set of sites, such as all metal atoms or all sites which adsorb a certain poison molecule, is the same as the set of sites which catalyzes a reaction. But unfortunately this assumption is not always justified.

Probably the most popular indirect methods of determining site density are the poison method and the method in which the amount of chemisorption of reactant is measured. The number of either the poison or the reactant molecules which chemisorb per unit area is then taken to be the site density. Some examples will be discussed.

Only a few years after the existence of active centers was first postulated, Dohse and coworkers in 1930 (68) determined the site density by what amounts to determining the number of chemisorbed reactant molecules when the reaction is zero order. For isopropanol dehydration over alumina they found a site density--assuming their alumina had a normal surface area--of about 10^{13} cm^{-2}. In 1937 Kubokawa (69) used a poisoning method, measuring the amount of mercuric ion which inhibited H_2O_2 decomposition over platinum black. Balandin and Vasserburg obtained in 1946 (70) a site density of the order of 10^{13} cm^{-2} for the dehydration of isopropanol over a mixed oxide of zinc and alumina.

The classic poisoning paper, by Mills, Boedecker, and Oblad, appeared in 1950 (71). They used several nitrogen-containing organic bases to quench cumene cracking over silica-alumina. The amount of base adsorbed when the reaction was just quenched was measured; the smallest area covered at that time, using a variety of bases, was taken to approximate the area active in cracking.

If one measures the amount of reactant which adsorbs, he may obtain a value equal to the site density or a value greater than the site density, since, after all, the reactant molecule might adsorb on non-catalytic sites. But in the 1958 paper of MacIver, Emmett, and Frank (72), in which they reported the amount of isobutane chemisorption on silica-alumina, there was an unexpected result. The amount adsorbed was only $3x10^{10}$ molecules per square centimeter, and therefore the site density for any silica-alumina catalyzed reaction of isobutane is no greater than $3x10^{10}$ cm^{-2}. Obviously, a very low upper limit to the site density is much more nearly correct than a high upper limit.

The poison method has been used more often than the reactant chemisorption method. Especially since the 1950's many Russian workers, such as V.P. Lebedev and coworkers (73-76), have used this method to determine the site density.

In 1970 Amenomiya and Cvetanovic (77) were able to use the reactant chemisorption method in the dimerization and hydrogenation of ethylene over silica-alumina catalysts; for ethylene chemisorption they found site densities of the order of 10^{11} to 10^{12} cm^{-2} for the catalysts.

A pulse poisoning technique to determine site densities has also been used in recent years. Here the site density is determined by increasing the size of the poison pulse until the reaction is quenched. Romanovskii described the method in 1967 (78) for the purpose of obtaining site distributions. In 1968 Murakami (79) not only used the method, but developed analytical means of determining just when the reaction was quenched as thiophene pulses poisoned benzene hydrogenation over nickel.

Two additional complications should be mentioned here: First, the poison technique can be complicated by the poisoning of the reaction by reaction products (80). Also, in some cases there has been a lack of correlation between the amount of reactant chemisorbed and the reaction rate; this was shown by Fukuda (81) in ethylene hydrogenation over WS_2.

Other indirect methods, those not involving poisoning or reactant chemisorption, have been used to determine site densities. Perhaps the most common of these methods is to count the surface metal atoms, with either pure metal crystals or metal supported on oxides, with the counting carried out by determination of the amount of O_2 or CO_2 which chemisorbs (82); the assumption has then sometimes been made that all such atoms are catalytically active.

Haag and Whitehurst had an interesting approach. In 1971 (83) they reported on the catalytic activity of $Pd(NH_3)_4^{2+}$ on CO addition to allyl chloride, both when the palladium species was in known positions in an ion exchange polymer and when this species was a homogeneous catalyst in solution. Presumably one could know accurately how many active sites this well-characterized solid possesses. With this assumption, they concluded that the palladium species was as active in the solid as in the liquid phase.

Many workers have been interested in determining site densities by using other chemisorption reactions. These efforts have generally been made during the last few decades; some examples follow:

In 1962 Hall (84) studied the chemisorption of organic cation radicals on silica-alumina and counted the number of adsorbed ions by means of ESR, obtaining about 10^{12} cm^{-2}. It was assumed that the sites so counted are Lewis acid cracking sites.

Some workers (85, 86) have titrated the surface of silica-alumina with nitrogen bases and related their results to the number of cracking sites. Dzisko and coworkers in 1966 (87) assumed the number of active sites for n-butanol decomposition on alumina could be counted by titrating with NaOH in the presence of an indicator.

In 1971 Bourne and coworkers in t-bultylbenzene cracking (already cited (56)) over silica-alumina determined the adsorption isotherm of benzene on the same material. A Langmuir plot was non-linear at one end; they suggested that this portion of the isotherm indicated that site heterogeneity exists. This part of the isotherm corresponds to a site density of about 10^{12} cm^{-2}, a value probably closer to the correct value than the value of 10^5 cm^{-2} cited above.

Other methods--indirect, but not utilizing chemisorption--have been used to determine site densities. In 1958 the magnetic moment of surface atoms was taken to be a measure of the number of catalytically active atoms (88). The number of surface free valencies was used for the same purpose in 1964 (89). In 1965 Mellor and coworkers (90) oxidized KI over silica-alumina; by chemical analysis the number of surface atoms capable of oxidation was determined and taken to be the site density. The number of hydrogen vibrations on the ZnO surface which catalyzed the dehydrogenation of isopropanol was used to calculate the site density (91).

Summary

This history of attempts to obtain site densities in catalytic reactions has been given taking one kind of approach at a time, observing where possible the chronological development of that approach. But no method which has ever been used is no longer used. Thus, it is not possible to say that one method has superseded another, and that the history of the problem is a history of successive approaches to solving the problem. This is another way of saying that as the science of catalysis has grown and consisted of more and more subdisciplines, so has the number of ways of finding the site density grown.

Literature Cited

1. Taylor, H.S. Proc. Royal Soc. (London), A 1925, 108, 105-111.
2. Taylor, H.S. Z. Elektrochem. 1929, 35, 542-549.
3. Boudart, M. Amer. Scientist 1969, 57, 97-111.
4. Taylor, H.S. J. Phys. Chem. 1926, 30, 145-171.
5. De Mourges, L.; Peyron, F.; Trambouze, Y.; Prettre, M. J. Catal.. 1967, 7, 117-125.
6. Vladyko, L.I.; Trokhimets, A.I.; Markevich, S.V. Dokl. Akad. Nauk Beloruss. SSR 1969, 13, 245-248.
7. Brunauer, S.; Emmett, P.H.; Teller, E. J. Am. Chem. Soc. 1938, 60, 309-319.
8. Kaliko, M.A.; Pervushina, M.N. Khim. i Tekhnol. Topliv i Masel 1959, 4, 35-49.
9. Sancier, K.M. J. Catal. 1971, 20, 106-109.
10. Sancier, K.M. J. Catal. 1971, 23, 404-405.
11. Vannice, M.A.; Neikam, W.C. J. Catal. 1971, 23. 401-403.
12. Bremer, H.; Steinberg, K.H.; Wendlandt, K.P. Z. Anorg. Allg. Chem. 1969, 366, 130-138.
13. Frost, A.V.; Shapiro, M.I. Compt. rend. acad. sci, URSS 1934, 2, 243-245.
14. Erbacher, O. Angew. Chem. 1950, 62A, 403-404
15. Thon, N.; Taylor, H.A. J. Am. Chem. Soc. 1953, 75, 2747-2750.
16. Eischens, R.P.; Pliskin, W.A. Adv. in Catal. and Related Subj. 1958, 10, 1-56
17. Kobozev, N.I. Uchenye Zapiski, Moskov. Gosudarst. Univ. im. M.V. Lomonosova 1955, 174, 17-51.
18. Ledeboer, D.; Post, E.; Bruxvoort, W.,; De Jong, R.; Maatman, R. J. Catal. 1965, 4, 480-484.
19. Schwab, G.M.; Rudolph, L. Z. Elektrochem. 1931, 37, 660-670.
20. Akhtar, M.; Tompkins, F.C. Trans. Faraday Soc. 1971, 67, 2454-2460.
21. Dabrowski, J.E.; Butt, J.B.; Bliss, H. J. Catal. 1970, 18 297-313.
22. Pignet, T.; Schmidt, L.D. J. Catal. 1970, 18, 297-313.
23. Maatman, R.W.; Leenstra, D.L.; Leenstra, A.; Blankespoor, R. L.; Rubingh, D.N. J. Catal. 1967, 7, 1-17
24. Maatman, R.; Friesema, C.; Mellema, R.; Maatman, J. J. Catal. 1977, 47, 62-68.
25. Maatman, R.W. J. Catal. 1976, 43, 1-17
26. Boudart, M.; Aldag, A.; Benson, J.E.; Dougharty, N.A.; Harkins, C.G. J. Catal. 1966, 6, 92-99.
27. Ashmead, D.R.; Eley, D.D.; Rudham, R. Trans. Faraday Soc. 1963, 59, 207-215.
28. Krylov, O.V.; Kushnerev, M.J.; Markova, Z.A.; Fokina, E.A. Proc. 3rd Int. Cong. Catal. 1965, 2, 1122-1134.
29. Krylov, O.V.; Livshits, V.S. Metody Issled. Katal. Reakts., Akad. Nauk SSSR, Sib. Otd. Inst. Katal. 1965, 3, 237-246.
30. Acres, G.J.K.; Eley, D.D.; Trillo, J.M. J. Catal. 1965, 4,

31. Harris, J.R.; Rossington, D.R. J. Catal. 1969, 14, 168-174.
32. Shigehara, Y.; Ozaki, A. J. Catal. 1968, 10, 183-187.
33. Carter, J.L.; Lucchesi, P.J.; Sinfelt, J.H.; Yates, D.J.C. Proc. 3rd Int. Congr. Catal. 1965, 1, 644-656.
34. Bett, J.A.S.; Hall, W.K. J. Catal. 1968, 10, 105-113.
35. Yermakov, Y.I.; Zakharov, V.A. Proc. 4th Int. Congr. Catal. 1969, 1, 276-292.
36. Gonzalez, F.; Munuera, G. An. Quim. 1969, 65, 849-863.
37. Criado, J.M.; Gonzalez, F.; Trillo, J.M. Rev. Chim. Miner. 1970, 7, 1041-1052.
38. Criado, J.M.; Gonzalez, F.; Trillo, J.M. J. Catal. 1971, 23, 11-18.
39. Franco, H.A.; Phillips, M.J. J. Catal. 1980, 63, 346-354.
40. Baddour, R.F.; Modell, M.; Goldsmith, R.L. J. Phys, Chem. 1970, 74, 1787-1796.
41. Glasstone, S.; Laidler, K.J.; Eyring, H. "The Theory of Rate Processes"; McGraw-Hill: New York, 1941.
42. Maatman, R.W. Adv. in Catal. 1980, 29, 97-150.
43. Aldag, A.W.; Lin, C.J.; Clark, A. J. Catal. 1978, 51, 278-285.
44. Simonyi, M.; Mayer, I. Acta Chim. Acad. Sci. Hung. 1975, 87, 15-32.
45. Laidler, K.J.; Tweedale, A. Adv. in Chem. Phys. 1971, 21, 113-125.
46. Solbakken, A.; Reyerson, L.H. J. Phys. Chem. 1960, 64, 1903-1907.
47. Galwey, A.K. Adv. in Catal. 1977, 26, 247-322.
48. Best, D.A.; Wojciechowski, B.W. J. Catal. 1977, 47, 343-357.
49. Horiuti, J.; Miyohara, K.; Toyoshima, I. J. Res. Inst. Catal., Hokkaido Univ. 1966, 14, 59-84.
50. Miyahara, K.; Kazusaka, A. J. Res. Inst. Catal., Hokkaido Univ. 1976, 24, 65-70
51. Fricke, A.L.; Altpeter, R.J. J. Catal. 1972, 25, 33-43
52. Maatman, R.; Mahaffy, P.; Mellema, R. J. Catal. 1974, 35, 44-53
53. Misra, D.N. J. Catal. 1969, 14, 34-42.
54. Cha, D.Y.; Parravano, G. J. Catal. 1970, 18, 200-211.
55. Hayward, D.O.; Herley, P.J.; Tompkins, F.C. Surface Sci. 1964, 2, 156-166.
56. Bourne, K.H.; Cannings, F.R.; Pitkethly, R.C. J. Phys. Chem. 1971, 75, 220-226.
57. Richardson, J.T. J. Catal. 1967, 9, 182-194
58. Maatman, R.W. J. Catal. 1970, 19, 64-73.
59. Brouwer, D.M. J. Catal. 1962, 1, 22-31.
60. Lawson, A. J. Catal. 1968, 11, 295-304.
61. Knozinger, H.; Buhl, H.; Kochloefl, K. J. Catal. 1972, 24, 57-68.
62. Sato, S.; Miyahara, K. J. Res. Inst. Catal., Hokkaido Univ. 1974, 22, 51-62.
63. Sato, S.; Miyahara, K. J. Res. Inst. Catal., Hokkaido Univ. 1975, 23, 1-16

64. Eley, D.D.; Russell, S.H. Proc. Royal Soc. (London), A 1974, 341, 31-44.
65. Maatman, R. J. Catal. 1981, 72, 31-36
66. Maatman, R.; Mahaffy, P.; Hoekstra, P.; Addink, C. J. Catal. 1971, 23, 105-117.
67. Maatman, R.; Ribbens, W.; Vonk, B. J. Catal. 1973, 31, 384-388.
68. Dohse, H.; Kalberer, W.; Schuster, C. Z. Elektrochem. 1930, 36, 677-679.
69. Kubokawa, M. Rev. Phys. Chem. Japan 1937, 11, 202-216.
70. Balandin, A.A.; Vasserburg, V. Acta Physicochim. U.R.S.S. 1946, 21, 678-688.
71. Mills, G.A.; Boedecker, E.R.; Oblad, A.G. J. Am. Chem. Soc. 1950, 72, 1554-1560.
72. Mac Iver, D.S.; Emmett, P.H.; Frank, H.S. J. Phys. Chem. 1958, 62, 935-942.
73. Kobozev, N.I.; Lebedev, V.P.; Maltsev, A.N. Z. physik. Chem. (Leipzig) 1961, 217, 1-41.
74. Lebedev, V.P. Probl. Kinetiki i Kataliza, Akad. Nauk SSSR 1966, 11, 122-131.
75. Filippov, Y.V.; Lebedev, V.P. Zh. Fiz. Khim. 1966, 40, 1846-1853.
76. Lebedev, V.P. Sci. Selec. Catal. 1968, 134-144.
77. Amenomiya, Y.; Cvetanovic, R.J. J. Catal. 1970, 18, 329-337.
78. Romanovskii, B.V. Kinet. Katal. 1967, 8, 927-930.
79. Murakami, Y. Kogyo Kagaku Zasshi 1968, 71, 779-783.
80. Constable, F.J. Nature 1943, 152, 135-136.
81. Fukuda, K. Trans. Faraday Soc. 1971, 67, 2158-2161.
82. Charcosset, H.; Barthomeuf, D.; Nicolova, R.; Revillon, A.; Tournayon, L.; Trambouze, Y. Bull. Soc. Chim. Fr. 1967, 4555-4558.
83. Haag, W.O.; Whitehurst, D.D. Proc. 2nd North Am. Catal. Soc. 1971, p. 16.
84. Hall, W.K. J. Catal. 1962, 1, 53-61.
85. Hirschler, A.E.; Hudson, J.O. J. Catal. 1964, 3, 239-251.
86. Parera, J.M.; Figoli, N.S. J. Catal. 1969, 14, 303-310.
87. Dzisko, V.A.; Kolovertnova, M.; Vinnikova, T.S.; Bulgakova, Y.O. Kinetika i Kataliz 1966, 7, 655-699.
88. Evdokimov, V.B. Kataliz v Vysshei Shkole, Min. Vysshego i Srednego Spets. Obrazov. SSSR 1958, 2, 142-152.
89. Strakhov, B.; Lebedev, V. Zh. Fiz. Khim. 1964, 38, 2235-2244.
90. Mellor, S.D.; Rooney, J.J.; Wells, P.B. J. Catal. 1965, 4, 632-634.
91. Kolboe, S. J. Catal. 1969, 13, 208-214.

RECEIVED November 29, 1982

MORE PIONEERS

32

Leonard C. Drake—His Contributions to the Development of Mercury Porosimetry for Catalyst Characterization

BURTRON H. DAVIS
University of Kentucky, Institute for Mining and Minerals Research, Lexington, KY 40512

Preparation of controlled porosity materials becomes an increasingly important function in catalyst research and preparation. Mercury porosimetry is a unique characterization technique since it permits measurement of the full range of pore diameters. The contributions that Dr. Leonard C. Drake made to the design and construction of the first instrument and in showing the utility of the technique are reviewed.

Diffusion plays an important role in heterogeneous catalysis. All too frequently the impact of diffusion on experimental results is not recognized, even by experts in catalysis. In 1939 Thiele gave a clear account of the role that heterogeneous catalyst particle size may play on the rate of catalyzed reactions (1). While this appears to be the first widely circulated publication, as well as the first one to make an important impact on the role of diffusion in heterogeneous catalysis, it was not the first one. Likewise, even though it was published prior to the development of mercury porosimetry, it did not play a direct role in initiating the development of the porosimeter.

The following, from the introduction to the first Ritter and Drake publication (2), provides an excellent introduction to mercury porosimetry (reference numbers changed to conform to the present manuscript):

> "Determination of total pore volume is a routine measurement in most laboratories dealing with porous materials. The value usually is calculated as the difference of two specific volumes (reciprocal density). Thus the internal pore volume is the difference between the reciprocals of real density and particle density; the intergranular (void) volume is the difference between the reciprocals of bulk density and particle density; and the sum of pore and void volumes is the difference between the reciprocals of bulk and real densities. [The nomenclature of McBrain (3) is followed in identifying the several densities,

0097–6156/83/0222–0451$06.00/0

assuming (with some error) that real and true densities are equal.] The total internal pore volume is then calculated from observations of the real and particle densities, determined, for example, by the usual pycnometric method using water and mercury, respectively, as the displacement liquids.

In processes involving diffusion rates and the availability of internal surface to large molecules, a knowledge of total pore volume is less important than a knowledge of the fraction of total pore volume contributed by pores in a given size range--i.e., of the distribution of pore sizes. It is convenient to classify the internal pores of porous materials roughly in two ranges. Present usage (4) applies the name "micropores" to those having radii smaller than 100 A.; "macropores" to those larger than 100 A. The division of the pore volume of a given porous material into micro- and macropores implies the existence of a distribution in size, yet little work has been done in the determination of such distribution functions.

Rabinowitsh and Fortunatow (5) have determined the respective fractions of micro- and macropores in a number of porous solids by means of the Kelvin equation...The adsorption equation of Brunauer, Deming, Deming and Teller (6) may be of some use in this connection, but is open to the criticism that it does not satisfactorily combine the simultaneous effects of adsorption and capillary condensation. Jellinek and Fankuchen (7) have used the scattering of x-rays at very small angles to evaluate pore size (or particle size) but assumed a constant average size. The unpublished work of Shull (8) on low-angle x-ray scattering takes into consideration a pore-size distribution; but, this method not only cannot conveniently be used for pores larger than perhaps 500 A. in radius, but the results in terms of pore size may also be open to question. This paper presents a method for determining the macropore-size distribution in a porous solid as well as the derived distributions for some typical porous materials.

Washburn (9) has pointed out the fact that surface tension opposes the entrance into a small pore of any liquid having an angle of contact greater than 90° (the common phenomenon of capillary depression); that this opposition may be overcome by the application of external pressure; and that the pressure required to fill a given pore is a measure of the size of the pore. Henderson, Ridgway, and Ross (10) have used this principle in a very limited way... The relation (quoted by Washburn) giving the pressure required to force liquid into a pore of given size is

$$pr = 2\sigma \cos\theta \qquad (1)$$

where p is the pressure, r the pore radius, σ the surface tension, and θ the contact angle. It may be derived as follows: In a pore of circular cross section, the surface tension acts along the circle of contact over a length equal to the perimeter of the circle. The force is $2\pi r\sigma$. Normal to the plane of the circle of contact, the force tending to squeeze the liquid out of the pore is $-2\pi r\sigma \cos\theta$. (The negative sign arises from the fact that the angle between the direction of action of the surface tension and the positive normal to the plane of contact is $\pi-\theta$ Since θ >90°, the term $-2\pi r\sigma \cos\theta$ is intrinsically positive.) Opposing this force is the applied pressure acting over the area of the circle of contact with a force equal to $\pi r^2 p$. At equilibrium these opposing forces are equal $-2\pi r\sigma \cos\theta = \pi r^2 p$, whence Equation 1 follows immediately.

From this relation it appears that a porous material under zero pressure will "absorb" none of any nonwetting liquid in which it is immersed. When the pressure is raised to some finite value, the liquid will penetrate and fill all pores having radii greater than that calculated from Equation 1...As the pressure is increased the amount of liquid "adsorbed" increases monotonically at a rate proportional to the differential pore volume due to pores of size corresponding to the instantaneous pressure. Thus, a given pore-size distribution gives rise to a unique pressuring curve; and, conversely, a given pressuring curve affords a unique determination of the pore-size distribution...Differentiation of Equation 1 and elimination of p give

$$\Delta r/r = -\Delta\theta \tan\theta \qquad (2)$$

as the fractional error incurred in calculated pore radius by an error of $\Delta\theta$ in contact angle. For θ in the neighborhood of 140°, $\Delta r/r$ for a 1° error in contact angle is only about 1.5%."

Reference 2 and a companion publication (11) defined the method and capabilities for mercury porosimetry. By 1949, Drake had increased the capacity of his equipment to 60,000 psi so that he was able to penetrate mercury into pores down to 18A in radius (12).

Graduate Studies

Dr. A. F. Benton received his Ph.D. at Princeton working under the direction of Prof. H. S. Taylor. Dr. P. H. Emmett recognizes Dr. Benton (Emmett's thesis director at Calif. Inst. Tech.) for his outstanding experimental techniques and his sound training in adsorption as well as his ability to teach these subjects (13). Benton was at the California Institute of Technology only briefly prior to joining the chemistry department at the University of Virginia. At Virgina, he quickly established a very productive research program emphasizing adsorption studies. Drake worked under Benton and received his Ph.D during this period. His thesis research included a detailed study of chemisorption of oxygen by silver and the stability of silver oxide. Thus, Drake left the University of Virginia at the height of the 1930's United States' depression with a Ph.D., a sound training in adsorption and experimental techniques, and no job.

Drake retained his youthful curiosity thought his graduate studies and employment. This is reflected in his drive to improve and to develop new experimental approaches. His research also reflected his independent and self-reliant character. Drake valued work, modesty in promoting accomplishments and, above all, honesty. These values strongly influenced his interactions with colleagues. For Drake, like many others, the challenge of the next problem was more important than the writing about completed work.

Early Years at Socony Vacuum

After a period of unemployment and several months with the U. S. Geological Survey in Washington, where he assisted in a survey of natural bleaching clays, Drake joined Socony Vacuum (now Mobil) at their Paulsboro, N. J. Research Laboratory where he was the lone Ph.D. chemist in a group of engineers. He worked for several years on solvent refining, dewaxing and wax processing problems. His first contact with the Houdry cracking process was in the application of molten salts as high temperature heat transfer agents for early commercial cracking units. In the late thirties he constructed low temperature physical adsorption equipment to measure the surface area of porous catalysts by the Point B method. Routine small glass cracking units were also installed to simulate Houdry commercial cracking units and to evaluate cracking catalysts.

Socony Vacuum wanted to use their recently developed moving bed (TCC) cracking process. This process permitted continuous movement of the catalyst from the cracking zone, purging and regeneration of the spent catalyst with air and reintroduction into the reactor. The initial development of this process began with 30/60 mesh catalyst but a need for larger particles was evident and more attrition resistant pellets were required. Fortunately a concurrent Socony Vacuum development (by Dr. Marisic) of synthetic silica alumina gel particles (beads) solved some of these problems. A large plant to manufacture these beads was built in Paulsboro and they replaced extruded and pelleted types. However, Marisic left Socony Vacuum before the bead catalyst was commercially developed. Drake was concerned with the laboratory evaluation and commercial development of this new catalyst form, using coke burning, surface area, gaseous diffusion and various attrition methods to measure hardness and abrasion resistance as well as catalytic activity.

During the first years of Drake's employment, the management of the Paulsboro Lab were quite restrictive in permitting discussions of laboratory research with people outside the lab. In this respect Socony Vacuum was no different than other labs. Restrictions and enforcement policies on exchange of research results and ideas appear to cycle between periods of utmost secrecy and periods where exchange and publication were encouraged. Brunauer and Emmett carried out and published much pioneering work on the application of physical and chemical adsorption to catalyst characterization during the 1930's culminating in 1938 with the publication of the BET equation for calculating surface areas. However, Drake honored the restrictions so that when he attended the ACS meeting in Boston in 1939, he avoided discussion of adsorption and catalyst characterization with both Brunauer and Emmett.

Shortly after the Boston meeting attitude at the laboratory changed. Roland Hansford had joined another group at the Paulsboro labs and had co-patented a method for thiophene production by reacting hydrocarbons and sulfur over catalysts. The development of this process was left to others and Hansford was made the head of Marisic's group. He introduced a program of more fundamental catalyst studies. Drake joined this group at its inception and assisted in setting up a laboratory for physical and chemical studies of porous catalysts. Hansford, as a former student of Professor Paul Emmett, was of course familiar with adsorption research. However, more important to Drake was Hansford's support and encouragement to pursue new ideas, both within and outside the laboratory, and Drake considers this research attitude as a major factor in developing the mercury porosimeter.

Hansford and Drake, in the early forties adapted the McBain-Bakr quartz fiber spring technique to low temperature nitrogen adsorption surface area measurements. This required at that time the construction of a spring winding device for making springs from quartz fibers and a traveling microscope reading device for multiple unit routine measurements. Improved models of this equipment are still in use at Mobil for surface area measurements. It must be realized that adsorption was an active research area in the thirties so that the six McBain-Bakr balances would be considered a "modern characterization instrument."

Mercury Porosimeter

Washburn had used liquid penetration to estimate the porosity of a range of ceramics; however, he employed liquids such as water or hydrocarbons and only used low pressures. Even so, he clearly recognized the advantage of mercury and published the theory for its use in 1921 (9). This paper, encountered by Drake in a search for new methods for measuring porosity, provided a basis for undertaking the experimental program. Drake, together with a recent employee (H. L. Ritter) and the encouragement of Hansford, started work on this in the early 1940's. Ritter remained at Socony Vacuum for only a brief period and had moved to the University of Wisconsin prior to publication of the two papers he authored with Drake (2, 11). While at Socony Vacuum Ritter also used small-angle x-ray scattering to measure pore size distribution in porous materials and published a paper with L. C. Rich on this in 1948 (14). He went to Miami University in Ohio after a few years at Wisconsin. After leaving Scocony-Vacuum he continued to use x-rays in his research but he did not again work on mercury porosimetry.

High pressure techniques were not commonly used at this time so that many problems were encountered. Precision-bore glass tubing was just becoming commercially available. Without this, a different method for measuring penetration would have been required. Even so, the precision-bore tubing was the component that limited the accurracy of the measurement.

The first bomb used by Ritter and Drake was limited to 2000 psi and used a commercial insulated electrical wire lead to the dilatometer they had developed. After publishing their first two papers, the authors learned that the Rusha Instrument Corp., Houston, Texas has marketed for a number of years a "high pressure porometer" which forced mercury by means of a tight piston into oil-well (12). However, the Rusha instrument only operated to 1000 psi (2100 A pores).

In the high pressure 60,000 psi unit, a P. W. Bridgeman (awarded a Nobel Prize for his high pressure work) designed insulated electrical lead, made in the Socony Vacuum shops, was at first limited to, at most, a few runs because of electrical shorting due to wire insulation failure (Figure 1). This leakage problem caused months of frustration. The substitution of Teflon coated copper wire allowed many runs to be made before eventual failure. This wire was made available by Dr. Wahl, a friend at the Wilmington duPont Research Laboratory, as a result of a discussion. This discussion, and the assistance that it provided, contrasts markedly with the situation a few years earlier.

A gas was used in the pressurization procedure and this, together with the limited experience of manufactures of high pressure vessel construction, provided an additional problem. "The first bomb used in this work split from end to end while at 50,000 pounds per square inch, throwing water and thermostatic equipment over the surrounding area. The sudden expansion of the 400cc of nitrogen (compressed from 2000 to 50,000 pounds per square inch) also lifted the relatively heavy bombshed roof several feet." (12) This is a rare mention of the experimental problems encountered in the development of this different

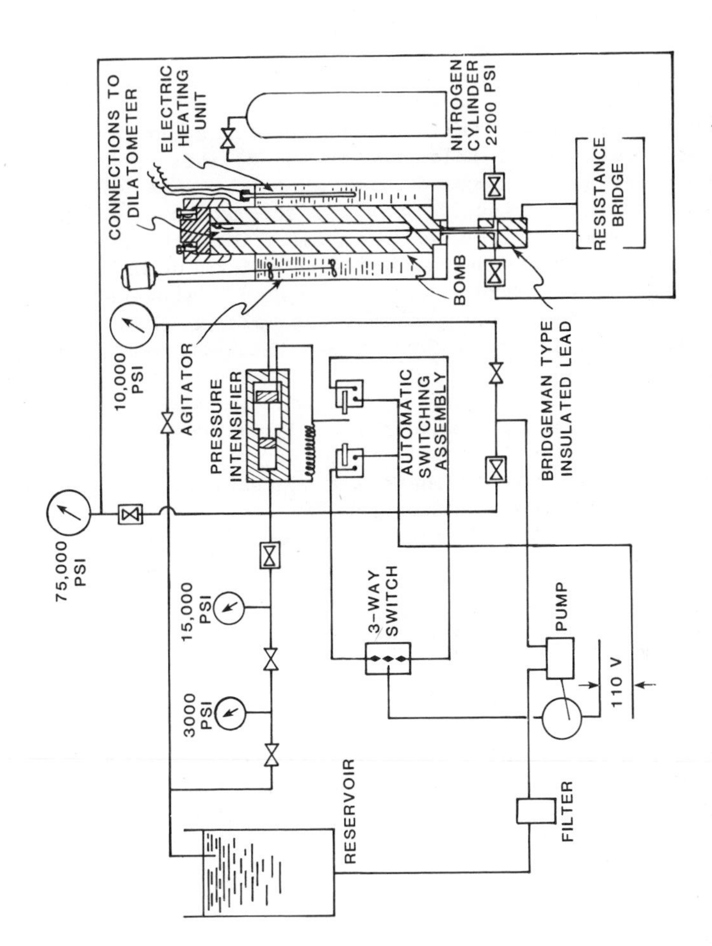

Figure 1. Schematic of the first mercury porosimeter. (Reproduced from Ref. 2. Copyright 1945, American Chemical Society.)

experimental procedure and even it was presented in the form of a safety precaution.

Never one to be satisified with a method, Drake tried many ways to improve the accuracy and ease of operation of the method. In the original dilatometer the penetration volume was obtained by measuring the resistance of a platinum wire in the calibrated bore (Figure 2). As the mercury level in the bore declined as pressure forced mercury into the catalyst pores the length of Pt wire in the mercury decreases; hence, the increase in resistance provided a measure of the penetration volume. In the original design, a gas was used to pressurize to 2,000 psi. In the high pressure unit, this gas was further compressed by an oil ram. Oil could not be tolerated in the calibrated bore area since it fouled the calibrated capillary bore. A sight was developed in order to visually measure the mercury levels but this was not considered to be an improvement over the resistance method. When this approach was not successful, Drake's effort to make visual measurements was terminated. However, this method was incorporated into a commercial unit so that Rootare, in a 1968 review (15), stated that it was the most common way to measure volume changes.

There is some uncertainty about the contact angle that mercury makes with catalyst materials. Drake used Woods Metal and sodium, heated above the melting point, to replace mercury in an attempt to learn the influence of contact angle. However, little difference was observed for the few results obtained with the two metals and the experimental difficulties were many.

Other Catalyst Characterization Techniques

While the mercury porosimeter was Drake's most noteable success during this period, he worked on many catalyst characterization methods that, while successful, were never published except in internal reports. One of the last catalyst characterization problems Drake worked on was with Pt-Al_2O_3 reforming catalysts. About 1950, Pt-Al_2O_3 was accepted as the preferred reforming catalyst. The need to measure the crystallite size of the low loading of Pt supported on alumina was apparent. X-ray line broadening was not possible for the dual component catalyst since the most intense Pt peak was a shoulder on an intense alumina peak; with small particle Pt and low concentrations there was insufficient Pt intensity to even make an estimate of the particle size. However, it was found that acetylacetone extraction selectively dissolved the alumina. The Pt could be recovered and analyzed for crystallite size by x-ray line broadening. It was also confirmed that they were not just recovering the large particles by the extraction. The Pt line for a supported large particle Pt formed by high temperature sintering was so sharp that it could be separated from the alumina peak. Thus, by determining that the x-ray peak area per gram of Pt was the same for the sintered, supported catalyst as for the extracted Pt, they concluded that the

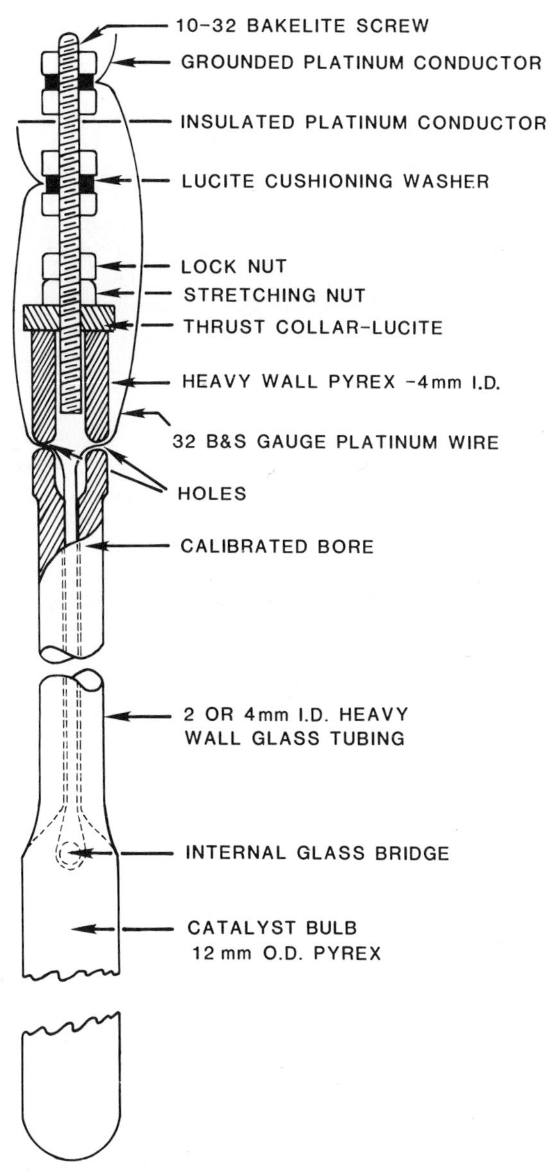

Figure 2. Schematic of the dilatometer used for the initial mercury porosimetry measurements.

extraction method recovered all of the reduced Pt initially on the catalyst. While the method was used at Mobil for years, it was never published in the open literature, but was disclosed to other outside catalyst groups by other Mobil researchers.

It is interesting to follow the development of the scientific thought from the discussion sections of the three Drake manuscripts (2,11 and 12). In the first paper (2) the concern was with establishing the validity of the experimental results and with showing that the results were consistent with geometric models for the pores. Deviations were in the direction expected for inkwell pores. In the companion paper (11) results for a wide range of porous materials are given as the authors established the scope of the method. Hysteresis was noted and a phenomenological discussion of it was provided. Only in the third paper (12) had the discussion attained the scientific maturity that one frequently observes after a few years intense work and thought by a pioneer.

Among the conclusions were:

a) the volume of a coke deposid on aged silica-alumina catalysts was greater than that of amorphous carbon and the carbon deposited rather uniformly throughout the pore size range during catalyst ageing,

b) pressurization, and the mercury retained by hysteresis, only slightly altered the original surface area showing that the porosity survived the pressurization of the measurement,

c) observed hysteresis, and the slow, nonequilibrium nature of the depressurization curve and noted that, at 25 psia on the depressurization curve, very little more mercury was retained after a 60,000 psi than for a 10,000 psi penetration.

d) verified mercury retention by weight of the catalyst after penetration as well as by resistance measurements, and

e) showed that the 60,000 psi data indicated that there was no reason to suspect the breakdown of the concept of surface tension even in pores with a diameter of 5 mercury atoms. To date, it does not appear that the last observation has been adequately recognized.

The experimental design of Ritter and Drake was the basis for the American Instrument Company developing and marketing a commercial instrument. Experimental difficulties and the time consuming operation required for a measurement delayed the wide usage of the technique. However, within the past years two U. S. companies have introduced two types of automated instruments that should make this method as popular as the widely used nitrogen adsorption method. Hopefully this account will help to introduce Leonard Drake to a new generation of scientists who, with improved commercial instruments, will participate in greatly expanding the use of the technique.

The author was fortunate to be associated with Leonard Drake during four years while employed at Mobil. His friendship was a valued acquisition during this period. Technical information on the development of the mercury porosimeter was provided by Leonard during a visit with the author during May, 1982 as well as by a review of the draft of this manuscript; however, the author is responsible for the final version as well as the conclusions. The photo for Figure 3, showing Leonard in the author's lab viewing a commercial porosimeter, was taken during the May, 1982 visit.

Figure 3. Picture (1981) of L. C. Drake with a commercial mercury porosimeter.

Literature Cited

1. Thiele, E. W.; Ind. Eng. Chem. 1939, 31, 916.
2. Ritter, H. L.; Drake, L. C.; Ind. Eng. Chem., Anal. Ed. 1945, 17, 782.
3. McBrain, J. W., "Sorption of Gases and Vapors by Solids"; George Routledge and Sons: London, England, 1932; p. 79.
4. Brunauer, S.; "Adsorption of Gases and Vapors," Princeton University Press; Princeton, N. J., 1943; p. 376.
5. Rabinowitsch, M.; Fortunatow, N.; Z. Angew, Chem. 1928, 41, 1222.
6. Brunauer, S.; Deming, L. S.; Deming, W. E.; Teller, E.; J. Am. Chem. Soc. 1940, 62, 1723.
7. Sellinek, M. H.; Fankuchen, I.; Ind. Eng. Chem. 1945, 37, 158.
8. Shull, C. G.; paper delivered at Gibson Island Conference, August, 1944.
9. Washburn, W. E.; Proc. Nat. Acad. Sci. 1921, 7, 115.
10. Henderson, L. M.; Ridgway, C. M.; Ross, W. B.; Refiner 1940, 19, 185.
11. Drake, L. C.; Ritter, H. L.; Ind. Eng. Chem., Anal. Ed., 1945, 17, 787.
12. Drake, L. C.; Ind. Eng. Chem., 1949, 41, 780.
13. Davis, B. H., J. Chem. Educ. 1978, 55, 249.
14. Ritter, H. L.; Erich, L. C.; Ind. Eng. Chem., Anal. Ed. 1948, 20, 665.
15. Rootare, H. M.; Aminco Laboratory News, 1968, 24, No. 3, pages 4A-4H.

RECEIVED November 29, 1982

Catalysis at Princeton University Chemistry Department

JOHN TURKEVICH

Princeton University, Chemistry Department, Princeton, NJ 08544

Catalysis started at Princeton in 1919 when Taylor(1) came from England to join the Chemistry Department. Turkevich came as a graduate student of Taylor's in 1931. Except for one year at Leipzig and Cambridge Universities, Turkevich remained at Princeton since that time. Taylor assumed the deanship of the Graduate School in 1948 and retired in 1958. Turkevich retired from teaching in 1975 and is still active in catalytic research. Steven Bernasek joined the faculty in 1976 and Jeffrey Schwartz in 1970. Both are active: one in surface chemistry and the other in catalysis. The work in the Chemistry Department has been complemented by that in the Department of Chemical Engineering by Richard Wilhelm (1934-65), Leon Lapidus (1954-1977), Michel Boudart (1953-1962) and David Ollis (1969-1980). For sixty-three years there has been a continuity of research in catalysis at Princeton University: continuity in basic ideas expressed by novel techniques. There was close association between the research activity and training of undergraduates, graduate students, postdoctoral research associates and visiting professors. The students carried out and are still carrying out the Princeton tradition throughout the world.

0097–6156/83/0222–0463$07.75/0

Hugh S. Taylor and Chemical Physics

The founder and leader for many decades of the Princeton School of catalysis was Sir Hugh Taylor. A student of H. Bassett, Jr. and F. W. G. Donnan, at Liverpool University, Taylor did research work with Svante Arrhenius at the Nobel Institute in Stockholm and with Max Bodenstein at the Technische Hochschule at Hanover, Germany. Bodenstein had a strong influence on Taylor. He spurred him to travel in kinetics along the path of catalysis and his steed was to be chemistry physics--the molecular structure approach to Chemistry. During World War I Hugh Taylor worked with Eric Rideal for the British Munitions Board on ammonia synthesis and the water gas shift reaction. During this period he wrote with Eric Rideal "Catalysis in Theory and Practice" (1919). On his return to Princeton in 1919 (where he had been for a short time in 1915) Taylor organized the first center of catalysis in the United States. Physical chemistry was in its infancy and was dominated by concepts of equilibrium. Thermodynamics was king and electrochemistry, queen. Taylor was influenced by his work on homogeneous reactions at the Arrhenius laboratory, by his research of gaseous kinetics in Bodenstein's institute and by his experience in heterogeneous catalysis in Britain. In Princeton he turned to kinetics of heterogeneous catalytic reactions with associated interests in photochemistry, discharge tube chemistry, alpha particle induced reactions. The initial approaches were macroscopic--characteristic of physical chemistry. Taylor however went beyond that, helping to establish the discipline of chemical physics. This motivated the Princeton school of catalysis for a half a century. Atomic and molecular theory of matter, quantum mechanics, statistical mechanics were in explosive development at the time. The Physics Department of the University and the Institute of Advanced Studies at Princeton were centers of this development. Albert Einstein, Niels Bohr, Wolfang Pauli, J. von Neumann, E. U. Condon, Eugene Wigner, P. A. M. Dirac were at Princeton at that time. On the experimental side spectroscopy was practiced at a high level by Alan Shenstone. A general-use infra-red spectrometer was built by Bowling Barnes and mass spectroscopy was developed and studied by Walker Bleakney. Chemical physics-investigation of diverse chemical species and their reactions at the atomic and molecular level, became the principal theme of the Princeton scene. Not only did Taylor use it as a bridge between the Chemistry and Physics Departments, but he also built up this discipline within the Chemistry Department. In 1927 he attracted G. Kistiakowsky from the Bodenstein Laboratory to Princeton. Unfortunately for Princeton Kistiakowsky moved to Harvard where he, together with another Princetonian, E. Bright Wilson, initiated a similar program of chemical physics. Taylor countered with the appointment of Henry Eyring as the first theoretical chemist. Interaction between Chemisty and Physics was further strengthened by frequent short visits to Princeton by Linus Pauling, Irving Langmuir, Harold Urey, Joseph Mayer, J. H. Van Vleck, J. Slater and longer stays of A. Tiselius, M. G. Evans, B. Topley, Lord Wynne-Jones and J. Kirkwood. Communal living at the Graduate School brought chemists in contact with such physicists as John Bardeen, Robert Hofstadter, Frederick Seitz, Joseph Hirschfelder. At the same time Taylor organized applied catalysis studies in the Engineering School. Richard Wilhelm came to Princeton in 1933 to carry out research on catalytic reactor design and mass and heat transfer in catalytic systems. The Princeton branch of the Rockefeller Institute for Medical Research served as a link with biochemistry. Such prominent personages as John Northrup, Wendel Stanley, M. Anson, M. Kuntiz with their colleagues, attended seminars in physical chemisty. One product of a post-seminar discussion of René Dubos, George Wald and Dean Burk was the work on hydrocarbon metabolizing bacteria by Turkevich, Goodale and F. Johnson.(2)

Taylor's Concepts- Activated Adsorption

Taylor developed three concepts which were made more precise by the work of his students in many countries and by his successor Turkevich at Princeton. The first was the concept of activated adsorption. It was known from the time of Michael Faraday, that heterogeneous catalytic reactions took place on the surface of the solid. The first attempt to explain the increase in reactivity by mere increase in concentration of reactants on the catalyst surface was abandoned. Physical absorption could not explain the specificity of catalytic activity. Physical adsorption was extensively studied during World War I on charcoal, silica and other high surface materials as part of the gas mask war effort. After World War I new catalyst materials came into industrial use in hydrogenation of fats, ammonia synthesis and methanol production. These catalysts consisted of metals, metals on supports, oxides of transition elements and zinc oxide. At the same time novel techniques were available for experimental studies: Pyrex glass handling systems, mechanical fore pumps, mercury high vacuum pumps, McCleod gauges, Dewar flasks with Dry Ice or liquid air. These Taylor and his students used to study the adsorption of gases on industrially important catalysis. It was found that there were at least two types of adsorption. One was a rapid low temperature adsorption--physical in type, associated with van der Waals forces and decreasing with rise in temperature. This was utilized effectively later by Brunauer, Emmett and Teller for the determination of the total surface area of a finely divided solid.(3) In addition, on many metals and oxides there was another slow adsorption process increased in extent with temperature, went through a maximum and then decreased. Since this was a slow process, involving a temperature coefficient, Taylor called it activated adsorption(4) and attempted to correlate it with the activation of the adsorbed reactants. This provoked considerable controversy--was it reaction with the surface, was it solubility in the solid, was it a diffusion controlled process, was it displacement of a poison from the surface? Many systems were studied by R. M. Burns,(5) A. W. Gauger,(6) R. A. Beebe,(7) W. A. Dew,(8) G. B. Kistiakowsky,(9) E. W. Flosdorf,(10) A. T. Williamson,(11-12) P. V. Kinney,(13) D.V. Sickman,(14) A. Sherman,(15) A. J. Gould,(16) J. Pace,(17) J.Turkevich,(18) R. L. Burwell, Jr.(19-20). A critical investigation was the adsorption of hydrogen on zinc oxide. In 1934 C. O. Strother(21) found two types of activated adsorption while in 1938 E. A. Smith and Taylor(22) showed that the reaction between hydrogen and deuterium which involved activation of hydrogen went at a much faster rate than either activated adsorption of hydrogen gas. Gradually the emphasis shifted from activated adsorption to chemisorption as a method of determining the number of reactive sites.(23) Elegant techniques for determining the number of active sites by chemisorption were developed by Emmett and his group and by students of Taylor: Benson and Boudart.(24) The relation between the numbers of centers which chemisorb reactants and the number of centers which are catalytically active came later from the Turkevich group.

Active Centers

Another important concept introduced by Taylor was that of heterogeneity of surface-active centers.(25-26) This stemmed from observation of R. N. Pease that minute amounts of carbon monoxide, much smaller than the amount necessary to cover the surface, were sufficient to poison the surface of a copper catalyst. Taylor proposed that there were active centers on the surface while others argued that nickel impurities segregated preferentially on the surface and acted as catalyst. The variation of the heats of adsorption with surface coverage as determined by R. Beebe was used as evidence supporting the concept of active centers. In spite of the contradictory interpretation of the same experimental data, the concept of active centers has been a fruitful one. It inspired imaginative research in the field of metal and oxide catalysis and has its present day expression in sophisticated surface physics studies. Subsequent work by coworkers of Turkevich at Princeton refined the nature of active centers in monodisperse metal particles and crystalline oxide catalysts.

Model Reactions

The third important concept introduced by Taylor was the use of model reactions, "yard sticks" to determine the mode of activation of molecules by surfaces. For hydrogen activation, Taylor(15) proposed the conversion of ortho to para hydrogen as a measure of the catalytic activity of a surface. This turned out to be more complicated than was first realized. A physical magnetic effect was also operative as was shown among others by Diamond and Taylor(27) for the case of rare earths and by Turkevich and Selwood.(25) Later Laroche and Turkevich(29) used magnetic resonance to quantify the catalytic effect of charcoal and to differentiate it from dissociative process. The discovery of deuterium opened up the use of isotope exchange reactions as delicate "model reactions" for elucidation of the activation of molecules. Immediately after H. Urey announced the discovery of heavy water in 1932, Taylor(30) realized its potential as a tool in catalytic research and engaged in a massive production in Princeton of heavy water. Soon hydrogen-deuterium gas exchange was studied on a variety of catalysts and also exchange between proteum and deuterium compounds. The methods of analysis were those using thermal conductivity and infra-red spectroscopy. Detailed distribution of deuterium among various molecules and different sites in the same molecule did not come until after World War II by Turkevich. Pioneer work was carried out by Taylor, Morikawa, and Benedict(32-35) on the hydrogenolysis of hydrocarbons as compared to deuterium exchange. It was shown that on nickel the exchange reaction took place at 373 to 403°K while hydrogenolysis to methane took place at 90° higher. Thus the activation of a hydrogen-carbon bond was more facile than that of the carbon-carbon bond. Morikawa went from Princeton to work for the Japanese government in Manchuria. After World War II he became Dean of the Tokyo Institute of Technology and facilitated the flow of Japanese post-doctorates to Princeton.

Dehydrocyclization

During the thirties, the pall of economic depression hung over the academic scene. There were few positions for graduate students and post doctorates in the chemical industry or the universities. Taylor went to New York to the M. W. Kellogg Company to obtain funds for an industrial fellowship. Its vice president P. D. Keith, Jr. said that money was available for the production of hydrogen from hydrocarbons. Taylor had intended this fellowship for one of the unemployed instructors,but in the meantime the latter had accepted a position with the DuPont Company. Taylor called in Turkevich to take on the project. Turkevich had studied in his doctoral thesis the adsorption of hydrocarbons on oxide catalysts, particularly chromium oxide. This oxide was chosen as the catalyst. However no pure hydrocarbons higher than C_3 were available at that time. Turkevich heard that pure normal heptane could be obtained as a sap from a pine tree growing on the shores of Lake Tahoe in California. This was chosen as the starting material. A special apparatus for a catalytic flow system in which the starting material was a liquid and the product liquid and gas was constructed. It consisted of a mercury leveling device whose bulb was lifted by an electric clock fitted with gears and containing n heptane in the other arm. The product collecting system had a liquid trap and two large carboys, one filled with salt water and the other empty. A mercury manometer with an electric contact was attached to the trap. As the pressure increased in the collecting trap, an electric relay allowed salt water to syphon from one carboy to the other collecting the gas evolved. When the reaction was carried out in the Spring of 1935, copious amounts of gas were produced at tmperatures above 400°C. This gas analyzed as 90% hydrogen. The liquid in the trap was toluene. Thus not only a fine source of hydrogen was discovered but a process was found for reforming in one step heptane of octane number one to toluene of octane number one hundred and ten. The material so produced was very useful in World War II in giving to the British Air Force a higher ceiling than the Luftwaffe and to the munitions works, toluene for production of TNT. Further this dehydrocyclization process became the first bridge between aliphatic and aromatic chemistry. Various forms of chromium oxide were tried and different types of alumina. The black form of chromium oxide was found to be active while the green form was "dead". Gamma alumina was a good support for chromia while alpha inactivated the chromia.(36) Turkevich spent the academic year of 1935-1936 at Cambridge University with Sir John Lennard Jones(37) and at Leipzig University with Karl Bonhoeffer working on quantum chemistry. This was a continuation of his theoretical studies with Henry Eyring(38) which resulted in a paper on one of the applications of group theory to the electron structure of symmetric molecules. During Turkevich's absence from Princeton, work on dehydrocyclization was carried out by S. Goldwasser(39), Harold Fehrer and Donovan Salley.(40-44) On his return from Europe Turkevich was appointed instructor in chemistry and resumed his studies on aromatization of saturated hydrocarbons on various molybdenum catalysts.(45) Molybdena on alumina, phosphomolybdic acid were found to be excellent aromatization catalysts while molybdena alone was poor, acting as a cracking catalyst. In addition, the relation between isomerization and aromatization was investigated by contrasting the aromatization behavior of 223 trimethyl pentane with that of normal heptane. An attempt was made to extend the aromatization process to the production of heterocyclic compounds. Charles H. Kline passed amyl amine over aromatizing catalysts in hope of making pyridine.(46) The disappointing product was amyl isocyanide, a highly toxic material. This approach was promptly abandoned and pyridine was synthesized catalytically from furfural. As a by-product of this investigation the infra-red spectra of pyridine was measured(47) and a group theoretical analysis(48) made of its normal vibrational modes.

War Research

War had flared up in the Far East. The Japanese had cut off the supply on natural rubber. The synthetic rubber program organized by the government required large amounts of butadiene. With the support of the U.S. Rubber Reserve Board, Turkevich organized a group consisting of A. Saffer, J. Arnold, W. McCarthy and R. Woodbridge to develop a steam insensitive catalyst for the thermodynamically difficult dehydrogenation of butane to butadiene. Temperatures of over 900K were required and steam was necessary to satisfy the thermodynamic conditions.

In the spring of 1941 Taylor enlisted Turkevich in the atomic weapon project. Hahn and Strassman had carried out their famous experiment of uranium fission the preceding year. Taylor, a British subject, was working for the British Canadian group before the American effort was organized into the Manhattan Project. The first task was to make deuterium by catalytic exchange with water. This process was to be used at the Trail, British Columbia plant where deuterium was to be a by-product of ammonia synthesis. This deuterium was to be used as a moderator in the Canadian pile. George Joris, a Belgium post-doctorate and Walter Moore, Jr. joined the group that studied this catalytic exchange process. The analysis for deuterium was the difficult aspect of the project. Joris and Turkevich built a mass spectrometer while Turkevich worked on the thermal conductivity method. Late in 1942, the Taylor group on deuterium production and Turkevich's butadiene project were incorporated into the newly organized Manhattan Project. The objective was the production of enriched uranium-235 from natural uranium by differential diffusion of highly reactive uranium hexafluoride through a porous membrane. A number of surface chemistry problems related to catalysis had to be solved. The work at Princeton was in support of the main effort at the SAM Laboratories of Columbia University. Aside from production of the diffusion barrier, the task was to characterize its pore distribution and stabilize it from corrosion. Taylor was associate director of the SAM Laboratories, spending his days in New York and evenings and weekends at Princeton. Turkevich, in addition to teaching both at Princeton and at Newark for the War Manpower Commission, brought the infra-red spectrometer back into working condition. It had been abandoned when Bowling Barnes left the Physics Department to join American Cyanamid Company. Together with P. C. Stevenson, he measured the infra-red spectrum of uranium hexafluoride(49) which with the Raman spectrum measured by Professor Duncan of Brown permitted the calculation of the thermodynamic properties of the uranium hexafluoride so necessary for the diffusion process.

In the meantime the RCA Laboratories moved to Princeton. Turkevich became their chemical consultant spending a half a day each week at their laboratories. He became particularly interested in the application of electron microscopy to catalysts and surfaces. With James Hillier he surveyed the texture of many typical catalysts, presenting a paper on their work at the Gibson Island Conference on Catalysis.(50) With the help of several engineers at RCA, he obtained replica electron-micrographs of the pores of the barrier materials. These gave a detailed view of the size, shape, surface condition of the pores which previously had been deduced indirectly by porosity measurements.

Taylor in Post War II Period

In the post World War II period Hugh Taylor became Dean of the Graduate School at Princeton retaining for a period his chairmanship of the Chemistry Department. He kept up his interest in catalytic research, particularly in hydrogenolysis of hydrodrocarbons, mechanism of ammonia synthesis and the application of semiconductor theory to catalysis. He was joined in these researches by J. Polanyi, C. Kemball(51) and F. S. Stone(52) from England and P. J. Fensham(53) from Australia, A. Ozaki(58), A. Amano (56) and K. Tamaru(55) from Japan, A. Cimino(57) and G. Parravano(58) from Italy and J. P. McGeer(59) and M. M. Wright(60) from Canada. Among the graduate students were A. Alei, H. Benesi, J. F. Black, L. C. Bostian, Michel Boudart,(61) G. D. Halsey,Jr.(62) E. J. Mikovsky,(63) Thor Rodin, H. Sadek(64) and S. C. Liang.(65)

Turkevich's Program in Post-War Period

Turkevich's research program during the post World War II period of thirty-five years followed the guide lines formulated by Hugh Taylor. What is the relation of adsorption to catalysis? What is the nature of active catalytic centers? In addition two new lines were established: What is the nature of activation of the adsorbed molecule by the active center and also what is the relation of catalysis in such diverse systems as gas phase, solutions, enzymatic reactions and medical treatment such as cancer therapy?

Turkevich's Training in Chemistry

Turkevich's training in chemistry and catalysis arose in the following way. He attended from 1919 to 1924, Columbia Grammar School in New York where he received a classical education with a strong blend of mathematics. Chemistry was introduced his senior year as the first science subject in this old secondary school once associated with Columbia College. In contrast to the other teachers, the chemistry teacher was primarily an athletic coach and was a poor chemistry teacher. He was so poor that Turkevich started to study chemistry on his own using college textbooks, particularly the one written by Deming. He became fascinated by the subject. On his matriculation at Dartmouth College where he obtained a scholarship and work for board due to the efforts of Frederick Alden, the headmaster, President Hopkins asked Turkevich what subject he would choose as a major. On hearing that it was chemistry, President Hopkins suggested that Turkevich see the Chemistry Professor, Leon B. Richardson. Not knowing that Richardson was called "cheerless" by the undergraduates, Turkevich went to get his advice, saying that he wanted to become a chemist. Richardson's retort was "there are too many bad chemists in the United States and there was no need for more." This remark further motivated Turkevich to obtain a perfect score in Richardson's General Chemistry course. It is interesting to note that during World War I Richardson worked with Arthur Lamb, subsequently a professor at Harvard, on adsorption of gases on charcoal. The teachers at Dartmouth: A. J. Scarlett, L. B. Richardson, F. Low, J. Amsden, C. Bolser, E. B. Hartshorn, were superb; attracting to chemistry in Turkevich's class such future chemists as G. Wheland of Chicago, Harry Scherp of Rochester and Alberto Thompson of Minnesota. On graduation, Turkevich was retained as an instructor for two years. He was in charge of the laboratory in Richardson's course and gave an occasional lecture. The only research activity in the department was that of Hartshorn, an organic student of Lauder Jones of Chicago. For his master's thesis under Hartshorn, Turkevich developed a method of using 30% hydrogen peroxide to make amine oxides. In the summer of 1930 Hartshorn obtained from his friend Dr. Peck, a summer job for Turkevich at the Standard Oil of New Jersey refinery in Bayway. The new research director, Per K. Frolich, an expert on catalytic synthesis of methanol, assigned Turkevich a catalytic process of converting methane using poisonous phosgene to acetyl chloride. Methane was plentiful and acetyl chloride was in high demand for "safe" photographic film. Because of the poisonous nature of the project, the work was set up in a shed outside the laboratory building. Monitoring tests for phosgene had to be developed and a flow system had to be constructed. By the end of the summer the system was operational and a small yield of acetyl chloride was obtained. Frolich suggested that Turkevich apply for graduate study at Princeton to work with Hugh Taylor on catalysis and then do postdoctorate work at MIT with Professor Lewis, the leading chemical engineer of that time and Frolich's mentor. After two years of teaching general chemistry at Dartmouth, Turkevich was asked to remain for another year to replace Hartshorn who was on sabbatical leave. Turkevich was in charge of the laboratory work in organic chemistry and taught advanced organic chemistry. One of his students was Frank Westheimer, now Professor of Organic Chemistry at Harvard, who did his first independent project in organic chemistry for Turkevich. Wheland, Scherp and Thompson and subsequently Westheimer went to Harvard to do graduate work with James Conant. The Dartmouth group wanted Turkevich to go to Yale to work in organic chemistry. However the summer with Frolich persuaded Turkevich to go in 1931 to Princeton to work with Taylor.

Adsorption Process

By the end of World War II it was well established that Van der Waals physical adsorption as determined by techniques of physical chemistry (volume-pressure measurements) could under BET conditions measure the extent of surface. This is a classical experiment in physical chemistry. J. Clarkson and Turkevich(65)showed how physical adsorption of oxygen on porous Vycor glass could be determined by spectroscopic methods of chemical physics. An oxygen gas molecule has a number of strong sharp lines in the electron paramagnetic resonance spectrum. The pressure of oxygen in the pores could be determined from the width of oxygen lines rather than from the manometer reading. This method was extended by E. Angelescu(67) from Rumania to the determination of chemisorbed O_2^- on palladium, gold and palladium gold alloys on silica. Oxygen adsorbed as O_2^- has a characteristic EPR spectrum while oxygen gas shows sharp lines in another region of EPR. Catalysts were treated with oxygen gas at various temperatures and then their temperature was lowered to that of Dry Ice. The excess oxygen was pumped off and then the catalyst was lowered to liquid nitrogen temperature. The intensity of the O_2^- signal was measured. It was shown that the number of centers increased on alloying palladium with gold and that chemisorption to produce O_2^- occurred. Furthermore when ethylene was admitted to a non-reacting palladium catalyst loaded with O_2^- the O_2^- signal disappeared and oxygen gas signals appeared. Thus ethylene displaced oxygen from its adsorbed state without reacting with it. On the other hand, on an active palladium-gold alloy addition of ethylene to an

O_2^- loaded surface, lowers the O_2^- signal but no oxygen gas signal appears. The oxygen does not desorb--it reacts.

Another example of the application of spectroscopic techniques to the study of chemisorption was the work of Sato(68) and Turkevich on the adsorption of oxygen on metallic sodium films. Pores of Vycor glass were filled with a solution of sodium in liquid ammonia. The solution was then evaporated to give a metallic film of sodium. The process was readily followed by EPR with the sharp line of sodium in liquid ammonia becoming the broad line characteristic of the conductivity electrons of sodium metal. On admission of oxygen gas the decrease in the conductivity electrons and the increase in the O_2^-signal could be measured independently as electrons were transferred from the surface to the adsorbed gas. Adsorbed CH_3I did not react with the sodium film. However on illumination with visible light facilitated electron transfer to produce a signal characteristic of the methyl radical.

Titration for Catalytic Sites

It has been tacitly assumed that the number of catalytically active sites was the number of chemisorption sites. An elegant extension of the method of determining the number of chemisorbed sites for hydrogen on platinum catalysts present in small quantities on supports was introduced by Benson and Boudart,(24) both students of Hugh Taylor. The Turkevich group developed a poisoning titration technique for determining the number of catalytically active centers.(69-74) In it the Emmett-Kokes microreactor was modified in the following way. A flow of a carrier gas (e.g. hydrogen) was passed over the catalyst, through a gas chromatographic column and into a detector. A pulse of an indicator substance (ethylene) was introduced into the gas stream and its conversion product (ethane) measured. A pulse of the titrant poison (carbon disulfide) was then introduced. This was adsorbed and was not detected. Another pulse of the indicator (ethylene) was introduced. The conversion was again noted. The cycle of indicator and titrant was repeated. After a certain number of pulses of the poison the conversion decreased and became zero. The number of active catalytic centers was then determined from the number of poison molecules just necessary to decrease the catalytic activity to zero. This technique was first used with cumene as an indicator for the cracking reaction and quinoline as a titrant for determination of the number of active centers on alumnia silica gel and zeolite cracking catalysts. The number of active centers was found to be proportional to the number of acid sites and in the case of zeolites with crystallographically identified aluminum atoms. In order to effectively extend this method to metals, to determine the number of active centers and the variation of activity of these centers with particle size, a program had to be instituted for preparing monodisperse particles of palladium, platinum, and alloys of palladium platinum and gold platinum. The description of this program will be presented in a later section of this account. Kim titrated monodisperse palladium(71) catalysts (of diameters greater than 75 Å) with carbon monoxide using ethylene hydrogenation as an indicator. The surface atoms of the catalyst were determined from electron microscopic size determination. Kim and Turkevich showed that all the atoms on the surface of the palladium particles of diameter greater than 74 A were equally active. Gonzales, Aika, Namba and Turkevich(72) used this method to determine the number of catalytic sites on monodisperse platinum of various sizes and found that in general all surface atoms were catalytically active for double bond hydrogenation in ethylene, propylene, butene, cyclohexene. It was the same for hydrogen-deuterium exchange. The number was distinctly smaller for the hydrogenation of benzene. Furthermore while the number of hydrogen chemisorbed centers is usually the same as that obtained from poison titration, the latter technique permits determination of centers for different hydrogenation reactions. The discrepancies between the two techniques found in certain cases give an experimental approach to hydrogen spill-over and carbon poisoning of the catalyst. In olefin hydrogenation over platinum, the turn-over number was found to be constant as the particle diameter decreased to 19 Å.

Enhanced activity of zeolites with respect to alumina silica gel type of catalysts was ascribed to the enhanced mobility of the protons in the crystalline zeolite bringing about increased acid strength.(73) Thus,though individual catalytic sites may be equivalent, their ensembles of different size would have different catalytic activity.(75) Furthermore C-13 nuclear magnetic resonance of adsorbed molecules showed that strong electric forces present in the regular pores of crystalline zeolites polarize and activate adsorbed molecules (155).

Laser Raman Spectroscopy was used by Buechler and Turkevich (76) to study the nature of adsorption of benzene and water vapor on porous Vycor glass. Unfortunately photocatalysis producing fluorescent materials limited the usefulness of this approach to catalytically-active surfaces. As we shall see later C-13 NMR was found to be an incisive tool for the study of adsorption of hydrocarbons on catalytic surfaces.

Use of Isotopes as Tracers

The discovery of deuterium in 1932 by Urey and tritium in 1939 by Alvarez and Corning opened up new techniques for catalytic studies. As previously stated Taylor organized the production of heavy water and used the deuterium so obtained to study the activation of molecules by catalysts. Exchange reactions of deuterium with hydrogen, deuterium with hydrocarbons, heavy water with benzene, heavy ammonia with light ammonia, were investigated. Overall exchange was determined using changes in index of refraction, thermal conductivity, and infra-red. There was limited use of mass spectroscopy. The detailed investigation of the isotope distribution between different positions in the molecule and among the different molecules was introduced and developed after the war by Turkevich.

Tritium as Tracer

Tritium was first used by Smith and Turkevich in studying the mechanism of double bond migration in the normal butenes. At that time analytical techniques using infra-red spectroscopy permitted analysis for the two isomers.(77) Phosphoric acid, sulfuric acid, alumina-silicas were good catalysts for this transformation. Tritium oxide obtained from E. O. Lawrence of the Radiation Laboratory at Berkeley, California was used to make tritiated phosphoric acid. When this was used to catalyze the double bond migration,(78) it was found that tritium introduction took place at the same rate (with correction for isotope effect) as the double bond migration. Since it was known that acids such as hydrochloric did not catalyze the double bond migration in the butene, the "hydrogen switch" mechanism was proposed: for catalysis active in double bond migration, a concerted atomic action must take place with hydrogen from the Bronsted acid of the catalyst adding to the first carbon atom of butene-1 and a hydrogen atom being removed from the third carbon atom to the Bronsted base of the catalyst molecule. The catalyst must have a Bronsted acid site (POH) and a Bronsted base site (P=O) separated by 3.2 Å. This condition is also satisfied by the alumina-silica catalysts. The hydrogen switch mechanism was extended to polymerization and cracking of olefins.(79)

Association with groups outside the University were invaluable for catalytic research at Princeton. As a result of successful experiment on dehydrocyclization Turkevich became a consultant to M. W. Kellogg Company attending for many years their staff meetings. During the War years catalytic cracking was the process of interest. Mass spectrometers graduated from "home made" apparatus to sophisticated equipment built by Consolidated Engineering Company. This was used at the Kellogg Company in the pioneering studies of determining position and number of tracer atoms in various isotopic isomers. As a result of the establishment of the RCA Research Laboratories at Princeton, Turkevich became their chemical consultant in electron microscopy, microwave spectroscopy and solid state physics (especially luminescence). Another fruitful association developed after the War, when the U.S. Atomic Energy Commission established the Brookhaven National Laboratory near his summer place on the North Shore of Long Island. Together with Professor R. W. Dodson of Columbia University, Turkevich organized, staffed and equipped the chemistry department of this laboratory.

Carbon-14 as Tracer

The first investigation carried out with the study with Francis Brown of the reaction of CO_2 with carbon to produce CO.(80) Radioactive C-14 carbon dioxide was allowed to circulate in a closed loop over a bed of heated charcoal. It was essential that the amount of carbon dioxide gas be smaller than the number of carbon atoms on the surface. The course of reaction was followed by change in pressure, by an internal Geiger Miller counter and by the ratio of CO_2 to CO. The reaction was found to procede in two steps--a rapid disappearance of all the CO_2 to produce CO with the same specific activity as that of the initial CO_2 and a slow build-up of the pressure of non-radiative CO. The mechanism proposed for this simple reaction was that in the first step, the CO_2 molecule on collision with the carbon surface lost one oxygen atom making CO and an oxygenated carbon surface. In the second step, a unimolecular decomposition of the oxygenated surface liberated the second molecule of non-radioactive carbon monoxide. This liberation took place by two exponential rates. Amick and Turkevich(81) extended this investigation to the reaction of steam with hot carbon to produce by an analogous process hydrogen first and then carbon monoxide. No further work was carried out on the use of radioactive C-14 as a tracer in catalytic work.

Deuterium as Tracer

Attention was then turned to the use of deuterium with special emphasis on where the deuterium was located in the molecule. Model deuterated compounds such as the methanes, ethanes, propanes, toluenes, isopropyl alcohols, were synthesized.(82-84) The first approach for analysis was the use of infrared spectra. The Brookhaven Chemistry Department had acquired an excellent spectrometer making the home-made one at Princeton obsolete. However it was decided that mass spectrometry offered a more incisive determination of the number of various isotopic molecules and the positions of the tracer atoms in these molecules. During the war years, the Chemistry Department at Princeton University had an all glass spectrometer made by Lee Harris, the glassblower of the Physics Department and placed into operation by G. Joris with the help of Turkevich. It was used primarily for hydrogen-deuterium analysis. It was fragile and its electronics fickle. During this period the Consolidated Engineering Company put on the market a rugged reliable but expensive mass spectrometer. It was used by petroleum companies to analyze hydrocarbon gas and liquids. Any given hydrocarbon gave under standardized conditions on electron impact, a characteristic spectrum of ion products whose masses and abundances could be determined. Turkevich decided to apply this technique to analysis of deuterated compounds. Together with Lewis Friedman, a number of deuterated hydrocarbons were synthesized and a new metallic spectrometer was built in Princeton. Turkevich enlisted Ernest Solomon and Francis Wrightson of the M. W. Kellogg Company to determine on their Consolidated mass spectrometer the electron impact spectra of those hydrocarbons. This collaboration resulted in a landmark paper on "Determination of Position of Tracer Atoms in a Molecule: Mass Spectra of Some Deuterated Hydrocarbons" published in 1948,(85) and another presented in 1951(86,87) at Discussions of the Farady Society. With Schissler and Irsa, Turkevich(88) made a detailed study of the interaction of ethylene and deuterium. With Thompson and Irsa the study of the Fischer Tropsch synthesis using deuterium gas,(89,90) with Lewis Friedman on reduction of acetone,(91,92) with G. C. Bond, the reaction of propylene(93) and cyclopropane(94) with deuterium. Using the ethylene-deuterium reaction as an example,(98) the mass spectroscopic approach permitted analysis of each of the five ethylenes and seven ethanes at the same time and showed, among other interesting data, that the first product of the reaction between deuterium and ethylene was an ethane containing no deuterium. Another example of the power of this method was its application to the Fischer Tropsch(90) synthesis. This reaction was first run with CO and H_2 until a steady state was attained. The reactive gases were flushed off with helium and deuterium was then introduced. The hydrocarbons produced were all completely deuterated suggesting that the active intermediate on the surface was carbon. The deuterium tracer work was carried out extensively in England by C. Kemball and G. C. Bond and in the United States by R. L. Burwell, Jr.

Catalyst Synthesis

In the pre-World War II days there was little work done at Princeton on synthesis of catalysts. Copper catalysts were made by reduction of Kahlbaum copper oxide, the iron ammonia catalysts were obtained from the Fixed Nitrogen Laboratory through the courtesy of Dr. Paul Emmett, the nickel on kieselguhr catalyst was obtained from DuPont. Platinum on asbestos was made in the laboratory by soaking asbestos with chloroplatinic acid and then igniting it, mixed chromite catalysts were precipitated and calcined and a study was made, particularly by R. L. Burwell, of the precipitation of an "active" chromium oxide gel. During the war and especially after the war, the synthesis of catalysts and their supports has been the preoccupation of the Turkevich group. The main diagnostic tool used in this work was the electron microscope. In 1942 the resolution of the microscope in the hands of James Hillier of the RCA Laboratories was 20 Å. Now in the hands of Joseph H. Wall of the Brookhaven National Laboratory it is 2.5 Å permitting visualization of the individual platinum atoms. A survey of catalysts made with the electron microscope in 1942(95) showed a diversity of size, shape and texture of catalytic substances. Many of the precious metals were large and consequently not very efficient--only a very small fraction of the atoms were available for surface reactions. However many of them were of colloidal size,(96) i.e. of one dimension at most of 2000Å. The usual method of making the catalyst was to soak the support with a solution of the salt of the precious metal and then subject it to thermal treatment. The complex topochemical reactions that take place are difficult to control to obtain monodisperse particles of optimum size. Two questions arose in the 40's and 50's. What is the dependence of catalytic activity on particle size? Is there a particle size below which there is no catalytic activity? It was proposed to synthesize the metal particles in solution in colloidal form; check their properties, both physical and chemical in solution; then mount them on a suitable support to study their activity in heterogeneous catalytic reactions. However, the colloidal chemistry of platinum and palladium was complex, poorly understood and difficult to reproduce.

Monodisperse Metal Studies

Stevenson, Hillier and Turkevich (97) turned to study the synthesis and properties of the most stable of colloidal metal systems--gold. An electron microscope examination was made of all classical preparations of this metal in the search of one producing the most uniform particles. The preparation based on the reduction of gold salts with sodium citrate was found to yield monodisperse particles. The distribution of particle size was correlated with the rate of nucleation, growth and coagulation of particles. Together with a "pure growth" reaction, monodisperse particles of diameters from 50 to 1200Å could be synthesized. The Mie theory of the dependence of color of colloidal gold on particle size was experimentally confirmed(98) as was also the Guinier theory of small angle X-ray scattering,(99-100) Smoluchowski kinetics for formation of dimers, trimers, etc., the Verwey Overbeek theory of fast and slow coagulation.(101-102) The latter permitted determination of the universal Van der Waals (Hamaker) constant.(103) Coagulation of small particles to flat crystalline plates with a growth spiral was also studied.(104) Furthermore the adherence of colloid gold particles to surface and supports was investigated.(105) A solid experimental and theoretical basis was produced for the preparation of monodisperse catalysts. Unfortunately gold is not a catalyst for most reactions of interest.

The knowledge of the colloidal chemistry of gold was used by Kim and Turkevich(106) to prepare monodisperse palladium particles in diameter greater than 75 Å. The number of surface atoms determined by electron microscopy was found to be equal to the number of catalytic centers determined by poison titration for the ethylene hydrogenation reaction. The surface of the palladium catalyst was homogenous. The velocity of the catalytic reaction was found to be proportional to the number of surface atoms. All surface atoms were active.

Application of these theoretical and experimental techniques to platinum catalysts had to await significant advances in electron microscopy. The resolution of the RCA electron microscope at the Chemistry Department of the University was only 20 A while the platinum particles produced under standard preparation conditions were at most 30 Å in diameter. However interesting findings were found even with this limited resolution. A smooth platinum foil was used as a catalyst in the Ostwald process of ammonia oxidation. It was known from optical observations that the surface became rough and tarnished. In 1953 Garton and Turkevich(107) applied the replica technique, developed during the war for the diffusion barrier, to observe with the electron microscope what was described as the "catalytic etch". During the course of ammonia oxidation, the amorphous Beilby layer produced on the platinum surface during manufacture of the foil rearranged to become a beautiful crystalline surface with terraces, corners, regular etch pits. The catalyst surface, even of high melting platinum, is not static during a catalytic reaction. While the surface area did not change appreciably, the number of edges, slip planes and corners increased enormously. Nevertheless the catalytic activity of the foil for hydrogen peroxide decomposition did not change after "catalytic etch". Thus, no special enhanced activity can be ascribed to various crystallographic loci--edges, corners, etch pits of an active catalyst.

Controlled synthesis of platinum particles was made possible in the late sixties and seventies by significant advances in electron microscopy. Lazlo L. Ban of the Cities Service Laboratories in nearby Hightstown was able to obtain lattice imaging with a resolution of 0.05 Å while Joseph Wall at the Brookhaven National Laboratory constructed a scanning transmission electron microscope with a resolution of 2.5 Å permitting visualization of individual platinum atoms. These techniques together with the use of the ultracentrifuge of Barbara

Bamman of the University Biology Department were applied to the platinum system. Particular assistance in this program was given by Seitero Namba and R. S. Miner, Jr.(72) The synthesis of monodisperse platinum(108,109,110),and also of its alloys with gold and palladium was studied.(112) It was found that only certain platinum compounds formed platinum particles. This served as an insight into the efficacy of certain platinum compounds as antitumor drugs in cancer therapy.(113) Electron microscopy was used to follow particle formation from atoms in the starting material. The lattice parameters of individual particles were determined. In solution it was found that a minimal diameter of 12 Å had to be exceeded in order to obtain catalytic activity for hydrogen peroxide decomposition. This limitation was not found for the hydrogenation of ethylene on platinum particles supported on alumina. All surface atoms of a 32 Å diameter particle were found to be active. Thus the main objectives of a synthesis program initiated three decades ago, were attained.

Alumina supports for metal particles were synthesized in the form of fibers and thin plates suitable both for catalyst studies and electron microscopic examination.(106) Light scattering, proton resonance, viscosity measurements, were used to study the formation of silica gels and monodisperse sols.(114) The same techniques were used to study synthesis of zeolites.

Magnetic Resonance. Interest at Princeton in magnetic phenomena related to catalysis goes back to 1931 when P. W. Selwood came to Princeton as a post-doctorate fellow from the University of Illinois where he worked with Professor Hopkins on the magnetic properties of rare earth. His help was invaluable to Taylor and Diamond in their work on the effect of rare earths on the ortho-para hydrogen conversion. At Princeton, Selwood built a Gouy balance for measuring static susceptibility using a magnet borrowed from the Physics Department. Turkevich, a new graduate student from Dartmouth College, where he taught organic chemistry suggested to Selwood that they measure the magnetism of the stable free radical alpha,alpha-diphenyl-beta-picryl hydrazyl (DPPH), which he made for this purpose by a four step synthesis. It was found to have the expected spin due to one unpaired electron.(115) DPPH was further used by Selwood and Turkevich as a catalyst for ortho-para hydrogen conversion. In addition C. P. Smyth, D. Oesper and Turkevich measured the electric susceptibility of DPPH and showed that the unpaired electron was not localized on the nitrogen but resonated to other parts of the molecule.(116) After Selwood left for Northwestern static magnetic susceptibility work was continued at Princeton on chromic oxide catalysts and on the relation of luminescence to the magnetic characteristics of manganese in zinc sulfide phosphors. During the summer of 1950 at lunch at the Brookhaven National Laboratory, Turkevich learned from

Charles H. Townes that he was using the very uniform magnetic field of the uncompleted cyclotron to construct an electron paramagnetic resonance (EPR) spectrometer. Turkevich persuaded Townes to measure the EPR of DPPH. Townes(117) found a very sharp resonance in the radical, so sharpened by exchange narrowing, that it has been used ever since its discovery, as a standard in EPR measurements. This opened up the application of EPR not only to free radical research(118) but also to the whole field of catalysis. Unfortunately the equipment for this work, particularly a homogeneous field magnet, was expensive. This limitation was circumvented when Pastor and Turkevich built at Princeton a rather inexpensive low magnetic field spectrometer using Helmholtz coils. It was suitable for measuring narrow lines.

Charcoal as Catalyst

On the basis of the Bonhoeffer work on the ortho-para hydrogen conversion, it was concluded that charcoal showed surface magnetism and was a suitable system for EPR studies.(119-120) Glucose was carbonized to various temperatures. A broad EPR signal appeared at 350°C, increased in intensity and decreased in width as the temperature of carbonization was raised. The signal reached its maximum intensity and minimum width on heat treatment at 605°C. At higher temperature treatment, the line broadened, decreased in intensity and the charcoal specimen became "lossy" due to electric conductivity. These observations were correlated by Laroche and Turkevich(29) with the catalysis of ortho-para conversion at liquid nitrogen temperature and hydrogen-deuterium equilibration at room temperature. On heating glucose, water and CO were evolved producing on the surface free radicals. As these increased in concentration, they showed exchange narrowing and intense magnetic fields. These fields catalyze the ortho-para hydrogen conversion at liquid nitrogen temperatures but did not catalyze hydrogen-deuterium equilibration at room temperatures. As the temperature is raised above 600°C hydrocarbons and CO are evolved from the charcoal and the spins begin to pair off to form anti-ferromagnetic domains, the surface magnetism drops and the ability of the surface to catalyze the ortho-para hydrogen conversion at liquid nitrogen is lowered. On the other hand the H_2-D_2 reaction and also the ortho-para reaction is catalyzed at room temperature. On further heat treatment, as antiferromagnetic domains grow into graphitic sheets, they lose their ability to catalyze the H_2-D_2 reaction. Previous studies(124) on EPR of potassium complexes of aromatic ring systems gave an insight into the size of pools of antiferromagnetic electrons which can serve as sources or recipients of electrons active in catalytic processes. Since that investigation EPR work was carried out on carbon black used for tire reinforcement, carbon from cigarettes, carbon in lungs of coal miners and city dwellers. The last investigation on carbon took place in 1980 when the late Professor S. Krzyzanowksi studied its formation in a ZSM-5 zeolite catalyst during petroleum synthesis from methanol.

Other Catalyst Systems

As previously noted Sato and Turkevich(68) studied the transfer of conductivity electrons from surface sodium films to adsorbed species. Clarkson and Turkevich(66) studied by EPR the physical adsorption of oxygen on silica, Angelescu and Turkevich(67) the transfer of electrons from metals to adsorbed oxygen and Vanderspurt, Che and Turkevich studied chemisorption of oxygen and ferrocene on silica observing not only O_2^- with its triplet gx, gy, gz, but also each having a super hyperfine structure of six lines corresponding to the five hydrogen atoms in the cyclopentadiene complex--a molecule-structural interpretation of the active adsorption center.

In the sixties the generous support from the Exxon Research Corporation and equipment grant from the U. S. Atomic Energy Commission permitted the purchase of commercially built EPR spectrometers at 9.3 and 35 G and a broad line nuclear magnetic spectrometer.

In 1964 a short investigation was made of EPR signals of platinum on alumina by F. Nozaki, D. Stamires and Turkevich.(70) The relation of catalytic activity of transition metal oxides to their EPR properties was studied. Thus in 1967 Kazanski investigated the chromium oxide on silica and its ability to carry out low temperature polymerization of ethylene. A series of papers were devoted to the TiO_2, its decomposition products and its interaction with oxygen. The results so obtained were compared with the behavior of Ti(III)(129) compound with hydrogen peroxide <u>in solution</u> using fast-mixing flow techniques. The active species was identified by EPR as an octahedral complex of Ti(IV) with OH, substrate and oxygen ligands. Zinc oxide, a material studied by Taylor and his students, an interesting semiconductor and active principle in xerography, was investigated by Turkevich. (130,131,132) Methyl radicals were produced on silica surfaces of porous Vycor glass by photolysis of methyl iodide.(75) The ESR characteristic of the proteum, deuterium and C-13 compound showed extensive mobility of the radical on the surface, its planar structure and its stability for days at room temperature. Its reactivity with gases was studied under diffusion control conditions. Thus the goal of the organic chemists of the middle nineteenth century was attained--under proper conditions methyl radicals can be produced and used as chemical reagents. Furthermore their stability at room temperature indicated that they can be important intermediates in catalytic hydrocarbon reactions. The structure and behavior of methyl radicals on surfaces was compared with that in aqueous solution. Methyl radicals were produced and characterized by EPR in a fast-mixing flow system of Ti(III) with tertiary butyl hydroperoxide.(133) As a side investigation, the reaction of methyl radicals with oxygen dissolved in water was monitored not only by EPR but by chemiluminescence using an electronic image intensifier.(134) Under appropriate experimental conditions a "cold flame" could be produced by this reaction in water.

Magnetic Resonance in Zeolites

For a number of years (1961-1981) the main thrust of the EPR work was the investigation of zeolites. The first investigation of these materials at Princeton was carried out in 1933 by the Swedish chemist Arne Tiselius(135) who brought with him samples which he personally collected in the Orkney Islands. C. O. Strother, a Princetonian who was the director of the Linde Research Laboratory of the Union Carbide Company at Tonowanda, New York sent samples of NaX zeolites for surface studies. Unfortunately the hydrogen (acid) form is not stable and the sodium form is inactive. However, the NaY form has a stable hydrogen form and this opened a comprehensive research program in the synthesis, physical examination and chemical reactions of zeolites.(70,136-144) These materials when free of iron impurities show no EPR signals. However after a variety of treatments, EPR signals are produced which are diagnostic of the molecular environment of the active acid centers: replacement of sodium ions by Cu(II) by Nikula, Stamires,(145) Ono, Soria;(146) by rare earths by L. Iton;(147) electron transfer to and from adsorbed molecules (trinitrobenzene and triphenylamine) from Lewis acids by Stamires, Nozaki and Ono(139-148) and trapping of electrons by Lewis acid sites on gamma irradiations.(138-150) The behavior of hydrogen atoms trapped in the lattice of calcium fluoride was compared to that of hydrogen atoms adsorbed in cavities of zeolites.(152) These researches were continued in Japan by Professors Y. Ono, Y. Fujita, S. Namba, I. Okura; in Spain by X. Soria, in France by M. Che and J.E. Vedrine; in the United States by R. Clarkson and L. Iton.

Proton resonance of charcoals prepared by heat treatment of dextrose at various temperatures was studied before and after hydrogen treatment by Turkevich, Mackey and Thomas.(136) This was followed by a preliminary study of double (electron-proton) resonance by Turkevich and Derouane(153) of benzene adsorbed on charcoal. This investigation was completed recently by M. Che and J. L. Vendrine.(154)

The status of water in pores of zeolites was examined by proton resonance by Mackey, Thomas and Turkevich.(136) The water hydrated the cation which moved freely in the cavity. On dehydration the cation was localized on aluminum centers in the cavity. L. Meecham studied water in glass which had uniform pores.[141] Different characteristics were found for water confined in pores of diameter less than 200 Å. The behavior of Al and Na in the zeolite as a function of water content was investigated by NMR of these nuclei.(141)

A thorough investigation of C-13 nuclear resonance of hydrocarbons adsorbed by zeolites was carried out in collaboration with Dorothy Denney of Rutgers University, S. Martikhin of the Institute of Catalysis, USSR, S. Namba of Tokyo Institute of Technology.(156) The chemical shift, the coupling constants, the relaxation times of the individual carbons in the adsorbed molecules indicated not only the presence of strong electric field in the pores of the zeolite but also permitted evaluation of the polarization of individual carbon atoms in the molecule.

Conclusion

Thus over the course of half a century, the basic concepts of activated adsorption and active centers were made more precise by application of novel techniques. In addition to the direct line work on catalysis, many scientific side lines were pursued: photochemistry, discharge tube chemistry, radiation induced transformation, solid state chemical physics, and chemotherapy of cancer. (113)

Acknowledgments

Support for these investigations of the Taylor and Turkevich groups came from the University Research Fund, the Higgins Fund, the M.W. Kellogg Company, the Chevron Research Fund, the Shell Company Grant, the U.S. Atomic Energy Commission, U.S. Deptartment of Energy and the National Science Foundation. These are gratefully acknowledged.

Literature Cited

1. Kemball, Charles, Biographical Memoires of Fellows of the Royal Society of London, 1975, 21, 517.
2. Johnson, F.H., Goodale, W. T., and Turkevich, J. J. Cell. and Comp. Physiol. 1942, 19, 163.
3. Brunauer, S., Emmett, P. H. and Teller, E. J. Amer. Chem. Soc., 1938, 60, 309.
4. Taylor, H. S., Proc. R. Soc. London, 1926, A113, 77.
5. Taylor, H. S. and Burns, R. M., J. Amer. Chem. Soc., 1921, 43, 1273.
6. Taylor, H. S. and Gauger, A. W., J. Amer. Chem. Soc.,1923, 45, 920.
7. Taylor, H. S. and Beebe, R. A., J. Amer. Chem. Soc., 1924, 46, 43.
8. Taylor, H. S. and Dew, W. A., J. Phys. Chem., 1927, 31, 277.
9. Taylor, J. S. and Kistiakowsky, G. B., Z. physik Chem., 1927, 125, 341.
10. Taylor, H. SD., Kistiakowsky, G. B. and Flosdorf, E.W., J. Amer. Chem. Soc., 1927, 49, 2200.
11. Taylor, H. S. and Williamson, A. T., J. Amer. Chem. Soc., 1931, 53, 813.
12. Taylor, H. S. and Williamson, A. T., J. Amer. Chem. Soc. 1931, 53, 2168.
13. Taylor, H.S. and McKinney, P. V., J. Amer. Chem. Soc., 1931, 53, 3604.
14. Taylor, H. S. and Sickman, D. V., J. Amer. Chem. Soc., 1931, 53, 3604.
15. Taylor, H.S. and Sherman, A., Trans. Faraday Soc., 1932, 28, 247.
16. Taylor, H. S. and Gould, A. J., J. Amer. Chem. Soc., 1934, 56, 1685.
17. Taylor, H. S. and Pace, J., J. Chem. Phys., 1934, 2, 578.
18. Taylor, H. S. and Turkevich, J., J. Amer. Chem. Soc., 1934, 56, 2254.
19. Taylor, H. S. and Burwell, R. L., Jr., J. Amer. Chem. Soc., 1936, 58, 1753.
20. Taylor, H. S. and Burwell, R. L., Jr., J. Amer. Chem. Soc., 1937, 59, 697.
21. Taylor, J. S. and Strother, C. O., J. Amer. Chem. Soc., 1934, 56, 586.
22. Taylor, H. S. and Smith, E. A., J. Amer. Chem. Soc., 1938, 60, 362.
23. Taylor, J. S., Activated Adsorption in Twelfth Report of Committee on Catalysis. National Research Council, Washington, D. C. 1940, Chapter III.
24. Benson, J. E. and Boudart, M., J. Catal.,1965, 14, 704.
25. Taylor, H. S., Z. Electrochem. Ang. Phys. Chem.,1929, 35, 542.
26. Taylor, H.S., Active Centers in Twelfth Report of Committee on Catalysis. National Research Council, Washington, D.C. Chapter IV.

27. Taylor, H. S. and Diamond, H., J. Amer. Chem. Soc., 1933, 55, 2613.
28. Turkevich, J. and Selwood, P. W., J. Amer. Chem. Soc., 1941, 63, 1077.
29. Turkevich, J. and Laroche, J. Zeit fur Physik Chemie,W.F. 1958, 15, 399.
30. Taylor, H. S., Science, 1934, 79, 303.
31. Taylor, H. S., J. Franklin Inst., 1934, 218, 1.
32. Taylor, H. S., Morikawa, K., and Benedict, W. S., J. Amer. Chem. Soc., 1935, 57, 592.
33. Taylor, H. S., Morikawa, K., and Benedict, W. S., J. Amer. Chem. Soc., 1935, 57, 2735.
34. Taylor, H. S., Morikawa, K., and Benedict, W. S., J. Amer. Chem. Soc., 1935, 58, 1445.
35. Taylor, H. S., Morikawa, K., and Benedict, W. S., J. Amer. Chem. Soc., 1935, 58, 1795.
36. Taylor, H. S., and Turkevich, J., Trans. Faraday Soc., 1939, 35, 921.
37. Lennard Jones, J.E. and Turkevich, J., Proc. Roy. Soc., London, 1937, 158, 297.
38. Eyring, H., Frost, A. A. and Turkevich, J., J. Chem. Phys.,1933, 1, 777.
39. Goldwasser, S. and Taylor, H. S., J. Amer. Chem. Soc., 1939, 61, 1766.
40. Taylor, H. S., Turkevich, J., Fehrer, H., J. Am. Chem. Soc., 1941, 63, 1129.
41. Taylor, H. S., Salley, D. J. and Fehrer, H., J. Amer. Soc.,1941, 63, 1131.
42. Taylor, H. S., Fehrer, H., J. Am. Chem. Soc., 1941, 63, 1385.
43. Taylor, H. S., Fehrer, J., J. Am. Chem. Soc., 1941, 63, 1387.
44. Taylor, H. S. and Briggs, R. A., J. Am. Chem. Soc., 63, 2500.
45. Turkevich, J. and Young, H. H., Jr., J. Amer. Chem. Soc., 1941, 63, 519.
46. Kline, C. H.,Jr., and Turkevich, J., Am. Chem. Soc. 1941, 66, 1710.
47. Turkevich, J. and Stevenson, P. C., J. Chem. Phys. 1943, 11, 328.
48. Kline, C. H.., Jr. and Turkevich, J., J. Chem. Phys. 1944, 7, 300.
49. Bigeleisen, J., Mayer, M. G., Stevenson, P. C., and Turkevich, J., J. Chem. Phys.,1948, 16, 442.
50. Turkevich, J., J. Chem. Phys., 1945, 13, 235.
51. Taylor, H. S. and Kemball, C., J. Am. Chem. Soc. 1948, 70, 345.
52. Taylor, H. S. and Stone, F. S., J. Chem. Phys., 1952, 26, 1339.

53. Taylor, H. S., Fensham, P. J., and Tamaru, K. and Boudart, M., J. Phys. Chem., 1955, 59, 806.
54. Taylor, H. A. Ozaki, A., and Boudart, M., Proc. R. Soc. London, 1955, A258, 47.
55. Taylor, H. S., Tamara, K. and Boudart, M., J. Phys. Chem., 1955, 59, 801.
56. Taylor, H. S. and Amano, A., J. Amer. Chem. Soc., 1954, 76, 4201.
57. Taylor, H. S., Cimino, A. and Boudart, M., J. Phys. Chem., 1954, 30, 796.
58. Taylor, H. S., Parravano, G. and Hamwell, E. F., J. Amer. Chem. Soc., 1948, 70, 2269.
59. Taylor, H. S. and McGeer, J. P., J. Amer. Chem. Soc., 1951, 73, 2743.
60. Taylor, J. S. and Wright, M. M., Canadian J. of Research, 1949, 27, 303.
61. Taylor, H. S. and Boudart, M., Res. Council of Israel, Special Pub.1, No. 1, Jerusalem 223.
62. Taylor, H. S. and Halsey, G. D., Jr., J. Chem. Phys., 1947, 15, 624.
63. Taylor, J. S., Mikovsky, R. J. and Boudart, M., J. Amer. Chem. Soc. 1954, 76, 3814.
64. Taylor, J. S., Sadek, H., J. Amer. Chem. Soc., 1950, 72, 1168.
65. Taylor, J. S., and Liang, S. C., J. Amer. Chem. Soc. 1947, 69, 2985.
66. Clarkson, R. B. and Turkevich, J., J. of Colloid. and Interface Science, 1972, 35, 165.
67. Turkevich, J. and Angelescu, E., unpublished.
68. Turkevich, J. and Sato, T., Proc. of Fifth International Congress of Catalysis. (J.W. Hightower, ed.) North Holland/American Elsevier, 1973, 1, 587.
69. Turkevich, J. Ichkawa, A., Ikawa, T., Abstracts, National Meeting, American Chemical Society, Chicago, Ill. Sept. 1961.
70. Turkevich, J., Nozaki, F., Stamires, D., Proc. Int. Congr., Catal. 3rd. 1964, 1965, 1, 556.
71. Turkevich, J. and Kim, G., Science, 1970, 169, 873.
72. Gonzalez-Tejuca, L., Aika, K., Namba, S. and Turkevich, J., J. Phys. Chem., 1977, 81, 1399.
73. Turkevich, J., Murakami, S., Nozaki, F., and Ciborowski, S., Chem. Eng. Prog., 1967, 63, 75.
74. Gonzalez-Tejuca, L. and Turkevich, J., J. Chem. Soc. Faraday Trans. 7, 1978, 74, 1064.
75. Turkevich, J. and Fujita, Y., Science, 1966, 152, 1619.
76. Buechler, E. and Turkevich, J., J. Phys. Chem., 1972, 76, 2325.
77. McCarthy, W. W. and Turkevich, J., J. Chem. Phys., 1944, 12, 405, and J. Chem. Phys., 1944, 12 , 461.

78. Turkevich, J., and Smith, R. K., J. Chem. Phys., 1948, 16, 466.
79. Turkevich, J. and Smith, R. K., Nature, 1946, 874.
80. Turkevich, J. and Bonner, F., J. Am. Chem. Soc., 1951, 73, 561.
81. Amick, G. A. and Turkevich, J., Brookhaven Nat. Lab. Quarterly Report. July 1950, 58-60.
82. Friedman, L. and Turkevich, J., J. Chem. Phys., 1949, 17, 1012.
83. Turkevich, J., McKenzie, H. A., Friedman, L. and Spurr, R., J. Am. Chem. Soc., 1949, 71, 4045.
84. Amick, J. and Turkevich, J., Brookhaven Nat. Lab. Quarterly Reports, July 1950, 60-65.
85. Turkevich, J., Friedman, L., Solomon, E. and Wrightson, F. M., J. Am. Chem. Soc., 1948, 70, 638.
86. Turkevich, J., Bonner, J., Schissler, D., Irsa, P., Trans. Faraday Soc., 1950, 8, 352.
87. Schissler, D. O., Thompson, S. O. and Turkevich, J., Disc. Faraday Soc., 1951, 10, 46.
88. Turkevich, J., Schissler, D. O., Irsa, P., J. Phys. and Colloid. Chem., 1951, 55 (6), 1075.
89. Thompson, S. O., Turkevich, J., Irsa, P., J. Am. Chem. Soc. 1951, 73, 5213.
90. Thompson, S. O., Turkevich, J., Irsa, P., J. Phys. Chem. Soc. 1952, 56, 243.
91. Friedman, L. and Turkevich, J., J. Am. Chem. Soc., 1952, 74, 1666.
92. Friedman, L. and Turkevich, J., J. Am. Chem. Soc., 1952, 74, 1669.
93. Bond, G. C. and Turkevich, J., Trans. Faraday Soc., 1953, 49 , 281.
94. Bond, G. C. and Turkevich, J., Trans. Faraday Soc., 1954, 50, 1335.
95. Turkevich, J., J. Chem. Phys.,1945, 13, 235.
96. Turkevich, J. and Hillier, J., Anal. Chem., 1949, 21, 475.
97. Turkevich, J., Stevenson, P. C. and Hillier, J., Disc. Faraday Soc.,1951, 11, 55-75.
98. Turkevich, J., Garton, G. and Stevenson, P. C., J. Colloid. Sci., 1954, Suppl. 1, 26.
99. Turkevich, J., Hubbell, H. H. and Hillier, J., Disc. Faraday Soc., 1950, 8, 348.
100. Turkevich, J. and Hubbell, H. H., J. Am. Chem. Soc., 1951, 73, 1.
101. Turkevich, J., American Scientist, 1959, 47, 97-119.
102. Enustun, B. V. and Turkevich, J., J. Am. Chem. Soc., 1963, 85, 3317.
103. Demirci, S., Enustun, B. V. and Turkevich, J., J. Phys. Chem., 1970, 82, 2710.
104. Chiang, Y.S. and Turkevich, J., J. Colloid. Sci., 1963, 18, 772-783.

105. Turkevich, J., Demirci, S., and Skvır, D. J., Croatica Chemica Acta, 1972, 45, 85.
106. Kim, G. and Turkevich, J., Science, 1970, 169, 873.
107. Garton, G. and Turkevich, J., J. Chem. Phys.,1954, 51,516.
108. Turkevich, J., Aika, K., Bann, L. L., Okura, I., Namba, S. J. Res. Inst. Catalysis, Hokkaido University, 1976, 24,34.
109. Turkevich, J., Bann, L. L. and Wall,, J. H., Perspectives in Catalysis (Larssen, ed.), 1980, 59.
110. Turkevich, J., Miner, R. S.,Jr., Okura, I., Namba, S., Zacharina, N., Perspectives in Catalysis (Larssen, ed.) 1980, 111.
111. Miner,R. S., Jr., Namba, S., and Turkevich, J., Proc. 7th. Int. Cong. Catal., Tokyo, 1980, 160.
112. Turkevich, J., Plenary Lecture,VIII, Ibero-American Symp. on Catalysis, La Rabida, Spain, July 1982.
113. Burchenal, J., Lokys, L., Turkevich, J., and Gale, G., in Cis Platin,(Prestayko, W. S., Crooke, S. T. and Carter, S. K., eds.) Academic Press, New York, 1951, 113.
114. Turkevich, J. and Bartholin, M., Proc. IV Simposio Ibero-Americano de Catalysis, Mexico City, 1976.
115. Turkevich, J. and Selwood, P. W., J. Am. Chem. Soc., 1941, 63, 1077.
116. Oesper, P., Smyth, C. P. and Turkevich, J., J. Am. Chem. Soc., 1942, 64, 1179.
117. Townes, C. H. and Turkevich, J., Phys. Rev., 1950, 77, 148.
118. Cohen, V. W., Kikuchi, and Turkevich, J., Phys. Rev., 1952, 85, 379.
119. Pastor, R. A., Weil, J. A., Brown, T. H. and Turkevich, J., Phys. Rev., 1956, 102, 918.
120. Pastor, R. C., Weil, J. A., Brown, T. H.,and Turkevich, J., Adv. in Catalysis, 1957, 9, 1078.
121. Brown, T. H. and Turkevich, J., J. Phys. Chem., 1957, 61, 1452.
122. Brown, T. H., and Turkevich, J., Proc. of 1957 Conf. on Carbon, 1957, Pergamon Press, London.
123. Turkevich, J., and Laroche, J., Zeit fur Physik Chemie,N.F., 1956, 15, 399.
124. Pastor, R. C. and Turkevich, J., J. Chem. Phys., 1955, 23, 1731.
125. Vanderspurt, T. H., Turkevich, J., Che, M., and Buchler, E. J. of Catalysis, 1974, 32, 127.
126. Kazanski, V. B., and Turkevich, J., J. of Catalysis, 1967, 8, 231.
127. Iyengar, R. D. Codell, M., Karra, J. S. and Turkevich, J., J. Am. Chem. Soc., 1967, 88, 5055.
128. Iyengar, R. D.., Codell, M. and Turkevich, J., J. of Catalysis, 1967, 9, 305.
129. Chiang, Y. S., Craddock, J., Mickewitch, D., and Turkevich, J., J. Phys. Chem., 1966, 70, 3509.

130. Larach, S. and Turkevich, J., J. Phys. Chem. Solids, 1968, 29, 1519.
131. Larach, S., and Turkevich, J., Applied Optics, 1968.
132. Larach, S., and Turkevich, J., Surface Science, 1970, 20, 192.
133. Mickewich, D. and Turkevich, J. J. Phys. Chem., 1968, 72, 2703.
134. Turkevich, J., Mickewich, D., and Reynolds, G. T., Zeit fur Phys. Chem., NF, 1972, 82, 185.
135. Tiselius, A. Zeit.Phys. Chem. A, 1935, 174, 401.
136. Turkevich, J., and Mackey, J. H.,Jr., Thomas, W. H., Acte du Deuxieme Congres International de Catalyse, Ed. Tech. , Paris, 1961, 1815.
137. Murakami, Y., Nozaki, F., and Turkevich, J., Shokubai, (J. Catalysis in Japan), 1963, 5, 262.
138. Turkevich, J., and Ciborowski, S., J. Phys. Chem.,1967, 71, 3208.
139. Turkevich, J., Murakami, Y., Nozaki, F. and Ciborowski, S., Chem. Eng. Progress, Symposium Series, 1967, 63, 75.
140. Turkevich, J. Catalysis Reviews, 1967, 1, 1.
141. Turkevich, J., and Ono, Y., Advances in Catalysis, Academic Press, New York, 1969, 20, 135.
142. Turkevich, J., and Ono, Y., Advances in Chemistry, 1971, 102, 315.
143. Buechler, E. and Turkevich, J., J. Phys. Chem., 1972, 76, 2325.
144. Miner, R. S., Ione, K. G., Namba, S., and Turkevich, J., J. Phys. Chem., 1978, 82, 214.
145. Nikula, A., Stamires, D. and Turkevich, J., J. Chem. Phys. 1965, 42, 3684; Rev. Roum. Phys., 1964, 9, 613.
146. Turkevich, J., Ono, Y., and Soria, X., J. of Catalysis, 1972, 25, 44.
147. Iton, L.E., and Turkevich, J., J. Phys. Chem., 1977, 81, 435.
148. Stamires, D. N. and Turkevich, J., J. Am. Chem. Soc., 1964, 86, 749.
149. Stamires, D. N. and Turkevich, J., J. Am. Chem. Soc., 1964, 86, 757.
150. Haul, R., Karra, J., and Turkevich, J., J. Am. Chem. Soc., 1965, 87, 2092.
151. Turkevich, J., Shokubai (J. of Catalysis in Japan), 1965, 7, 328.
152. Iton, L. E. and Turkevich, J., J. Phys. Chem., 1978, 82, 200.
153. Derouane, E., and Turkevich, J., unpublished.
154. Che, M. and Vendrine, J. C., in press.
155. Denney, D., Mastikhin, V. M., Namba, S., and Turkevich, J., J. Phys. Chem., 1978, 82, 1752.

RECEIVED February 1, 1983

34

Murray Raney—Pioneer Catalyst Producer

RAYMOND B. SEYMOUR
University of Southern Mississippi, Department of Polymer Science, Hattiesburg, MS 39406
STEWART R. MONTGOMERY
W. R. Grace and Company, Davison Chemical Division, Columbia, MD 21044

A catalyst prepared by the reduction of nickel oxide, was used as a hydrogenation catalyst by Sabatier and Senderens almost a century ago. (1) While many modifications of this versatile catalyst (2, 3) and the use of high pressure nickel catalyzed hydrogenations have been described, (4, 5, 6) none of these modifications was as effective as the nickel alloys developed by Murray Raney in the early 1920's.

Since Sabatier was a Nobel laureate (1912) and Ipatieff and Adkins were known worldwide for their chemical expertise, one may wonder how a nonchemist, such as Raney, was able to develop his superior nickel catalyst. Actually, there is nothing in his education or early endeavors to suggest the "flash of genius" displayed by this lone inventor.

One of us (R.B.S.) had the pleasure of knowing Murray Raney in the 1940's and to learn of the early years of this Kentuckian first hand. As described in the only three available biographies of Raney, (7, 8, 9) this inventor's precollege education ended at the eighth grade. He was born on October 14, 1885 in Carrollton, KY. Instead of working for his father, who was a contractor, Murray took a job as a polisher in a furniture factory earning twenty-five cents a day. After realizing that such a job did not have a very promising future, he enrolled in a correspondence course in order to obtain his high school diploma. In 1905, after a summer course at the University of Kentucky he enrolled as a full-time student even though he had not finished high school. He was interested in automotive design and therefore pursued the field of mechanical engineering. During his college career he received financial assistance from the electrical engineering department where he served as an assistant. He received his baccalaureate degree in mechanical engineering in 1909.

His first professional position was an instructor in Eastern Kentucky Normal School at a salary which was slightly greater than the 25 cents a day earned previously as a furniture polisher. After a tenure of one year, he accepted a more

0097–6156/83/0222–0491$06.00/0

Figure 1. Murray Raney.

lucrative position at the A. L. Ide Engine Company in Springfield, IL. Two years later he accepted a position as a sales engineer for the Chattanooga Railway Light and Power Company.

After arriving in Chattanooga, he had a good fortune to meet Mercer Reynolds, president of Lookout Oil and Refinery Company and Paul Kreusi, president of both the America Lava Corp. and Southern Ferro Alloy Corp. These entrepreneurs and their plants provided the stage for Murray Raney's invention. It was my (R.B.S.) good fortune to know both of these industrialists who along with Murray Raney were members of the Advisory Board of the Industrial Research Institute, with which I was associated, at the University of Chattanooga in the 1940's.

As a result of having designed a cottonseed oil hydrogenation plant for Mercer Reynolds, Murray Raney became intrigued with catalytic hydrogenation. He demonstrated his unusual technical ability by first developing a test for catalyst efficiency and then showing that the efficiency of a nickel catalyst prepared by the reduction of nickel oxide could be improved by heating with cane sugar. The important feature of this empirical discovery was that the ratio of nickel oxide to nickel was just the inverse of that previously considered to be the optimum by the leaders in the field of catalysis.

The nickel catalyst then in use for the hydrogenation of fats was prepared by the hydrogen reduction of a supported nickel oxide. The catalyst was generally prepared by the vegetable oil processor and variations in catalytic activity were somewhat of a problem. Mr. Raney became fascinated with the concept of a catalyst. He likened it to that of a hammer driving nails into a board. When the job is done, he said, the hammer remains unchanged. He believed that through a careful control of the nickel oxide reduction step, the nickel catalyst then in use could be made consistently more active. To pursue this concept, he formed his own research organization in 1915 (The Chattanooga Research Co.) through the financial support of this employer, Mr. Mercer Reynolds. He installed a laboratory in his home in Chattanooga where he carried out research studies on the catalyst for about five to six years. His studies were directed toward controlling the ratio of nickel oxide to nickel in the catalyst with the belief that this would allow him to prepare a catalyst with consistently high activity. He was not able to achieve this goal to his satisfaction. Finally, after about six years of work, he gave some thought to other reactions that might also yield a nickel catalyst.

Mr. Raney had already demonstrated that the sale of electrolytically-produced oxygen could support the entire facility for the production of hydrogen. Also, because of his familiarity with the commerical production of hydrogen, he was aware of an earlier European process in which a ferrosilicon alloy was treated with aqueous sodium hydroxide. This process produced finely divided iron oxide in addition to hydrogen. He speculated that a nickel-silicon alloy might also give hydrogen and a finely divided nickel oxide since both metals appeared in the same group of the periodic table. His suspicion that he would obtain nickel oxide and not nickel caused him to delay trying the experiment for about two years. Finally in June of 1924, he prepared about 200 grams of a 50% nickel-silicon alloy in Paul Kreusi's foundry and took it to his home laboratory for treatment with aqueous sodium hydroxide. By 1924, he had moved from his house on Oak Street in Chattanooga to a new house on top of Lookout Mountain, a rather pleasant suburb of Chattanooga. His new house also had a chemistry laboratory which was built as a wing adjacent to the kitchen (Figure 2). The laboratory, approximately 25 feet square, was equipped with a work bench, two hoods and a sink as shown in Figure 3 and 4. Not only was the first Raney catalyst prepared and tested in this laboratory, but all those that followed during the four decades that he continued to experiment with the process were prepared here.

In his first preparation from the nickel-silicon alloy he observed that the reaction was quite vigorous and that a greyish metallic solid, quite pyrophoric when dried and exposed to the air, settled out after the reaction subsided. Because of it pyrophoricity he concluded that pure nickel, not the inactive nickel oxide, had formed. On testing this material for catalytic activity for the hydrogenation of cottonseed oil in his rather simple laboratory test equipment (10), he reported that it was five times as active as the best nickel catalyst then in use. He applied for his first patent (11) in the United States on September 20, 1924 which issued December 1, 1925. Probably because of his enthusiasm over his discovery, he often remarked that he was never able to prepare a catalyst as good as the first one and he attributed this to a welding rod used to break up the upper crust of the molten alloy. Over the years he made numerous attempts to substantiate this belief by trying to exactly duplicate the first experiment. He often stated that he never found the answer to this question.

With the successful preparation of a catalyst from the nickel-silicon alloy he gave thought to finding other alloys that might also work. He considered aluminum a good choice because it is readily dissolved by aqueous sodium hydroxide. But because of its high cost at the time, he thought that its use would never be practical. His concern for the economics of the nickel-aluminum alloy caused him to delay, for almost two years, experimentation involving the alloy. In later years he often remarked that he was surprised that some enterprising chemist, after seeing his first patent, didn't try the nickel-aluminum system and obtain, what later proved to be his most valuable composition of matter patent, during the two years it took him to carry out his second experiment. In the spring of 1926 he prepared a 50% Ni-Al alloy (a crank case from an old truck was

Figure 2. Raney's home laboratory - exterior.

Figure 3. Raney's home laboratory - interior.

Figure 4. Raney's home laboratory - interior.

his source of the aluminum) which after treatment with aqueous sodium hydroxide again yielded a pyrophoric nickel that was also catalytically active. He applied for his second patent on May 14, 1926 which issued May 10, 1927 (12). It is of interest to note that the 50% Ni-Al alloy first used by Mr. Raney is still the most important Raney alloy in use today from an industrial standpoint. Years later when asked why he chose the 50-50 composition, he remarked that he wanted to give both metals an equal chance to react.

In his search to prepare a hydrogenation catalyst by a chemical reaction that produces hydrogen, he was convinced that such a catalyst should have excellent activity, because it would contain "nacent" hydrogen. The concept of "nacent" hydrogen intrigued him when he first learned about it in freshman chemistry. The action of hydrochloric acid on zinc (the Marsh test for arsenic) was thought to involve "nacent" hydrogen. This notion that the presence of hydrogen in the catalyst would improve its activity has turned out to be more or less correct. The excellent low-temperature activity of Raney nickel has often been explained on the basis of the hydrogen content of the catalyst.

In 1925, Mr. Raney joined the Gilman Paint and Varnish Co. of Chattanooga as general manager. Later advanced to secretary, then vice president and finally president in the 1930's. He retired from this company in 1950 at the age of 65. However, he maintained control of his patents while they were in effect and later registered the name Raney as a trademark for his catalyst and alloy powders. He sold small amounts of his alloy powder to academic and industrial research laboratories as early as 1930. These powders were prepared for him by the Aluminum Corporation of America in their Maryville, Tennessee plant. Because of his relationship with the Gilman Paint and Varnish Co., these powders were sold as products of this Company.

Shortly after its discovery, Raney nickel was used by the vegetable oil processing industry as a hydrogenation catalyst. Facilities to produce the active catalyst were installed by his former employer, The Lookout Oil and Refining Co., for this purpose. Drawings showing the equipment used were published by Mr. Raney in 1940 (10). Pictures of the alloying crucible and the reaction vessel used to prepare the catalyst starting about 1928 are shown by Figures 5 and 6, respectively. From time to time Mr. Raney sold small amounts of active catalyst that were prepared using this equipment.

Until 1932, Raney nickel was considered only as a very active hydrogenation catalyst for the hydrogenation of vegetable oils. However, in the October 1932 issue of the Journal of the American Chemical Society, Professor Homer Adkins (13) of the University of Wisconsin announced to the scientific community that a new nickel catalyst "preferable to the nickel on Kieselguhr catalyst" was available from Murray Raney of

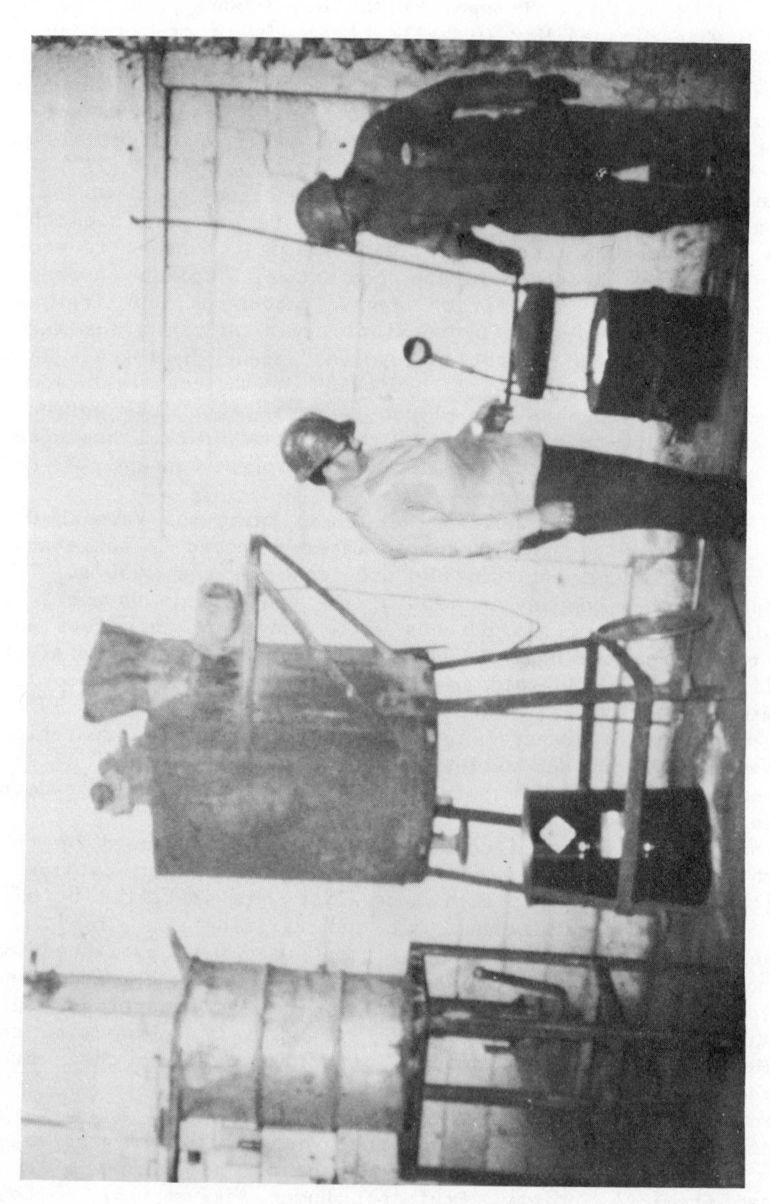

Figure 5. Raney's original alloying crucible.

Figure 6. Raney's original tank used to prepare catalyst.

Chattanooga, Tennessee. In his note to the editor, he referred to the catalyst as the Raney catalyst. This identification of the catalyst has been used almost universally thoughout the world ever since. In the years to follow, Professor Adkins and his students experimented on ways to prepare the catalyst and described a number of catalysts designated as the W-type Raney nickel. Numerous applications of the catalyst for the hydrogenation of unsaturated organic compounds were also reported by Adkins and his students.

For about two decades after its discovery, Raney nickel was used almost exclusively in the laboratory by organic chemists. Gradually, industrial hydrogenation processes evolved based on the use of the catalyst. Practically all of the early industrial users of the catalyst had their own facilities for activating the Alloy (caustic treatment). Both the powder and the granular forms were activated by the user. Some of the first industrial processes to use the catalyst were the conversions of ethylene oxide to ethylene glycol, the conversion of dextrose to sorbitol and the conversion of benzene to cyclohexane.

By 1950, the use of Raney nickel for vegetable oil processing had fallen off considerably, but at the same time the need for the catalyst by the organic chemicals industry had risen sharply. It was in that year that Mr. Raney retired at the age of 65 from the Gilman Paint and Varnish Co. to start his own company, The Raney Catalyst Co., to produce the catalyst. A plant was designed and built by Mr. Raney in south Pittsburg, TN, which is 25 miles southwest of Chattanooga. Commercial production of the catalyst started in June of 1952. As the demand for the catalyst increased, Mr. Raney met the challenge with necessary plant expansions. In 1959 he incorporated his business. At that time he was producing active catalysts from nickel, copper, cobalt, and iron as well as a chromium promoted nickel catalyst which had been shown by the French chemist, Raymond Paul (14), to have excellent catalytic properties. In 1963, he sold his business to W. R. Grace & Co. and remained active in the business as a Grace consultant. Raney catalysts are now produced and marketed by Davison Specialty Chemical Co., a subsidiary of W. R. Grace & Co. Shortly after his second retirement (1966) he was killed in an automobile accident in Chattanooga. He died at the age of 80.

Mr. Raney continued to research on his catalyst from the time it was discovered until he died. He was granted a total of six U. S. Patents and five foreign patents. In addition to the two basic patents that issued prior to 1930, he obtained a patent for the preparation of granular catalysts made from aluminum and silicon alloys containing metals such as iron, cobalt, copper and nickel (15), a method for reclaiming spent catalyst (16), the production of a nonpyrophoric catalyst (17) and a catalyst prepared from a mixture of nickel and a nickel-aluminum alloy (18). A list of his inventions is given in Table 1. Over the

TABLE I

Catalyst Developments Patented by Raney

Year	Development
1925	Ni Catalyst from Ni-Si Alloy
1927	Ni Catalyst from Ni-Al or Ni-Al-Si Alloy
1933	Granular Catalysts from Ni, Fe, Cu, Co Alloys of Al or Si
1938	Reclamining Spent Ni Catalyst from Oxide by Aluminothermic Reaction with Al
1949	Nonpyrophoric Ni Catalyst by CO_2 Treatment
1961	Ni Catalyst from Powdered Ni and Ni-Al Alloy

years he experimented with the preparation of Raney catalysts from other metals such as molybdenum, silver and platinum. He was quite discouraged with his results, since he found them less active than nickel for the hydrogenation of cottonseed oil. He never explored other uses for these catalysts. He also prepared catalysts from ternary and quarternary alloys as shown in Table 2.

TABLE II

Other Catalysts Prepared by Raney

One Metal	
Cobalt	Molybdenum
Copper	Silver
Iron	Platinum
Nickel With Other Metals	
Boron	Molybdenum
Boron-Iron	Tantalum
Cerium	Thorium
Chromium-Copper	Titanium
Chromium-Iron	Vanadium
Columbium	Zirconium

In recognition of his achievement he was awarded a honorary D.Sc. from the University of Kentucky in 1951, although he never used the title. Numerous other marks of recognition were bestowed on him by chemical and engineering societies. While he was grateful for these awards, he stated that he received the most satisfaction from knowing that his catalyst was important enough to make his business a commerical success.

In his pursuit to find a better vegetable oil hydrogenation catalyst, Mr. Raney accomplished something more significant than he realized at the time. Later, when he became aware of the impact his discovery had made in the field of organic chemistry, he modestly said that it was all accomplished by chemists. Since he claimed he wasn't a chemist, he believed he shouldn't take credit for their discoveries. While, Murray Raney was not a chemist and never professed to be one, presumably in self defense, he joined the American Chemical Society in 1940 and was an ardent supporter of the Southeastern Tennessee Section which became the Chattanooga Section of the ACS in 1945. He summarized his investigations of catalysts in an article in Industrial and Engineering Chemistry in 1940 (10) and in addresses to the Memphis (1948) and Pittsburgh Sections of the American Chemical Society (1964).

Literature Cited

1. Sabatier, P.; Senderens, J. B. Compt. Rendu 1897, 124, 1358.
2. Sabatier, P. "Catalysts in Organic Chemistry" (translated by Reid, E. E.); Van Nostrand: New York, NY, 1922.
3. Fraser, O. B. J. Trans. Electrochem. Soc., 1937, 71, 425.
4. Ipatieff, V. I. "Catalytic Reactions at High Pressure and Temperatures"; MacMillan: New York, NY, 1937.
5. Adkins, H. "Reactions of Hydrogen"; University of Wisconsin: Madison, WI, 1937.
6. Ipatieff, V. I. "The Life of a Chemist" (translated by Haensel, V.; Lusher, R. H.); Stanford University Press: Standford, CA, 1946.
7. Seymour, R. B. Chem. Eng. News 1947, 32, 2628.
8. Tarbell, D. S.; Tarbell, A. T. J. Chem. Ed. 1977, 54, 26.
9. Biographical Account, in "The National Encyclopedia of American Biography", J. T. White & Co.: New York, NY, 1969; p. 443.
10. Raney, M. Ind. Eng. Chem. 1940, 32, 1199.
11. Raney, M., U. S. Patent 1,563,587 (Dec. 1, 1925).
12. Raney, M., U. S. Patent 1,628,190 (May 10, 1927).
13. Covert, L. C.; Adkins, H. J. Amer. Chem. Soc., 1932, 54, 4116.
14. Paul, R. Bull. Soc. Chem., Fr. 1946, 13, 208.
15. Raney, M. U. S. Patent 1,915,473 (June 27, 1933).
16. Raney, M. U. S. Patent 2,139,602 (Dec. 6, 1938).
17. Raney, M. U. S. Patent 2,461,396 (Feb. 8, 1949).
18. Raney, M. U. S. Patent 2,977,327 (March 28, 1961).

RECEIVED November 17, 1982

Frank G. Ciapetta—Pioneer in Catalytic Chemistry

W. P. HETTINGER, JR.
Ashland Petroleum Company, Ashland, KY 41101

Frank G. Ciapetta was born in Philadelphia in 1915. He earned his B.A. in Chemistry at Temple University in 1937, his M.A. in 1939, and his Ph.D. in 1947 from the University of Pennsylvania. His career involved responsible positions with Atlantic Refining Company, Socony-Vacuum and the Davison Division of W. R. Grace & Co. and endeavors including research involving homogeneous catalysis, the discovery of a new general class isomerization catalyst and concepts of dual-size catalysis. During later years, efforts in developing a reforming catalyst making possible the Atlantic "Catforming Process" culminated in his receiving in 1955 the Precision Scientific Company Award of the American Chemical Society.

In every phase of human activity - business, industry, technology, science, politics, law, medicine, the arts, athletics - there are "real pros." These are the stand-outs who have won the admiration and respect of their most important group of critics - their peers.

In catalysis, "real pro" is an appelation that fits Dr. Frank G. Ciapetta. Frank was born in Philadelphia in 1915. With the aid of a senatorial scholarship, he enrolled at Temple University where he received a B.A. in Chemistry in 1937 and an M.A. in 1939. Upon graduation, he started work as a laboratory assistant in Research and Development Department of Atlantic Refining. At the same time, he began graduate study at the University of Pennsylvania and was awarded a Ph.D. in 1947; a remarkable achievement in its own right and a tribute to his dedication to science. Homogeneous catalysis was the subject of both his M.A. and his Ph.D. theses. Figure #1 is a picture of Frank taken in 1947 just after graduation.

0097-6156/83/0222-0505$06.00/0

Figure 1. A photograph of Frank G. Ciapetta taken in 1947 just after his graduation from the University of Pennsylvania.

In 1948, he was placed in charge of Atlantic's catalytic process group and in 1950 was made principal chemist. There, discovery of a new general class of isomerization catalysts, consisting of a hydrogenation-dehydrogenation agent such as nickel or platinum, in combination with an acidic cracking catalyst such as silica-alumina, became his major contribution. These catalysts possessed high activity and selectivity for the isomerization of unsaturated hydrocarbons. This most important contribution to catalysts occurred in the late 40's at Atlantic Refining Company Research while he was working on isomerization of normal paraffins.

At that time, unbeknown to each other, Dr. Vladmir Haensel was also involved in his work on platinum on alumina reforming catalyst promoted with fluoride, which work ultimately led to the platforming process of UOP and for which work Dr. Haensel ultimately received the President's Medal of Science. Unbeknown to either man, Frank independently began to explore the use of nickel and platinum supported on silica-alumina and discovered that by the proper treatment of silica-alumina catalyst by high temperature steam treatment, he could couple or harness the activity of nickel and later platinum to the acidity of the catalyst so as to provide a very interesting material for reforming of hydrocarbons. This work eventually led to the commercialization of the "Catforming Process" of Atlantic. He later went on to explore this reaction in great detail and published a number of important papers related to this work. His work and that of Dr. Haensel led to the realization, and later proof by other researchers as well, that two different catalytic sites were required in isomerization namely both hydrogenation-dehydrogenation sites and acidic sites. This work ultimately led to the concept of dual-site catalysis. Since that time, dual-site catalysis has been studied in great detail by many catalyst investigators around the world.

Previous knowledge of the isomerization of hydrocarbons had been limited primarily to low temperatures and low molecular weight hydrocarbons, using catalysts such as aluminum chloride. With the dual-functional catalysts, it was now possible to isomerize hydrocarbons of much higher molecular weight and at considerably higher temperatures.

In 1951, Frank joined the Research and Development Department of Socony-Vacuum and became Research Associate in change of catalyst research. At Socony-Vacuum, Ciapetta's research was focused on the chemical and physical factors that influence the activity and selectivity of metal-acidic oxide catalysts. His study of isomerization-dehydrogenation catalysts led to Socony-Vacuum's introduction of the Sova-Forming Process used commercially in several refineries.

This was a most successful eight-year research campaign on the applications of catalysis to hydrocarbon chemistry, half of which was spent at Atlantic and half at Socony as a Research Associate. His major area of activity with both oil companies was in refinery catalytic processes - cracking, reforming, hydrocracking, isomerization and alkylation, and petroleum conversion generally, and with all types of bed concepts - fixed, moving, and fluidized. During his latter years at Atlantic, he worked closely with Davison scientists on developing the reforming catalyst that subsequently made possible the Atlantic "Catforming Process."

This work culminated in his receiving the coveted Precision Scientific Company Award of the American Chemical Society in Petroleum Chemistry in 1955.

Figure #2 is from the front page of the Chemical and Engineering News, April 11, 1955, when Frank received the Precision Scientific Award. Following is an abstract of his address which was presented before the Division of Petroleum Chemistry at the American Chemical Society Meeting in Cincinnati, April 4-7, 1955.

THE REACTIONS OF PURE HYDROCARBONS IN THE PRESENCE OF DUAL-FUNCTIONAL CATALYSTS

BY
FRANK G. CIAPETTA
(A DIVISION OF SOCONY-VACUUM OIL COMPANY, INC.)
RESEARCH AND DEVELOPMENT DEPARTMENT
PAULSBORO, NEW JERSEY

A major factor in the rapid commercial utilization of catalytic reforming processing for upgrading low octane naphthas, and the production of aromatics from petroleum sources, has been the development of more active and selective dual-functional catalysts. These catalysts contain a very active hydrogenation-dehydrogenation agent such as platinum, in combination with an acidic oxide support such as alumina or silica-alumina.

Figure 2. Frank Ciapetta in 1955 receiving the Precision Scientific Award of the American Chemical Society in Petroleum Chemistry.

At hydrogen pressures of approximately 500 p.s.i.g., these catalysts maintain their activity for reforming virgin naphthas for long periods of time.

An extensive study of the reactions of pure hydrocarbons, including alkanes, cycloalkanes, alkenes and aromatics has been made in the presence of hydrogen, these catalysts are active for (1) isomerization of alkanes, cycloalkanes, and alkylaromatics, (2) hydroisomerization of alkylcyclopentanes to aromatics, and (4) dehydrogenation of alkanes and cyclohexanes to aromatics. The activity and selectivity of these catalysts are dependent primarily on (1) the hydrogenation-dehydrogenation component used and its concentration, (2) the acidity of the oxide support, and (3) the reaction conditions. The effect of these factors on the conversion of pure hydrocarbons as well as the mechanism of these reactions will be discussed.

Due to their high activity and selectivity of isomerization and dehydrogenation reactions, dual-functional catalysts produce large quantities of high purity hydrogen as a by-product in the reforming of naphthas. Utilization of this hydrogen for commercial processes such as hydrorefining of cracked naphthas, distillate fuel oils and lube oils, and ammonia production is rapidly expanding throughout the petroleum industry.

At that time he also made the following observations and comments which were quite perceptive. "During the past few years our petroleum industry has rapidly increased the octane number of its total gasoline pool by adding catalytic reforming processes to upgrade the quality of straight-run and thermally cracked naphthas. Reforming capacity has grown at a phenomenal, rate from less than 80,000 barrels per day in 1950 to 200,000 barrels at the beginning of 1955. By the end of this year, reforming capacity will pass the 1 million barrel-per-day mark, an amount almost equivalent to the volume of straight-run naphtha now produced.

Large amounts of high purity hydrogen (80 to 95%) are obtained during reforming. If we assume 500 cubic feet of hydrogen per barrel of straight-run naphtha as an average yield then refineries will be turning out 500 million cubic feet of hydrogen per day by the end of 1955. Many companies are installing new processes to utilize this output; to reduce sulfur and unsaturation in reformer charge stocks; to reduce sulfur and mercaptan contents, and to improve the color, storage ability, and combustion characteristics of light distillate fuel oil; to reduce sulfur and carbon residue, and to improve the color, odor, and stability of lube oils. By year-end almost 11% of the ammonia produced in the United States will utilize hydrogen from catalytic reforming."

In 1955, he began his association with Davison; first at the Research Facilities at Curtis Bay, Maryland, and later in 1958 at the Washington Research Center near Clarksville, Maryland. Frank joined W. R. Grace & Company as Research Director of the Davison Chemical Division.

Frank's enthusiasm for, and success in, matters catalytic continued at Davison and led to some further remarkable developments in cracking, especially the new molecular sieve based catalysts which appeared in the early 60's, as well as auto exhaust and polymerization catalysts.

In 1968 he was named Vice President for Research, Davison Division of W. R. Grace in recognition of his many contributions not only to catalysis but to Davison as well. In those later years he saw the dedication of a research center devoted to catalysis at the Clarksville Research Laboratories in Maryland. For the first time a large research facility devoted to catalyst research was constructed in order to support work in industrial catalyst research.

Many patents, publications, special honors and awards came to him. He was the author of more than 27 scientific publications and his name appears on 43 United States and many foreign patents on cracking, reforming, and hydrocracking catalysts; on reforming and hydrocracking catalytic processes; and on oxidation catalysts for exhaust gases.

The entire field of catalysis was greatly enriched by his many-faceted contributions. In addition to his active memberships in the American Chemical Society, the American Petroleum Institute, and the American Association for the Advancement of Science, he devoted many hours to the technical groups concerned specifically with catalysis and participated actively in many related scientific and technical organizations.

He was the first president and a founding director of the

Catalysis Society, an evolutionary development from what was originally the International Congress on Catalysis, Inc.

In 1949 a group of Philadelphia catalytic chemists (including Dr. Ciapetta) organized the first catalyst club, the Catalysis Club of Philadelphia. Then, a group of these men, also including Dr. Ciapetta, organized and held, in 1956, the world's first International Congress on Catlaysis. Some 800 academic and industrial catalytic experts came from 17 countries to Philadelphia for the meetings.

In 1960 the locale of the meeting moved to Paris; and to Amsterdam in 1964, each Congress showing strong gains in attendance. The Congress is now organized as a recognized international group of some 14 nations (under IUPAC). Frank was president of the International Congress on Catalysis and Chairman of the Petroleum Research Fund Advisory Board when he died.

One of Dr. Ciapetta's last accolades was his selection to present the major review paper on trends in petroleum refining catalysis at the Seventh World Petroleum Congress in Mexico City in 1967. This was one of 18 papers covering major phases of petroleum technology, delivered by leading scientists from all over the world. Frank was one of only four Americans on the program. This was a natural consequence of a virtual lifetime devoted to work in the field of catalysis - academically, industrially, and in professional and scientific groups.

A quiet, unassuming man, he was respected by all who knew him. He was loved by many, both for his professional activities as well as for his warm, friendly and charming personality. He was a good friend and a warm individual to know and to be associated with. He had a love of life and the fun of friendship as well as sporting activities. In his earlier days, he was an enthusiastic tennis player and in later years transferred that enthusiasm from tennis to golf; which he entered into with the same enthusiasm that he had shown for all of his professional interests. Figure #3 shows Frank participating in his favorite sport at an annual Davison golf outing.

Frank was also a great family man and very proud of his son and daughter. He was especially proud and traveled extensively with his sweet and charming wife, Carolyn. Wherever there was a professional catalysis meeting, Frank would be in attendance and Carolyn was almost always there to add to the enjoyment of

Figure 3. Frank Ciapetta enjoying the annual Davison golf outing.

his company. She, too, became well known throughout the catalyst research world and was always a charming hostess.

On February 7, 1972, on a hot summer day (summer for Argentina), while attending a barbecue picnic in the hot Argentina sun where he was touring and lecturing, Frank experienced an unpleasant feeling and late that afternoon and succumbed to a heart attack. He was in Argentina at that time lecturing on his favorite subject, catalysis, and as usual was having a tremendous time traveling. Frank loved to travel as did Carolyn. Fortunately she was at his side when this occurred.

All of his friends and associates will long remember, with warmest feelings, being associated with him and wish that he were here today to lend his charm, warmth, wit and enthusiasm to this symposium. We all know we have lost a wonderful friend.

Attached is a list of Frank's patents and publications.

Literature Cited

1. Tomlinson, Hazel M.; Ciapetta, Frank G. Ind. Eng. Chem. Anal. Ed., 1941, 13, 539-540.
2. Ciapetta, Frank G. Ind. End. Chem., 1945, 37, 1210-1216.
3. Leum, Leonard M.; Ciapetta, Frank G. U.S. Patent 2 390 951, 1945.
4. Ciapetta, Frank G. Ph.D. Dissertation, University of Pennsylvania, Philadelphia, Pa., 1947.
5. Ciapetta, Frank G. U.S. Patent 2 430 803, 1947.
6. Ciapetta, Frank G. U.S. Patent 2 434 833, 1948.
7. Ciapetta, Frank G.; Kilpatrick, Martin. J. Am. Chem. Soc., 1948, 70, 639-646.
8. Macuga, Stephen J.; Ciapetta, Frank G.; Heiligmann, Randall G. U.S. Patent 2 447 762, 1948.
9. Ciapetta, Frank G.; Macuga, Stephen J.; Leum, Leonard M. Anal. Chem., 1948, 20, 699-704.
10. Ciapetta, Frank G.; Macuga, Stephen J.; Leum, Leonard M. Ind. Eng. Chem., 1948, 40, 2091-2099.
11. Ciapetta, Frank G. U.S. Patent 2550 531, 1951.
12. Ciapetta, Frank G.; Tomlinson, H.M. J. Phys. Colloid Chem., 1951, 55, 429-439.
13. Ciapetta, Frank G.; Pitts, Paul M.; Leum, Leonard M. Adv. Chem. Ser., 1951, 5, 76-82.
14. Ciapetta, Frank G.; Buck, William H. U.S. Patent 2 589 189, 1952.
15. Ciapetta, Frank G.; Hunter, James B. Ind. Eng. Chem., 1953, 45, 147-155.

16. Ciapetta, Frank G.; Hunter, James B. Ind. Eng. Chem., 1953, 45, 155-159.
17. Ciapetta, Frank G. Ind. Eng. Chem., 1953, 45, 159-162.
18. Ciapetta, Frank G. Ind. Eng. Chem., 1953, 45, 162-165.
19. Ciapetta, Frank G. U.S. Patent 2 651 598, 1953.
20. Ciapetta, Frank G.; Plank, C. J. In "Catalysis 1, Fundamental Principles"; Ries, Herman E.; Reinhold: New York, 1954; Part 1, p. 315-352.
21. Conner, J. E., Jr.; Ciapetta, Frank G.; Leum, Leonard M.; Fowle, M. J. Ind. Eng. Chem., 1955, 47, 152-156.
22. Ciapetta, Frank G. General Papers, ACS, Div. of Pet. Chem., 33, 167, 1955.
23. Ciapetta, Frank G.; Hunter, James B. U.S. Patent 2 721 226, 1955.
24. Meisel, Seymour L; Koft, Emil Jr., Ciapetta, Frank G. Prepr., Div. Pet. Chem., Am. Chem. Soc., 1957, 2 (4), A45-A51.
25. Baker, Richard W.; Ciapetta, Frank G.; Dobres, Robert M.
U.S. Patent 2 849 378, 1958.
26. Ciapetta, Frank G.; Dobres, Robert M.; Baker, R. W. In "Catalysis-Alkylation Isomerization, Polymerization, Cracking, and Hydroforming", Emmett, Paul H., Ed., Reinhold: New York, 1958, Vol. 6, p. 495-692.
27. Ciapetta, Frank G.; Koft, Emil Jr.; Meisel, Seymour L. U.S. Patent 2 876 268, 1959.
28. Myers, C. G.; Sibbett, D. J.; Ciapetta, Frank G. J. Phys. Chem., 1959, 63, 1032-1041.
29. Ciapetta, Frank G. U.S. Patent 2 914 461, 1959.
30. Coonradt, Harry L.; Ciapetta, Frank G.; Garwood, William E.; Leaman, W. K.; Miale, J. N. Prepr., Div. Pet. Chem., Am. Chem. Soc., 1960, 5 (4), B35-B51.
31. Pitts, Paul M., Jr.; Ciapetta, Frank G. U.S. Patent 2 942 041, 1960.
32. Ciapetta, Frank G.; Coonradt, Harry L.; Garwood, William E. U.S. Patent 2 945 800, 1960.
33. Ciapetta, Frank G.; Coonradt, Harry L.; Garwood, William E. U.S. Patent 2 945 802, 1960.
34. Ciapetta, Frank G.; Coonradt, Harry L.; Garwood, William E. U.S. Patent 2 945 805, 1960.
35. Ciapetta, Frank G. U.S. Patent 2 945 806, 1960.
36. Ciapetta, Frank G. Petro/Chem. Eng., 1961, 33 (5), C19-C31.
37. Coonradt, Harry L.; Ciapetta, Frank G.; Garwood, William E.; Leaman, W. K.; Miale, J.N. Ind. Eng. Chem., 1961, 53, 727-732.
38. Ciapetta, Frank G.; Coonradt, Harry L.; Garwood, William E. U.S. Patent 3 015 549, 1962.
39. Ciapetta, Frank G.; Coonradt, Harry L.; Garwood, William E. U.S. Patent 3 023 159, 1962.

40. Ciapetta, Frank G.; Coonradt, Harry L.; Garwood, William E. U.S. Patent 3 046 219, 1962.
41. Wilson, Charles P., Jr.; Ciapetta, Frank G.; Elston, James W. U.S. Patent 3 055 737, 1962.
42. Wilson, Charles P., Jr.; Ciapetta, Frank G.; Elston, James W. U.S. Patent 3 066 012, 1962.
43. Ciapetta, Frank G.; Coonradt, Harry L.; Garwood, William E. U.S. Patent 3 111 482, 1963.
44. Ciapetta, Frank G.; Coonradt, Harry L.; Garwood, William E. U.S. Patent 3 115 456, 1963.
45. Wilson, Charles P., Jr.; Jones, John B.; Ciapetta, Frank G. U.S. Patent 3 124 540, 1964.
46. Wilson, Charles P., Jr.; Ciapetta, Frank G.; Jones, John B. U.S. Patent 3 124 541, 1964.
47. Baker, R. W., Blazek, J. J.; Maher, P. K.; Ciapetta, Frank G.; Evans, R. E. The Oil and Gas Journal, 1964, 62 (18), 78-84.
48. Baker, R. W.; Ciapetta, Frank G. Proc. Am. Pet. Inst., Section III, 1964, 44, 452-461.
49. Ciapetta, Frank G.; Elston, James W., Jr.; Wilson, Charles P., Jr. U.S. Patent 3 129 189, 1964.
50. Wilson, Charles P., Jr.; Ciapetta, Frank G.; Elston, James W., Jr. U.S. Patent 3 131 156, 1964.
51. Ciapetta, Frank G.; Coonradt, Harry L.; Garwood, William E. U.S. Patent 3 144 401, 1964.
52. Ciapetta, Frank G.; Coonradt, Harry L.; Garwood, William E. U.S. Patent 3 150 071, 1964.
53. Dobres, R. M.; Rheaume, Leo; Ciapetta, Frank G. Ind. Eng. Chem. Prod. Res. Dev., 1966, 5 (2), 174-182.
54. Ciapetta, Frank G.; Jones, John B.; wilson, Charles P., Jr. U.S. Patent 3 267 044, 1966.
55. Baker, Richard W.; Blazek, James J.; Ciapetta, Frank G. U.S. Patent 3 271 300, 1966.
56. Magee, John S.; Briggs, Warren S.; Ciapetta, Frank G. U.S. Patent 3 325 247, 1967.
57. Ciapetta, Frank G.; Henderson, Donald S. The Oil and Gas Journal, 1967, 65 (42), 88-93.
58. Henderson, Donald S.; Ciapetta, Frank G. Prepr., Div. Pet. Chem., Am. Chem. Soc., 1967, 12 (3), 83-99.
59. Carr, Brownell; Ciapetta, Frank G.; Wilson, Charles P., Jr. U.S. Patent 3 352 636, 1967.
60. Baker, Richard W.; Ciapetta, Frank G.; Wilson, Charles P., Jr. U.S. Patent 3 404 097, 1968.
61. Ciapetta, Frank G. Chem. Ind., 1969, 51, 1173-1183.
62. Baker, Richard W.; Ciapetta, Frank G.; Wilson, Charles P., Jr. U.S. Patent 3 425 956, 1969.
63. Baust, George A.; Ciapetta, Frank G.; Henderson, Donald S. U.S. Patent 3 471 270, 1969.
64. Magee, John S., Jr.; Surland, George J.; Briggs, Warren S.; Ciapetta, Frank G. U.S. Patent 3 501 418, 1970.

65. Carr, Brownell; Ciapetta, Frank G.; Wilson, Charles P., Jr. U.S. Patent 3 520 654, 1970.
66. Ciapetta, Frank G.; Parthasarathy, R. U.S. Patent 3 557 199, 1971.
67. Ciapetta, Frank G. Catalysis Reviews, 1971, 5 (1), 67-158.
68. Parthasarathy, R.; Warthen, John L.; Ciapetta, Frank G. U.S. Patent 3 830 847, 1974.
69. Briggs, W.; Ciapetta, Frank G.; Warthen, John L. U.S. Patent 3 853 789, 1974.

RECEIVED November 29, 1982

A Society of Catalytic Chemists and Engineers

HEINZ HEINEMANN
University of California, Lawrence Berkeley Laboratory, Berkeley, CA 94720

While catalysis is an old art and a fairly old science, until the second quarter of this century, it has functioned essentially as a part of various branches of chemistry (physical, organic, inorganic), physics, and engineering (reactor engineering, modeling).

The first international meeting primarily devoted to catalytic research was held by the Faraday Society in 1949. Annual Research Conferences on Catalysis in the U.S.A. started in 1943 as one of the Gibson Island Conferences, and moved to Colby Junior College in New Hampshire in 1947 as one of the Gordon Research Conferences. A whole generation of catalytic scientists has attended these prestigious and pleasant meetings restricted to about 100 persons.

In 1949, a group of seven scientists gathered in Philadelphia to discuss the possibility of holding regular local meetings in the field of catalysis. They were A. Farkas, F. G. Ciapetta, R. Hansford, H. Heinemann, R. Kennedy, K. A. Krieger, and A. G. Oblad. These men started the Catalysis Club of Philadelphia, which from that time on held monthly meetings. All of the founders served as chairmen of the Club in its early years. Early meetings were held at the University of Pennsylvania, and attendance at the meetings rapidly increased from about 15 to 80-100 persons. Men prominent in the field like D. A. Dowden, P. H. Emmett, and Otto Beeck, were among the first speakers. Beginning in 1951 an annual all-day symposium was held--a custom that still prevails. The first symposia were on Saturdays at Swarthmore College, then at the DuPont Country Club in Wilmington, and later in Princeton, New Jersey, and in Chester, Pennsylvania. Membership of the Philadelphia Club rapidly increased to above 200 and has remained at that level. A Chicago Catalysis Club was founded a few years later, followed by the New York Club--whose founders were charter members of the Philadelphia Club. Within the next 20 years, catalysis clubs were formed in California, New England (Boston), Pittsburgh,

0097-6156/83/0222-0519$06.00/0

Houston (Southwest Catalysis Club), Michigan, and also in Canada. In 1979, the Tri-State Catalysis Society joined the group.

In 1954, Rudolf Brill and Heinz Heinemann suggested to the Philadelphia Club that an International Congress on Catalysis should be held, and the Club under its 1954 Chairman, G. A. Mills, enthusiastically agreed to sponsor the First International Congress on Catalysis in Philadelphia in 1956. In 1956, Charles A. Plank was Chairman of the Philadelphia Club. P. H. Emmett, Eric Rideal, and Eugene Houdry were honorary chairmen, and Heinz Heinemann was executive secretary of the First International Congress on Catalysis with responsibility for the overall organization. Numerous committee chairmen worked hard to make the Congress a success and they, as well as the sponsors who contributed $43,000 to the Congress, are listed in Volume 9 of ADVANCES IN CATALYSIS that contains the proceedings of the meeting. The Congress was attended by over 600 persons, many of whom came from abroad with support furnished by part of the funds raised for the event. During preparations for the meeting, the "International Congress on Catalysis" was incorporated in the U.S., and at the meeting it was decided to hold such Congresses once every four years in a country offering an invitation. Heinz Heinemann was elected president of the organization for the period 1956-1960 and through the second Congress held in Paris in 1960. He was succeeded as president by A. Farkas (1960-1964), F. G. Ciapetta (1964-1968), and V. Haensel (1968-1972). The Third Congress was held in Amsterdam in 1964, the fourth in Moscow in 1968, the fifth in Palm Beach, Florida in 1972, the sixth in London in 1976, and the seventh in 1980 in Tokyo. The eight is scheduled for Berlin in 1984. All Congresses after the first five were attended by at least 1,000 participants, and proceedings have been published for all of them. The Second Congress was cosponsored by IUPAC, and subsequent ones were joint enterprises. After the success of the Second Congress, a truly international organization was established with a board comprising two representatives from each member country. Local organizing committees did most of the work.

By 1965, the individual Catalysis Clubs in the United States felt the need for closer affiliation, and the Catalysis Society of North America was established, which comprised the membership of all Catalysis Clubs that were renamed "Catalysis Society of California," etc. The organization of the Catalysis Society and its first national meeting in Atlantic City in 1969 were financed with funds remaining from the First International Congress on Catalysis. National meetings of the Society have been held every two years in odd-numbered years in Atlantic City, Houston, San Francisco, Toronto, Pittsburgh, Chicago, and Boston. All have been attended by hundreds of participants and parallel scientific sessions have been required. The Society is administered by a

Board of Directors elected by the member societies and the membership at-large. The Board in turn elects a president, or vice-president, secretary, and treasurer.

In 1967, the Catalysis Society of North America proposed the establishment of two prestigious awards to be given once every two years. A committee comprised of A. H. Weiss and Heinz Heinemann obtained sponsorship for the "Paul H. Emmett Award" in fundamental catalysis and the "Eugene J. Houdry Award" in applied catalysis--the former to carry a $2,000 prize, and the latter a $2,500 prize. Emmett Award has been sponsored since its inception by the Davison Division of W. R. Grace & Company; the Houdry Award was originally sponsored by the Houdry Process Corporation, Sun Oil Company, Oxycat Corporation, and the Houdry family. More recently, Air Products and Chemicals Corporation has taken over the sponsorship. Award winners for the first ten years were:

Emmett Award	Houdry Award
R. Kokes	H. Bloch
J. H. Sinfelt	C. Plank
J. Lunsford	H. Heinemann
G. A. Somorjai	V. Haensel
G. Ertl	A. Farkas
P. Jacobs	H. Pines

Catalysis Clubs have also prospered in other countries. There are now such groups in Hungary, Japan, and Holland. The Gordon Conferences found a European equivalent in the Roermond Conferences held in Holland once every four years.

RECEIVED October 29, 1982

INDEX

INDEX

A

Acetylenic complexes ... 162, 165
Acid catalysis
  Beeck's involvement ... 165
  and catalytic reactions, similarities ... 237
Acidic components, typical ... 294
Acidic sites, required in isomerization 507
Activation of hydrogen in catalytic reactions of hydrocarbons, 1939 Faraday Society Discussion ... 108
Active catalysts, choice in automotive exhaust catalysts ... 423
Active sites
  catalytic, spectroscopic studies ... 472
  measurements ... 435
  number related to reaction rate ... 437
  isolation, discussion ... 334–42
Adipic acid, catalytic production ... 318
Adsorbates, electronegative, adsorption on zinc oxide ... 188
Adsorbed polymer weight, method of determining site density ... 438
Adsorption
  activated
    concept of catalytic reactions ... 465
    controversies ... 466
  and desorption rates, nitrogen, Brunauer, Love, and Keenan ... 206–8, 208*f*
  hydrocarbons on oxide catalysts ... 468
  isotherms, equation ... 202
Alkene to alkane conversion ... 10
Alkylation
  aromatic ... 318
  isoparaffins and cycloparaffins with olefins ... 79
Alkylbenzene disproportionation, hydrocracking of aromatics ... 298
Alumina
  intrinsic acidity and catalytic activity ... 84
  pellet supports, automotive exhaust catalysts ... 423
  synthesis of supports for metal particles ... 480
Aluminosilicate catalyst, Superfiltrol ... 242
Aluminum halide catalysts, investigation by Stevenson ... 166
Ammonia
  catalysts, basic study ... 200
  chemisorbed, differentiation in cracking catalyst acidity ... 184
  kinetics of decomposition and synthesis ... 204
  nitrides as intermediates in synthesis 209
  related to methanol synthesis ... 351, 352*f*, 353
  synthesis
    American process ... 196–98
    equilibrium concentration, various temperatures ... 197
    and Hugh S. Taylor ... 464
Ammoxidation process, Sohio ... 331
Analysis methods, various, Princeton University catalytic research ... 467
Anderson, carbon chain growth network, FTS ... 390*t*, 391
Anisotropic motion ... 382
Anisotropy, chemical and surface behavior ... 126
Aromatics, hydrocracking reactions ... 298, 299*f*
Arrhenius equation, exponential factor related to site density ... 439
Arsenic poisoning, platforming process ... 150
Associative mechanism vs. dissociative mechanism, Polanyi ... 112
Automobile exhaust
  catalyst development ... 415, 429
  composition and conditions ... 416
  emission standards
    1959–1981 ... 418, 422, 428
    delays in enforcement ... 425

B

Banks, Robert L., olefin metathesis ... 403
Bardeen, John, and Albert Einstein ... 223
Base metals, active automotive exhaust catalyst component ... 424
Base titrations, to determine site density ... 444
Beacon Laboratory, IR studies of catalyst surfaces ... 183
Beeck, Otto
  1943–1950 ... 153
  achievements ... 155

Beeck, Otto—*Continued*
early history ... 156
final publications on metal films ... 164
BET equation
development ... 200–202, 465
history by Edward Teller ... 227
and Paul H. Emmett ... 49
required changes ... 230
Blanding, testing cracking catalysts ... 255, 256*f*
Bloch, initial observation of NMR ... 377
Boedecker, method of determining site density ... 442
Bohr, Niels, catalysis research at Princeton University ... 464
Bond energy, metal-oxygen, discussion ... 334–42
Bond number as a function of temperature ... 137*t*
Bonhoeffer–Farkas mechanisms
confirmation by Rideal and Trapnell ... 111
criticisms ... 110
proposal ... 11
Bonhoeffer–Harteck, ortho- and para-hydrogen ... 91
Boudart, Michel, catalysis research at Princeton University ... 463
Bragg, William Lawrence, work on crystal structure ... 20
Bray, W. C. ... 196
Bridgeman, P. W., insulated electrical lead ... 456, 457*f*
Brunauer, Stephen ... 218, 227
and Albert Einstein, 1943 ... 217
Bureau of Mines, catalysis, 1930–1960 ... 400
Bureau of Ordnance, Navy, 1942 ... 218
n-Butane, observations of silica-alumina catalytic cracking ... 249
2-Butanol, dehydration ... 440
Butene-isobutane feed mixture, olefin metathesis ... 404
n-Butene, metathesis reaction ... 405

C

Campbell, D. L., development of fluid catalytic cracking ... 273
Carbon-14
CO_2 with carbon ... 476
technique development applied to catalysis, Paul H. Emmett ... 53–55
Carbon-13 NMR
adsorption of hydrocarbons on catalytic surfaces ... 474
study of heterogeneous systems ... 384
Carbon chain growth, Fischer–Tropsch synthesis ... 389
Carbon formation, platforming process ... 145
Carbon number distribution, carbon chain growth, FTS ... 391, 392*t*
Carbon dioxide, mechanism of CO production ... 476
Carbon monoxide
adsorbed, dipole-dipole interactions 188
first pollution standards ... 418
IR studies of iron chemisorption ... 185
poisoning of copper catalyst ... 466
Carbonium ion
discussion ... 168, 239
intermediate concept applied to catalytic cracking ... 237–39
mechanism
of activation of paraffins ... 251
catalytic cracking ... 235
strong acid catalysts ... 243
Carboxylate formation, factor in catalyst deactivation ... 189–91
Casale recirculation unit, methanol synthesis ... 363, 364*f*, 365*f*
Catalysis
application of EPR ... 481
Bureau of Mines, 1930–1960 ... 400
catalyst, and catalytic force, terms first used ... 6
developments in 1920s ... 465
first use without metal ... 67
flow system ... 468
metaphysical theories ... 4
petroleum industry reforming processes ... 510
Princeton University, 63 years ... 463
production of organic chemicals ... 318, 319*f*, 320, 321*t*
refinery processes ... 508
solid phase studies, Fixed Nitrogen Laboratory ... 207
state of the art in 1938 ... 278*t*
Catalysis Society of North America, established, 1965 ... 520
Catalysts
application of electron microscopy .. 470
applications of the NMR chemical shift ... 382–85
automotive exhaust
degradation and deactivation studies ... 420–423
effect of California legislation ... 419, 420, 423
review of technical aspects ... 430
bead
development at Mobil Oil Corp. 248
synthetic silica-alumina gel particles ... 454

Catalysts—*Continued*
calculations of site densities, various systems 437–44
characterization techniques, Leonard C. Drake 458
chrome bead, development 249
conversion, automotive exhaust development 415
development
Fixed Nitrogen Laboratory 198
magnetic resonance techniques 375, 380–85
nickel alloys 494–500
the R series 152
selective ammoxidation and oxidation 331, 332*f*
doubly promoted, development in the 1920s 198, 199*t*
dual-function, reactions of pure hydrocarbons 508, 510
early surface research 179, 183
effect of matrix
on activity 270*t*
on stability 269*t*
effect of pore structure on behavior 180, 181*f*
efficiency test by Murray Raney 493
EPR studies, various 480–84
ether formation, nickel containing nickel oxide 85
evaluation in olefin metathesis of various olefins 406, 408*f*
Fischer and Tropsch 359
fused magnetite iron nitrides and FTS 393
general discussion of nature of active sites 436–37
heterogeneity of surface active centers 466
high-temperature metathesis 407
homogeneous and heterogeneous, similarities, metathesis reaction 411, 413
Lazier, Lewis, and Frolich 360
major cause of deactivation 257
measuring the number of active centers 435
metathesis reactions, various 406
methanol synthesis life, discussion 367, 370*f*
Mittasch and Schneider 358
Mittasch, Pier, and Winkler 359, 360
monolithic, automobile exhaust emissions 418
optical rotatory properties as method of site-density determination 438
poisoning of silica-based catalysts 407, 408*f*

Catalysts—*Continued*
possible origin of acidity 250
requirements for platforming process 145
selective oxidation, discussion 334–42
sodium/zeolite/matrix, discussion 267–70
steam-hydrocarbon reforming modification 357
supports, various, chromium oxide catalysts 468
synthesis at Princeton University 477
three essential characteristics 260
vanadium oxide, hydrocarbon conversion 417
Catalytic activity
effect of coke formation 255
effect of particle size 478
first related to rate 8
transition metal oxides, related to EPR properties 482
Catalytic converters
1962 417
automotive
1968 GM study 421–22
impetus for development 416
effect of fuel sulfur on base metal catalysts 424
effect of relaxation of Federal standards 425–26
Environmental Protection Agency test procedure 425
initial tests 416
vehicle particulate emission studies 426–28
Catalytic cracking
acidic nature 238
carbonium ion mechanism 235
commercial development 61, 70, 239, 247
development by Houdry 66, 69
discussion of mechanism 236–38
durabead type catalysts 264*t*
early equipment 284*f*, 285*f*
early research at Universal Oil Products Co. 241
economic considerations of new zeolite catalysts 265, 266*f*
first synthetic silica-alumina 74
fluid, development 273, 274, 278, 279*t*
fluidized bed
essential elements 282
first commercial units 286, 287*f*
Houdry adiabatic process 74
mechanism, nature of catalysts, Paul H. Emmett 53–55
metal acid faujasites 263*t*
moving bed process, Socony Vacuum 454
nitrogen base effect 238

Catalytic cracking—*Continued*
pure hydrocarbons, compared with thermal cracking 243
role of olefins 251
silica-alumina, of *n*-butane 249
similarities between low-temperature acid-catalyzed reactions .. 237
the war effort 247–48
zeolite, improvements 267, 268*f*
Catalytic decomposition, paraffins, Otto Beeck 157
Catalytic etch, platinum foil 479
Catalytic experiments with wires, flaw 96
Catalytic oxidation on platinum 6
Catalytic Research Associates 274
Catalytic research, Fixed Nitrogen Laboratory, 1926–1940 ....
Catalytic society, development, 1949-present time 519–21
CAT-C test, discussion 257, 258*f*, 259*f*
Catforming process, reforming of hydrocarbons 507–8
Cations, importance to isomerization .. 79
Ceramic support, monolithic automotive exhaust catalysts 423
Charcoal
active catalyst for para-H_2 conversion, discussion 95
catalyst studies at Princeton University 481
Chemical feedstocks, future trends 342, 346*f*
Chemisorption
application of spectroscopic techniques 472
benzene dissociation, effect of temperature 137
to determine site density 444
first observations 8
hydrogen on tungsten, methods by J. K. Roberts 110
selective, method development by Paul H. Emmett 50
Chevron Research, development of hydrocracking 293
Chicago Tribune headline, January 4, 1983 430
Chromite, catalyst synthesis at Princeton University 477
Chromium oxide
adsorption of hydrocarbons 468
on silica, low temperature polymerization of ethylene 482
static magnetic susceptibility 480
Ciapetta, Frank G.
association with Davison, 1955 511
catalytic research, 1937–1972 505
patents and publications 515–17
Claude type methanol unit 361, 362*f*
Clean Air Act Amendment of 1970 .. 422
Coal hydrogenation, early work 400
Coke
deposition on aged silica-alumina catalysts 460
formation effect on catalysis 255, 256*f*
Collective paramagnetic particles, discussion 136
Compressibility, various gases 197
Condon, E. U., catalysis research at Princeton University 464
Conjunct polymerization 30
Herman Pines 78
Contact angle, mercury with and catalyst materials 458
Contact catalysis, first observation by Ipatieff 25
Conversion of ortho-hydrogen to para- hydrogen, 1929 91
Copper, catalysts by reduction of Kahlbaum copper oxide 477
Cottonseed oil hydrogenation plant, Murray Raney 493
Cottrell, F. G. 196, 197
Cracking catalysts
acidity, Lewis or Brönsted acid sites 184
sample preparation, chemisorbed ammonia studies 184
petroleum, activated clay 67–69
Crain, D. L., mechanism of olefin metathesis 407, 408*f*
Crystal faces, exposed, determination by IR studies 188–91
Crystal lattice deformation, as catalytic active sites 436
Cyanamid catalyst, Ford report, 1967 421
Cyclohexane, catalytic production 318
Cycloparaffins, hydrocracking reactions 300, 301*f*, 302*f*

D

Davy, Edmund and Humphrey 5
Deacon process 7
Dehydrocyclization
first bridge between aliphatic and aromatic chemistry 468
research at Princeton University ..467–69
Dehydrogenation 318
Dehydrogenation activity (DA) test, Charles J. Plank 255
Deuterium
and Adalbert Farkas 100, 104
catalytic studies 469, 476
chemical separation, 1934 101
various interactions 108
Diffusion
barrier production and characterization 469
role in heterogeneous catalysis 451

Dilatometer, mercury porosimeter 458, 459f
αα-Diphenyl-β-picrylhydrazyl—
See DPPH
Dipole-dipole interactions,
adsorbed CO ... 188
Dirac, P. A. M., catalysis research
at Princeton University ... 464
Dobereiner, selective oxidation ... 5
Downflow operation, fluidized bed
catalytic cracking ... 286, 288f
DPPH
catalyst for ortho to para hydrogen
conversion ... 480
EPR measurements ... 481
Drake, Leonard C.
catalytic bead testing ... 455
development of mercury
porosimetry ... 451
graduate studies ... 454
at Mobil Oil Corp. ... 249
Dual-site catalysis ... 507

E

Egloff, Gustav, 1935 ... 141
Einstein, Albert
catalysis research at Princeton
University ... 464
correspondence ... 220, 221
discussion ... 217
involvement with war effort ... 219, 223
personal views ... 224, 225
Eischens, R. P., IR studies of
catalyst surfaces ... 183, 436
Electric quadrupole moments ... 378
Electric susceptibility, DPPH ... 480
Electron microscopy
monodisperse platinum particles 479, 480
use in catalyst synthesis ... 478
Electron paramagnetic resonance—
See EPR
Ellipsometry technique, nondestructive studies of the rate of growth
of oxide films ... 126
Emission standards, automobile
exhaust ... 418, 422
Emmett, Paul H.
accomplishments, 1922 to
present time ... 45, 47
ammonia synthesis ... 195
development of the BET equation ... 49
educational background ... 46
effects of high-energy radiation
on catalysts ... 381
Emmett–Kokes microreactor,
modifications ... 473
Emmett Award, Catalysis Society of
North America ... 521
Entropy loss, various surface species ... 441
Environmental protection agency test
procedure, catalytic converters ... 425
Epoxidation, heterogeneous
catalysis ... 322, 323
EPR
catalytic activity of charcoal ... 379
initial experiments ... 376
Princeton University, various
studies ... 480–84
ESR, to determine site density ... 444
Ethylene
dissociative mechanism of
deuterium exchange ... 108
inhibition, first observation ... 6
hydrogenation
on nickel, comprehensive
studies ... 158, 159–65
site density calculations ... 438
interaction with deuterium ... 477
rate of removal by hydrogenation ... 164
reactions with Group VI
hexacarbonyls ... 405
Eyring, Henry ... 44, 439
and Albert Einstein ... 223

F

Faraday, M., 1834 ... 7
Faraday Society Discussion of 1932 ... 94
Farkas, Adalbert
and Ladislas ... 108
1933–1934 ... 98, 102
catalysis research with para-
hydrogen ... 89
deuterium experiments ... 100, 101
Father of petrochemistry, Ipatieff ... 29
Faujasite
base exchanged ... 262t
durabead type catalysts ... 264t
metal-acid cracking ... 263t
x-form, initial experiments ... 260
Federal Clean Air Act, 1967 ... 420
Feldspar, as catalyst ... 241, 242
First International Congress on
Catalysis, Philadelphia, Pa.,
1956 ... 378, 520
Fischer, methanol synthesis ... 359
Fischer–Tropsch synthesis
carbon chain growth and iron
nitrides ... 389
deuterium tracer work ... 477
iron catalyzed ... 183
mechanism, Paul H. Emmett ... 53–55
Fixed bed reactor, commercial
development ... 71
Fixed Nitrogen Research Laboratory,
ammonia synthesis ... 195
Fleet and Life testing, catalytic
converters ... 418

Fluidized-bed cracking process, history 236
Fluidized solids
inventors 282, 283f
technique 273, 290t
Fluorided catalyst, first development 144, 145
Free radical research, application of EPR 481
Frohlich, methanol synthesis 360
FTS—*See* Fischer–Tropsch synthesis
Fuel
effect of automotive catalytic converters on economy 429
technology status in 1922 62
synthetic, 1948 183

G

Gas, adsorption 179, 228–230, 465–66
Gas oil conversion, reasons for development 296, 297
Gas purity, methanol synthesis 371, 372f
Gasoline
desulfurized, over a platinum catalyst 143, 144
effect of unleaded on catalytic converter operation 422
General Motors, sulfuric acid emission studies of catalytic converters 428
Glucose carbonization, EPR studies 481
Gold, studies of synthesis and properties of colloidal metal system 478
Gombers, Moses 24
Griest, Edward, acidity of catalyst beads 248
Gwathmey, Allan T., 1903–1963 121

H

Haagen-Smit, A. J., studies of automobile exhaust pollutants 415, 416
Haber, Fritz 89
Haensel, Vladimir, development of the platforming process 141
Hall, Keith, application of NMR to real catalyst systems 381
Hansen, initial observation of NMR 377
Hansford, Roland
method for thiophene production 455
with Paul H. Emmett 52
Harris, David, management of UOP in 1945 146
Harteck and Bonhoeffer, ortho and para- hydrogen 91
Hayes, John 152
Heat of adsorption
fractions of a monolayer on evaporated metal films 162
Heat of Adsorption—*Continued*
variation with surface coverage 466
Health impact
development of automotive catalytic converters 429
sulfuric acid content automatic catalytic emission 426
Heavy hydrogen, Royal Society Discussion, 1934 99
Heavy water, production at Princeton University 467
Heckelsberg, L. F., new catalyst compositions 406
Henry, 1824 6
Henry, William, work on crystal structure 20
Herington, carbon chain growth probability, FTS, 1946 389
Heterogeneous catalysis
before American contributions, brief survey 3
contributions of olefin metathesis studies 411, 412f
description 13
diffusion 173
first technological application 6
research leading to metathesis discovery 403
fundamental laws, Langmuir 19
two basic catalyst requirements 334
Hexacarbonyl, Group VI transition metal 404, 405
n-Hexadecane reactions, comparison 305f
Hexamethylcyclohexane, hydrocracking products 301f
High-pressure catalytic technique, introduction by Ipatieff 25
Hillier, monodisperse metal studies at Princeton University 278–80
Horiuti–Polanyi, mechanistic proposal 11
Houdry, Eugene J.
associates 74
catalytic work 64, 69, 70, 247
cracking process 72, 235, 276
development of the lignite process 64–66
innovations 75
Houdry Award, Catalysis Society of North America 521
Hydrocarbons
adsorption on oxide catalysts 468, 484
conversion
catalysts, silicophosphoric acid 29
catalysts, vanadium oxide 417
light olefinic 403
effect on steam-hydrocarbon reforming 357
hydrogenolysis 467
intermolecular exchanges 167

Hydrocarbons—*Continued*
origin of metabolizing bacteria 465
reforming
and Frank G. Ciapettia 507
heptane to toluene 468
selective oxidation by heterogeneous catalysis 317
Hydrocracking
definition 294
development 293
discussion of various reactions 297–308
growth 398–11
Hydroforming catalyst deactivation, study 163
Hydroformylation reaction 401
Hydrogel, silica-alumina 248
Hydrogen
adsorption, sintered nickel films 163
chemisorption, research development 202
conversion
effect of rare earths 480
measure of catalytic activity in hydrogen activation 467
mechanisms 92, 93, 111
ortho to para, method of determining site density 438
exchange reactions
butanes in concentrated sulfuric acid 167–69
hydrocarbons with acid catalysts 166
double bond migration switch mechanism 475
and hydrogen sulfide, magnetization-volume isotherms 137, 138*f*
ortho and para 89, 91
production from hydrocarbons 467–69, 510
Hydrogen sulfide inhibition, first observation 6
Hydrogenation-dehydrogenation sites, required in isomerization 507
Hydrogenation components, typical 294
Hysteresis, platinum catalyst characterization 460

I

Incorporation, carbon chain growth .. 391
Interim emission standards, California 1975 425
Ion exchange capacity, silica-alumina hydrogel 248
Ipatieff, Vladimir Nicholaevitch 24, 27, 30, 142
IR
deuterium location studies 476
early studies, discussion 191
Iron catalysis
chemisorption 202
promoted 209
surface studies 209, 211*f* 212*f*, 213, 214
poisoning by water vapor 213
Fixed Nitrogen Research Laboratory 196
Iron nitrides
carbonitrides, and carbides, reaction sequences 393
catalysts in Fischer–Tropsch synthesis 389, 393–98*f*
Isoamyl alcohol, catalysis through iron tube 25
Isocracking, Chevron Research 295
Isomerization
activity, double-bond metathesis process 410
catalysts, nickel or platinum hydrogenation agent 507
n-butane to isobutane 164
hydrocarbons with acid catalysts 166
and hydrogen exchange, important difference 167
of olefins, mechanisms 82–84
Pines and Ipatieff 79
saturated hydrocarbons, mechanism 79
Isotope
elucidation of molecular activation 467
tracers in catalytic studies 474–77
Isotopic hydrogen exchange, liquid benzene and heavy hydrogen 112
Isotropic motion 382

J

Jahnis, C. E., development of fluid catalytic cracking 273

K

Kaiser Wilhelm Institute, para-hydrogen and heavy hydrogen 89
Kehl, application of EPR to real catalyst systems 380
Kerosenes, hydrocracking 143
Kinetics
ammonia, decomposition and synthesis 204
and activation energy, effects of catalyst porosity 159
carbon chain growth scheme, FTS .. 390
in catalysis, first introduced 8
determining site density, discussion 437–39
Langmuir theory 15
Kirchoff, hydrolysis of starch to glucose, 1814 5
Kunsman, C. H. 196

L

Lamb, Arthur B. 196
Langmuir, Irving, contributions 13, 15
Lapidus, Leon, catalysis research at Princeton University 463
Laroche, charcoal catalysis of ortho to para conversion 481
Lazier, methanol synthesis 360
Lead, effect on catalytic converters 420, 421
Lewis, W. K.
  catalyst slippage in upflow pipes 281
  chemical engineering application to petroleum industry 278–90f
  fluid bed velocities 276
  methanol synthesis 360
Lewis acid cracking sites, chemisorption of organic cation radicals 444
Lignite process, development 64
Lind, S. C. 196
Lomax, Universal Oil Products Co. 295

M

Maatman, Russel W., attempts to measure the number of active sites 435
MacIver, application of EPR to real catalyst systems 380
Magnetic dipole moments 378
Magnetic resonance
  application development in catalytic research 375
  catalysis research at Princeton University 480
  to quantify catalytic effect of charcoal 467
  review of the literature 385
  zeolites 483
Manganese, magnetic characteristics in zinc sulfide phosphors 480
Manhattan Project, Princeton University 469
Marisic, Milton, bead catalyst 248
Martin, H. Z., development of fluid catalytic cracking 273
Mass spectrometry, deuterium studies 476
McBain–Bakr quartz fiber spring technique, adaptation 455
Mercury poisoning, to determine site density 442
Mercury porosimetry
  catalyst characterization 451
  method development 455–59f, 460
Metal carbonyl compounds, supported on a high surface area substrate .. 404
Metal oxides, catalytic properties ..405, 406
Metallic single crystals, application to studies of surface phenomena 125
Metallocarbene mechanism, homogeneous and heterogeneous catalyzed metathesis reactions 413
Methane, conversion to acetyl chloride 471
Methanol
  prices 371,373f
  production 318, 349, 350f, 351, 353 371, 373f
  problems from synthesis gas 353
  synthesis, unit designs and types, discussion 91, 358–69f
  various gas feedstock sources 354–56
  various uses 371
Methyl radicals, EPR studies 482, 483
Microconductivity method for deuterium analysis 98
Microreactor, Emmett–Kokes, modifications 473
Mills, method of determining site density 442
Mittasch, methanol synthesis 91, 358–60
Model reactions, to determine mode of activation of molecules by surfaces 467
Molecular chemisorption, to determine site density 442
Molecular desorption, to determine site density 438
Molecular structure approach, chemistry physics 464
Molecular templates, Dickey and Pauling 260
Molybdenum
  aromatization of saturated hydrocarbons 468
  hexacarbonyl olefin metathesis 404
Molybdenum oxide, catalyst in olefin metathesis 406
Monodisperse metal studies at Princeton University 478–80
Montmorillonite, IR study of drying .. 191
Motor Vehicle Pollution Control Board, California 418
Muskie Bill, effect on automotive exhaust catalysts 420

N

Naphtha, reforming process production 510
Nickel
  alloy
    catalytic development 494, 497
    Murray Raney 491
  films
    sintering 161
    various, studies by Al Smith 160–61

Nickel—*Continued*
in nickel oxide
catalyst for ether formation 85
magnetic moment 132, 133*f*
Nickel oxide, hydrogenation catalyst .. 491
Nitrided iron catalysts, discussion 397, 398*f*
Nitrogen
adsorption on iron 179, 200–202
base effect in catalytic cracking 238
chemisorption, research development 202
curves 203*f*
kinetics of adsorption and desorption 206
Nitrogen oxides (NO_x), first pollution emission standards 422
NMR
chemical shift application to catalytic systems 382–85
effects of supported paramagnetic catalysts 377
initial experiments 376
study of adsorption on metal oxides 379
Noble metals, use in automotive catalysts 422, 424
Nuclear magnetic resonance—*See* NMR
Nuclear relaxation, description of several aspects 378

O

Oblad, method of determining site density 442
Oil commercial feeds, hydrocracking reactions 304, 305*f*, 306*f*
Oil injection, fluidized bed catalysis 277
Olefins
allylic oxidation and expoxidation .. 320
in carbonium ion mechanism 251
cyclic, metathesis reactions 410, 412*f*
detergent range, metathesis reaction 410, 411
metathesis
discovery and development 403
potential uses 410, 411
various contributions to catalysis 411
potential relationship of reactions .. 413
synthesis 403
Olefin disproportionation—*See* olefin metathesis
Optical rotatory properties as method of site-density determination 438
O'Reilly, D. E., first review of magnetic resonance applied to catalysis 380
Origination of catalysis, 1814 5
Ostwald, Wilhelm 3, 8
Overhauser effect, EPR transitions in a paramagnetic material 382
Oxidation
allylic, heterogeneous catalysis 327
aromatic, heterogeneous catalysis .. 324
discussion of selective and non-selective processes 333, 334
heterogeneous 318
homogeneous 318
hydrocarbon partial combustion reaction 358
mechanism of bismuth molybdate catalyzed 334, 342, 343*f*, 344*f*
paraffin, heterogeneous catalysis 325
selective, heterogeneous catalysis 320, 321*t*
Oxygen adsorption, application of spectroscopic techniques 472

P

Packard, initial observation of NMR .. 377
Palladium
asbestos catalyst 4
monodisperse, determining number of active sites 474
properties of monodisperse particles 479
Paraffins, hydrocracking reactions 300, 303*f*, 304, 305*f*
Paramagnetic para-hydrogen conversion 96
Paramagnetic susceptibility, related to magnetic moment 131, 132
Particle size, effect on catalytic activity 478
Particulate emission studies, catalytic converters, vehicle 426–28
Pauli, Wolfgang, catalysis research at Princeton University 464
Pentanes and hexanes, iso-to-normal ratio, hydrocracking catalysis 300, 303*f*
Perfume musk synthesis 411, 412*f*
Petroleum fractions catalytic cracking, discussion 235
Petroleum industry, during the 1930s 278
Philips, Peregrine, 1831 6
Phillips neohexane process, commercial application of olefin metathesis 411
Philosophers' stone 4
Phosgene, catalytic conversion of methane 471
Phthalic anhydride, production by heterogeneous catalysis 324
Physisorption, first observations 8
Pier, methanol synthesis 359, 360
Pines, Herman
awards 85

Pines, Herman—*Continued*
with Ipatieff at Northwestern University ... 80
1930s–1980s ... 77, 78
Plank, C. J.
invention of catalytic cracking ... 253
at Mobil Oil Corp. ... 248
Platformate and platforming, origin of terms ... 147
Platforming process
1960 to present day ... 146
arsenic poisoning ... 150
initial use ... 145, 148
Platinum
active automotive exhaust catalyst ... 424, 425
on alumina, EPR signals ... 482
characterization by Leonard C. Drake ... 458
determining chemisorbed sites for hydrogen ... 473
monodisperse, determining number of active sites ... 474
properties ... 5, 479
synthesis ... 477, 479
use ... 429
Pliskin, 1950s IR spectrum of adsorbed molecules ... 436
Poisoning
first example ... 6
method to determine site density ... 442, 473, 474
preferential, hydrocracking catalysis ... 307*t*
Polanyi and coworkers, kinetic equations for associative and dissociative mechanisms ... 112
Pollitzer, Ernie ... 151
Pollutants, automobile exhaust ... 415, 416
Pollution control
automotive devices ... 419
legislative action, effect on catalyst development ... 420
Polymerization and cracking of olefins, hydrogen switch mechanism 475
Poole, application of EPR to real catalyst systems ... 380
Pore structure, role in determining catalyst performance ... 180, 181*f*
Pore volume, discussion ... 451–53
Pound, initial observation of NMR ... 376
Precision scientific award, Frank G. Ciapetta ... 508
Pressure bomb, high, first use by Drake ... 456
Pressure, operating, methanol synthesis, discussion ... 371, 372*f*
Production capacity, methanol synthesis ... 374
Propylene
ammoxidation ... 328, 342, 345*t*, 346*t*
metathesis products ... 407, 409*f*
oxidation to acrolein ... 331*f*
Pulse poisoning, technique to determine site density ... 443
Purcell, initial observation of NMR ... 376

R

Raney catalyst, applications ... 497, 500
Raney, Murray
alloying crucible and reaction vessel 497
patents ... 500, 501*t*
research on nickel catalysts ... 493–500
various catalysts ... 502*t*
Reactant chemisorption, method to determine site density ... 443
Reactor types, various, methanol synthesis ... 363, 366*f*–369*f*
Reductive dehydroxymethylation of primary alcohols, Pines and Ipatieff ... 84, 85
Refinery catalytic processes ... 508
Reforming
catalytic ... 143
steam-hydrocarbon synthesis gas ... 354–58
Reid, E. Emmet, work with Paul H. Emmett ... 52
Rhenium oxide, catalyst in olefin metathesis ... 406
Rhodium, active automotive exhaust catalysts ... 424, 425
Rideal, Eric K., 1933 ... 98
Ritter, Howard, at Mobil Oil Corp. ... 249
Rosinski, E. J.
early zeolite work ... 257
invention of catalytic cracking ... 253
Rotational motion, surface species ... 441
Royal Society Discussion on heavy hydrogen, 1934 ... 99
Rubber, synthetic, development ... 469
Rusha Instrument Corp., high pressure porometer ... 456

S

Sabatier and Ipatieff, 1900 ... 10
Schneider, first patent for methanol synthesis ... 358
Scott, John W., development of hydrocracking ... 293
Selectivity phenomenon, metathesis of butenes ... 407
Selwood, P. W.
application of magnetic methods to catalytic systems ... 131, 377
Gouy balance for measuring static susceptibility ... 480

Silica
catalyst support effect on metathesis 407
high surface area and nonporous, as support for IR studies 185, 186
hydrogel starting material for cracking catalysts 242
produced by burning silicon chloride 185
Silica–alumina
catalyst bead development 454
explanation of acidity 239
high-temperature acid 243
hydrosol 248
observations of catalytic *n*-butane cracking 249
various site densities 441, 443
Silicophosphoric acid, catalyst in copolymerization of butenes 29
Site density
calculations, determination of mechanism validity 440
definition 435
effect of temperature 438
indirect method determination 442–44
various, direct, non-TST determinations 437–39
various, direct, TST determinations 439–42
Site synergism 436
Smith, Albert E.
nickel film studies 160–61
and Otto Beeck, 1936 158
Society of catalytic chemists and engineers, development, 1949–present time 519–21
Sodium aluminosilicates, amorphous, zeolites 242
SOVA-Forming process, isomerization-dehydrogenation catalysts 508
Spin-echo method, NMR, various applications 378, 382
Spin lattice relaxation processes, description 378
Standpipe concept, fluidized bed catalysis 276, 277
Steam-hydrocarbon reforming
methanol synthesis gas 354, 356
observation 8
synthesis gas purification 356–58
Stevenson, monodisperse metal studies at Princeton University 478–80
Sticking coefficient of gas molecules, method of determining site density 438
Stohman, Z., interpretation of catalysis, 1894 3
Storch, Henry H., developments in Fischer–Tropsch synthesis, 1945–1960 389
Structural imperfections in copper single crystals, studies by F. W. Young, Jr. 128
Styrene, catalytic production 318
Sulfur, effect on automotive exhaust catalysts 424
Sulfuric acid, catalytic converter exhaust emissions 426–28
Sullivan, Richard F., development of hydrocracking 293
Superfiltrol, acid-treated clay 241
Support configuration, automotive exhaust catalyst development 423
Surface area, measurement by Leonard C. Drake 454
Surface atoms, metal, to determine site density 443
Surface attraction, role in catalysis 14
Surface energy, related to use of TST 440
Surface geometry
importance in catalytic reactions 126
related to activity of different crystal faces, Allan T. Gwathmey 124
Surface species, effect of degrees of freedom on density determination 441
Susceptibility isotherm for chromium in chromium oxide 134
Synthesis gas purification, methanol synthesis 356
Synthetic Fuels Act, 1944 401

T

Tamele, concept of origin of catalyst acidity 250
Taylor, Hugh S.
1890–1974 33, 38, 377
atomic weapon project 469
catalysis research at Princeton University 464
post World War II period 470
Teller, Edward, history of the BET equation 227
Temperature, effect on charcoal catalysis 481
Terpenes, studies by Pines and Ipatieff 80
Tetramethylcyclohexane, hydrocracking products 302*f*
Thenard, 1818 6
Thermal-diffusion effects in equilibrium measurements, Paul H. Emmett 49, 50, 51*f*
Thiele, Ernest W. 173–175
Thiele modulus, discussion 173
Thomas, C. L.
carbonium ion mechanism 238
catalytic cracking research at UOP 241
Tolman, R. C. 196

Torrey, initial observation of NMR 376
Transition metal carbonyls, early catalyst developments 405
Transition metals, catalytic and thermochemical measurements .. 162
Transition state theory 439–42
Translational motion, effect on surface density determinations 441
Transmutations, various 4
Triolefin process
 commercial applications 410
 olefin metathesis technology 406
Tritium, mechanism of double bond migration 475
Tropsch, methanol synthesis 359
Tungsten oxide, olefin metathesis 406
Turkevich, John
 atomic weapon project 469
 catalytic research using magnetic resonance 379, 480
 charcoal catalysis of ortho to para conversion 481
 chemistry training 471
 development of poisoning titration technique 473
 Princeton University catalysis research 463, 478–80
 production of hydrogen from hydrocarbons 467–69

U

Unicracking, Union Oil Co. 295
Universal Oil Products Company, early catalytic cracking research 241
Unsaturated chemical bonds 14
Upflow operation, fluidized bed catalytic cracking 286, 289*f*
Uranium hexafluoride
 calculation of thermodynamic properties 470
 differential diffusion through a porous membrane 469
Urey, Harold, 1934 102

V

Valence inductivity approach, studies of supported manganese oxides .. 132
Vanadium oxide, hydrocarbon conversion 417
van der Waals forces, multilayer adsorption 231
von Neumann, J., catalysis research at Princeton University 464

W

Water, effect on catalysis with platinum foils 163
Water gas shift
 catalysts 200
 equilibrium constant, discussion 214
 Hugh S. Taylor 464
Waugh-type pulse sequences, to characterize adsorption 384
Weiss constant 134, 135*f*
Wheeler, Ahlborn
 and Otto Beeck, 1937 159
 research on catalyst surfaces 179
Whitmore, Frank, carbonium ion theory 251
Wigner, Eugene, catalysis research at Princeton University 464
Wilhelm, Richard, catalysis research at Princeton University 463
Wilstater, Richard 24
Winkler
 fluidized bed limitation 281, 282
 methanol synthesis 359, 360

Z

Zeolite
 catalyst 260, 262–65, 474
 Courtald model of a faujasite 260, 261*f*
 discussion of matrix synergism 267–70
 effect of coke formation 263
 silica-alumina hydrogel 248
 sodium aluminosilicates, amorphous 242
 various EPR studies 483, 484
Zinc oxide
 adsorption of electronegative adsorbates 188
 EPR studies 482
 IR spectrum 186, 187*f*, 188

Jacket design by Kathleen Schaner
Editing and production by Susan Robinson and Frances Reed

Elements typeset by Service Composition Co., Baltimore, MD
Printed and bound by Maple Press Co., York, PA